NETWORK ECONOMICS OF MARINE ECOSYSTEMS AND THEIR EXPLOITATION

Network Economics of Marine Ecosystems and their Exploitation

Christian Mullon
Institut de Recherche pour le Développement UMR212
Exploited Marine Ecosystems
Avenue Jean Monnet, 34200, Sète
France

CRC Press
Taylor & Francis Group
Boca Raton London New York

CRC Press is an imprint of the
Taylor & Francis Group, an **informa** business
A SCIENCE PUBLISHERS BOOK

CRC Press
Taylor & Francis Group
6000 Broken Sound Parkway NW, Suite 300
Boca Raton, FL 33487-2742

Printed in the United States of America on acid-free paper

International Standard Book Number: 978-1-4665-9005-2 (Hardback)

Visit the Taylor & Francis Web site at
http://www.taylorandfrancis.com

CRC Press Web site at
http://www.crcpress.com

Science Publishers Web site at
http://www.scipub.net

Preface

Studies about sustainable development imply to jointly consider, at the time horizon of a generation, systems with processes of diverse kinds, relevant to diverse scientific disciplines, at diverse scales (from the small ecosystem to the whole planet).

A common way to perform a necessary integration is the coupling of models that have already been formulated and implemented in the different disciplines. In some cases, the intrinsic complexity of the models to couple, some theoretical incoherencies between them, or the number of parameters to jointly estimate make it a difficult way.

Here I propose an alternative way, less ambitious. Having looked for common features in ecological and economic processes that are involved in the exploitation of marine ecosystems, I suggest a common model to represent them and I analyze its effective implementation in several case studies.

The approach I propose is based on:

(1) a focus on equilibration processes in ecological and economic processes, on their interplay, the identification of possible synergies,
(2) the formalism of networks and therefore the emphasis on structural constraints, rather than on deterministic relationships,
(3) the use of several tools from the modern mathematical theory of optimization, mainly the variational inequality and complementarity problems.

The origin of the questions I address in this book lies in the many scientific projects of IRD I have been involved in. IRD is a French research institute for developing countries. Almost all of its programs involve several scientific disciplines (oceanography, fisheries biology, economics, sociology, etc...), a high diversity of approaches, of scales, of processes to represent. All of them imply to bring together a dedicated team of scientists, from developing and developed countries, to realize field studies and surveys. In these programs, the use of models is often crucial and may be problematic. For some scientists they are a tool, for others they are a goal. When they are seen as a tool, they can be an integration tool (between searchers) or a communication tool (from searchers to other stakeholders). There are specific models in all disciplines; a common feature is often the use of mathematics and it happens that the construction of an integrated model is attributed to a mathematician. This is the end of a story of this kind I try to tell here.

This book is organized into four parts.
Part I introduces the issues. Chapter 1 briefly discusses the question of marine ecosystems subject to double exposure. Chapter 2 provides a short introduction to network theories. It pays particular attention to: (a) the diversity of formalisms, terms, and applications, (b) the relationship between structure and function, and (c) the arbitrariness of graphical representations. The network concept is useful because it can be applied equally to mathematics, physics, material sciences, biology, social science, and traffic studies. We briefly discuss this point and give some illustrative examples of network uses in various research fields, with an emphasis on those that are most relevant to network economics. Chapter 3 discusses the equilibrium versus dynamics issue with network economics principles. It aims to place some of the principles of network economics in the context of theoretical debates in economics and ecology.

Part II presents the mathematical tools that form the background to the network economics approach. Although the approach is elementary and illustrative the presentation is mathematical, i.e., with definitions, propositions, and demonstrations. Chapter 4 introduces principles and notations. Chapter 5 is an introduction to convex optimization, ending with the Karush-Kuhn-Tucker theorem. Chapter 6 is dedicated to the main ideas of network economics, namely variational inequality and complementarity. The final section of this chapter describes a practical tool that links modeling and simulations based on real data. In an ending digression, we discuss the issue of equilibrium in ecology and economics in more detail.

Part III presents the way to apply network economics to ecological and economic systems. Chapter 7 is dedicated to the equilibrium of trophic systems. Chapter 8 is dedicated to economic networks: spatial economics, supply chains; an annex relates it to the general equilibrium theory. Chapter 9 introduces to the formulation of migrations problems.

Part IV consists of several studies that illustrate the application of network economics principles and tools to ecological and economic problems. They show the kind of results that can be obtained, highlight changes in perspective, and the scenarios that can be modeled and analyzed. Some studies are purely speculative and involve abstract models; others are based on the simulation of complex real-world systems. Chapter 10 insists on the necessity of a specific way to present complex modelling experiments.

In Chapter 11 network economics are used to throw a new light on Gause's exclusion principle and underline the importance of intra-specific competition as a factor in biological diversity.

Chapter 12 applies the network economics approach to Humboldt's productivity paradox and suggests that a focus on the accessibility of primary production is a main factor in fish abundance.

Chapter 13 shows how the equilibrium found in a network model of the Chilean coastal ecosystem is related to ecological controls and the identification of key species in an upwelling ecosystem. The research described in chapters 11, 12 and 13 was carried out in collaboration with Yunne Shine and Philippe Cury.

Chapter 14 presents a model of the global tuna supply chain. Unlike the models described in previous chapters this is a dynamic model, based on the coupling of economic equilibrium and deterministic functions for fish stock production and fisheries investment. This work was carried out in collaboration with Jeanne Fortilus.

Chapter 15 presents a model of the worldwide fish supply chain, including data collection, model calibration, and scenario analysis. It demonstrates the combined effects of climate change and economic globalization on the network equilibrium of this system. This chapter summarizes ongoing work with Gorka Merino and Fabien Steinmetz.

Chapter 16 discusses how complementarity may help to reassess how fisheries are financed.

Chapter 17 presents a simple model, using a variational inequality approach to migration, of the movements of small and big fish in a water column (an example of a pursuit evasion game). This work was carried out with Anna Nagurney.

Chapter 18 analyzes the spatial behavior of Atlantic Bluefin tuna. It seeks to explain changes in migration patterns as a trade-off between the availability of food in feeding areas and the energetic cost of migration. This chapter is a summary of ongoing work with Patrizio Mariani and Vlastimil Krivan.

This work is part of the program of the Exploited Marine Ecosystems research unit led by Philippe Cury. I sincerely thank him for giving me availability and guidance all along the writing and the publishing of this work. I have benefited from discussions with my colleagues in this research unit: Frédéric Ménard, Jean-Marc Fromentin, Nicolas Bez, Patrice Guillotreau, Christian Chaboud, and even Olivier Maury. I especially thank my accomplice Pierre Fréon who has heavily contributed to this research for several years: commenting, inspiring, encouraging. This work has been supported by the EUROCEANS consortium through the organization of two summer schools in 2006 (Dragerup-Denmark) and 2009 (Sète, France). I thank Pierre-François Baisnée and the courageous attendants of the presentations I gave at these occasions

The works dedicated to a model of the worldwide fish supply chain and to the migrations of the Bluefin tuna, have been supported by the European program BASIN led by Mike Saint-John and Manuel Barange.

The part of this work dedicated to a model of the global tuna supply chain has been supported by the program MACROES led by Olivier

Aumont and funded by the ANR. Jeanne Fortilus was supported by the program AMPED led by David Kaplan and funded by ANR.

The origin of the idea of applying network economics to ecological problems has been the reading of Anna Nagurney's book: Network Economics, a Variational Inequality approach. I have found there the level of abstraction, the level of complexity, the mathematical tools I was seeking for several years to model real marine systems. I acknowledge this influence and I thank her for the collaboration we have initiated since.

As a modeler of complicated systems, I have known a period before Volker Grimm and Steve Railsback and a period after them. And I know how much the methodological principles they suggest have positively changed the modeling practice in ecology. Volker has always been reactive to my questions and his outspoken comments about a preliminary version have been very useful.

I thank Ferenc Jordan and Charles Mullon for reading a previous version, comments and suggestions.

Jeanne Fortilus, Micharl Mansoor, Fabien Steinmetz, Gorka Merino, Jose Fernandez, Patrizio Mariani have been involved in the different case studies that are presented in this book. I thank them for their important contribution.

I have benefitted, in the interdisciplinary works I have been associated to inside IRD, from the enthusiasm and the wisdom of Philippe Couty, Joël Bonnemaison, Gaston Pichon, Jacques Weber, Francois Bousquet, Larry Hutchings, Marcel Kuper, Yveline Poncet, Pierre Morand, John Field, Carolina Parada, Sergio Neira, Lynne Shannon, Claude Roy, Christophe Lett, Olivier Thébaud, Benjamin Planque. I am grateful to Jean-Pierre Treuil and Pascal Renaud for our collaboration in a previous life.

I thank my colleagues of the international research unit Resiliences, Jean Luc Dubois, Mama Ouattara, Allioune Kane, Awa Niang, Marie-Hélène Durand, Catherine Pasquier, Pierre Morand, Isabelle Droy, Marie Piron, Jacques Quensière for hospitality and stimulating debates.

I thank Elaine Seery for reviewing the English of the manuscript.

This book is dedicated to my wife Dominique, to my children Charles, Rachel, Justine and to my parents, Jane and André.

Christian Mullon
Avenue Jean Monnet, 34200, Sète
France

Contents

Part I

Introduction

Chapter 1

Double Exposure and Network Economics

1.1 Double Exposure

This book shows how to apply the principles and tools of network economics, such as variational inequalities and complementarity problems (Nagurney, 1999; Facchinei and Pang, 2003), to study large exploited natural systems, especially the marine systems, at a global, continental or regional level. These systems are exposed to both climate change and economic globalization; a phenomenon that has been defined by Leichenko and O'Brien (2008) as "double exposure"[1]. Many of them are endangered in the present context.

A representative example is the supply chain that links small pelagic fisheries and fishmeal and fish oil markets shown in figure 1.1. The importance of this chain is underlined by the fact that it accounts for 30% of global marine catches. The variability of the underlying natural processes, an increasing global demand (mainly due to the development of aquaculture), and the expansion of trade in the context of World Trade Organization agreements are other important factors. Climate change affects both the supply side (e.g. the productivity of fisheries) and the demand side (e.g. the functioning of marine farms), while globalization affects demand (e.g. aquaculture, fishmeal, and fish oil prices, and fishing, shipping, and fuel costs), and commercial relationships (e.g. import taxes). This global supply chain was studied with network economics tools by Mullon et al. (2009a).

[1]See also Perry and Ommer (2010).

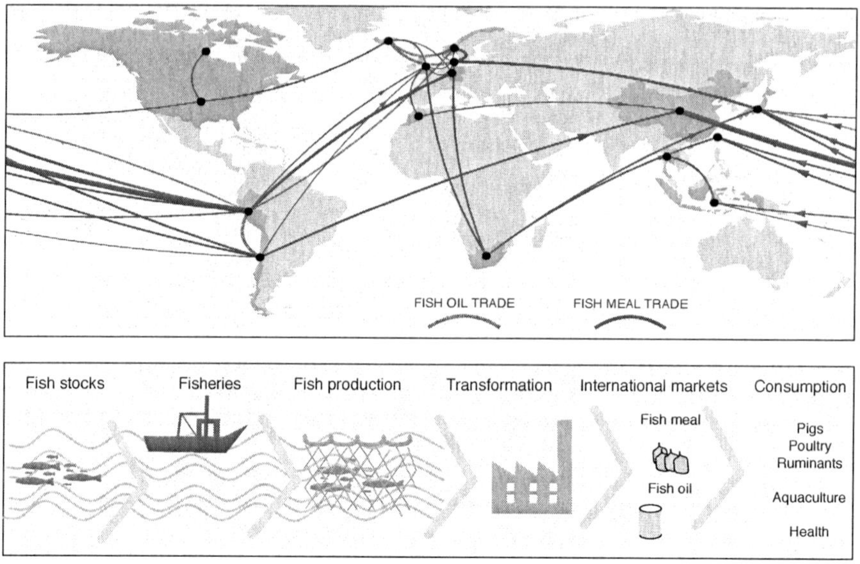

Figure 1.1: A global marine system. The spatial structure of the global network relating producers and consumers of fishmeal and fish-oil. Courtesy of Gorka Merino.

In their exploration of double exposure, Leichenko and O'Brien (2008) addressed the following questions: How can the different levels (local, regional, national, continental, worldwide) and time horizons (now, in the near future, in the long term) at which double exposure operates be distinguished? And how, for both types of exposure, at all levels and in all time horizons, to distinguish between winners and losers? The use of the terms winners and losers should be noted as it suggests that issues related to double exposure can be formulated using a game theory framework. For example, when Harris (2011) mention the needs of having, on the processes involved in double exposure, a "cosmopolitan" worldview, i.e., less international, more global, this implicitly supposes changing the roles of the players, i.e., the stakeholders.

1.2 Network Economics

1.2.1 Network Structure

Figure 1.2 shows the network structure of a marine system subject to double exposure. This example is typical of the systems that are examined and modeled here. Nodes are diverse: ecosystems, fish stocks, fleets, traders, and consumers. Links show various types of flows, corresponding to predation, catches, and trade. We can use figure 1.2 as an example to investigate

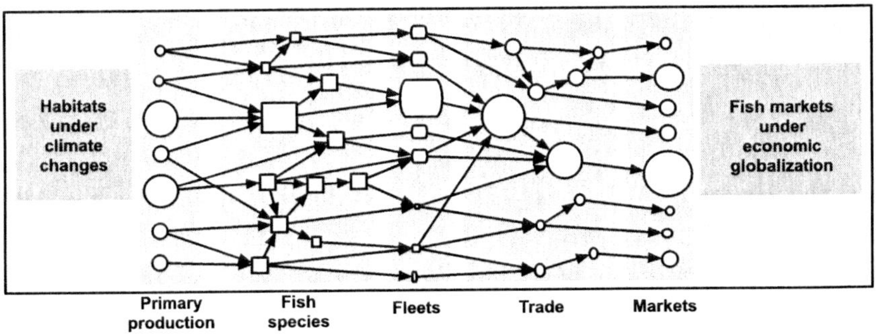

Figure 1.2: The network structure of a fish supply chain and the double exposure.

questions such as what would happen if there was a change in demand (right-hand side of the figure). How do its effects propagate through the network? Will a new equilibrium be reached?

1.2.2 Alternative Approaches

There are already many approaches that can be used to model these systems and address such questions: examples include network theory, system dynamics, input-output modeling and network games theory.

With network theory (1) the structure of the network (topological features, organization of nodes and links) is defined, and (2) its functional characteristics (stability, robustness, weak points) are linked to its structure. Chapter 2 gives an overview of network theory, with a particular focus on how network economics operate.

With system dynamics (Forrester, 1973), (1) a state function is defined; it links the state of a node at a given time to its own state and the states of the nodes it was linked to in the network at an earlier time, (2) the resulting dynamic system is expressed and simulated, and (3) the dynamics (stationary points, attractors) of this system are related to the structural characteristics of the network.

With an input-output analysis (Miller and Blair, 1985), (1) input and output nodes are distinguished, (2) relationships between inflows and outflows are identified on intermediate nodes. This is usually a linear relationship: a specified amount of a given input is necessary to provide a unit of output; (3) the desired output is defined for output nodes, and (4) the necessary input is defined for input nodes and consequently all network flows can be determined using elementary linear algebra.

With the theory of network games (Abramson and Kuperman, 2001; Galeotti et al., 2010; Easley and Kleinberg, 2010; Watts, 2001), (1) nodes are considered as players with their own strategy, whose income depends on their strategy and that of the players they are linked to in the network, (2) the resulting game equilibrium is analyzed, and (3) it is related to the network structure.

1.2.3 The Network Economics Approach

We will use a network economics approach (Nagurney, 1993) to model a network such as the one shown in figure 1.2. It can be seen as a mixture of the approaches described above. In network economics: (1) interactions between nodes occur on links, and interactions between links occur on nodes; conservation equations represent physical constraints occurring on nodes (e.g. inflow must be greater than outflow); equations of an other kind, said complementarity equations, represent economic constraints occurring on links (e.g. when there are positive flows on two links with common extremities, transportation costs are equal), (2) the solutions of these equations, considered as equilibrium states, are computed using generic tools from variational inequality theories or complementarity problems, and (3) dependencies of the resulting equilibriums and network topology dynamics on the organization of nodes and links are analyzed.

1.2.4 Network Economics and Other Approaches

Network economics is one component of network theory because it takes account of specific network features such as topology and structure. However, it focuses on stock and flow systems, which is somewhat unusual in the context of network theory, e.g. in comparison to social network theory, which underlines the existence of links rather than their strength.

Network economics differs from system dynamics (Randers, 1980) in several aspects. In system dynamics theory there is: (1) a set of rules for expressing knowledge of the system, stock and flow diagrams and causal loops, and (2) tools to translate these diagrams into differential or finite difference equations and carry out simulations. Network economics makes intense use of stock and flow diagrams but requires complementarity relationships, which are not part of system dynamics theory. Systems theory represents dynamics using ordinary differential equations, while network economics focuses on equilibrium states and switches between these equilibrium states. Network economics represents dynamics using projected dynamical systems theory that implements a particular type of differential equations where the trajectories of solutions are constrained to stay in a subset of the set of states.

Input-output modeling is based on the idea of rigid structural relationships. Technical coefficients, i.e. the ratio between outflow and inflow for a sector are assumed to be constant, at least in the near future. Rather than assuming a deterministic relationship (an equality) between coefficients relating inflow and outflow, network economics represents constraints between outflow and inflow on a node as inequalities. Then, in the set defined by these constraints, some specific states are characterized as economic equilibrium. Network economics has two advantages over input-output modeling, namely robustness and representativity: robustness because it circumvents the problems of input-output modeling that are due to the fact that solutions are obtained by an inverting a potentially ill-conditioned matrix; representativity because it takes into account elastic processes. Network economics makes it possible to represent, in a more natural way than input-output modeling, features of economic systems such as switching and collapses.

Network games can be easily implemented using network economics tools. However, on a network not all nodes are optimizing agents; some agents are reactive and behave deterministically. Network economics suggests that a system does not consist of one or several optimizers that exchange commodities, but rather of commodity exchanges between agents. The rules of exchanges are based on the complementarity principle: there is an exchange on a link if there is no better, i. e., lest costly, alternative.

1.2.5 Key Features

The main features of the application of network economics to the study of ecological and economic systems can be summarized as follows.

It provides generic tools for very different problems. It can also be applied to domains as diverse as structural mechanics, spatial economics, ecology, traffic studies, migration studies, supply chain analysis, trophic systems, exploited trophic systems, and non-cooperative games.

It takes into account the network structure of stock and flow models that are commonly used in economics and ecology; as is usual in these modes it considers stocks as nodes and flows as links.

It makes it possible to tackle theoretical issues in ecology, economics, and the interaction between them. It emphasizes the importance of constraints in economic and ecological systems. It provides an original perspective on the relationship between equilibrium and optimization. Through a focus on the fact that there are flows on a path if they are necessary, it stimulates discussion on the idea of rational agents, utility, and utility optimization.

It takes into account similarities between economic and ecological systems. Ideas from economics can be applied in an ecological framework and vice-versa. Results obtained with network economics to real world applications may improve the quality of controversial ecological debates between very diverse actors: scientists from different disciplines, politicians, fishery managers, and the general public.

It makes it possible to revisit the main paradigms of ecological and economic management. For example, when control theory is applied to the management of an ecological system, it set up a scenario with actors and roles. This has been analyzed by Fath (2004a): in general terms, this scenario describes a manager who has a (quantified) goal, means (controls), and knowledge of how these means transform the state of the system (state functions). This is how things should be, and there is a tendency to adopt this point of view. However, in real-life ecological and economic studies it is not always easy to identify who is the manager, whether there is only one, what his means and goals are, or how to parameterize state functions. Network economics makes it possible to envisage in a different way the stakes and stakeholders involved in the management of an ecological system, and can provide a different perspective on management issues. This is a commonly the case when a new approach (rather than a paradigmatic one) is taken.

1.3 Marine Systems

The example given in figure 1.1 illustrates the issues surrounding the future of marine systems. Such systems are the subject of this book and the following paragraphs summarize some of their key features[2].

1.3.1 A Natural System

Marine systems have a physical component. Their dynamics are based on flows of water, heat, and organic matter that can be described by physical and chemical laws. A key point is the diversity of scales at which these processes can be analyzed: from the global (planet) scale to a microscopic (micro-organism) scale. From a physical point of view, climate change can have an effect at both the large scale (streams) and the small scale (hydrodynamics processes such as stratification, vertical mixing, and turbulence). From a chemical point of view, climate change can impact

[2]This is only an introduction. To a reader wanting to go further, we suggest (1) for all subjects, the encyclopaedia of Ocean Sciences in six volumes (Steele et al., 2009), (2) for the physical oceanography point of view a classical book such as Pickard and Emery (1990) or more recent, Denny (2011), (3) for the biological point of view, references books such as Mann and Lazier (1996); Kaiser et al. (2005), see also Platt et al. (2007); (4) for the fisheries point of view: the annual State of World Fisheries and Aquaculture from FAO (FAO, 2012); see also Jennings et al. (2001). An interesting review about the different components has been given in Mann (1993). We recommend also for the reasons of pertinence Margalef (1997); Sinclair (1988); Pauly (2010).

ocean temperature (although not uniformly) and produce changes in acidity or oxygen distribution in water columns.

Marine systems have a biological component. They represent by far the major source of biological production of our planet. For several decades we are analyzing their most important processes: from primary production (photosynthesis and primary production) to top-level predation (including fisheries). Here again, the diversity of scales (spatial, time) is a major issue. Another important characteristic of marine systems is their third dimension: depth and its relationship with light and oxygen. Most of the ocean cannot be seen. Climate change can impact plankton productivity (e.g. by changing stratification), and species distribution as some species prefer to live in a specific temperature range.

1.3.2 A Globally Exploited Natural System

Fisheries are an important economic sector. Worldwide, there are 38 million fishermen and 16 million fish-farmers. At the beginning of XXIst century, the average annual value of all United States marine fisheries landings was about 4 billion dollars per year. Fisheries are part of an important trading network. The worldwide ratio between exports (at national level) and production is around 40%. This means that a large part of the fish consumed in one country has been caught by a fishery in another country. Fisheries are an important supplier of protein: fish supplies 18.8 kg per person per year out of a total of 110 kg per person per year. The fishing sector is heterogeneous: there are vast differences between the productivity of a fisherman or a fish farmer in Europe or North America (20 tons per year), South America (7 tons per year), and Asia or Africa (2 tons per year). About 60% of African fishing boats do not have an engine, compared to 30% in Asia, and less than 5% in Europe and North America.

1.3.3 A Changing Exploited Natural System

In recent decades fisheries have changed. Wild fish catches have stabilized at a level of about 90 million tons per year for the last 20 years (FAO, 2012), while aquaculture has seen enormous growth from 10 millions tons per year to 60 millions tons during the last 20 years; today one third of caught fish is forage fish, notably used for the feeding of farm fish. Globally, between 1990 and 2010 there has been a decrease in fish prices from 3 to 2 US dollars per kilogram (FAO, 2012), excluding fishmeal and fish oil. Moreover, in a context where coastal areas are subject to demographic densification (Curran et al., 2002), it is increasingly important to take into account competition between different uses of the sea such as tourism, recreational fishing, oil drilling, and shipping (Cooke and Cowx, 2004).

It is becoming more and more apparent that fisheries are having a dramatic impact on marine ecosystems: 1,500 different fish species are exploited (this is the number of items in the FAO FishStat data base (FAO, 2013)), 90 million tons of fish are extracted and fish stocks are declining. Fifty percent of stocks are fully exploited, while 25% are overexploited. Further details about the level of exploitation of world fish stocks can be found in the FAO annual report (FAO, 2012).

At the same time, biological transformations due to climate change are affecting plankton productivity (Hays et al., 2005), fish abundance (Brander, 2007) and species distribution (Perry et al., 2005; Poulard and Blanchard, 2005). The result of these changes has been efforts to manage fisheries using an ecosystem approach (Pikitch et al., 2004; Link, 2010).

1.3.4 A Scientifically Managed Exploited Natural System

Fish are one of the last wild resources to be consumed globally (Cury and Cayré, 2001), and in the wild they are a communal asset. Almost always and almost everywhere, ways have been found to share marine resources (Caddy and Cochrane, 2001; Berkes et al., 2000; Dietz et al., 2003). From the beginning of the twentieth century, the management of fisheries has been based on the principle that there are levels of exploitation (maximum sustainable yield) that preserve the resource in the long term and produce the highest income (Hjort, 1926; Gordon, 1954; Beverton and Holt, 1957; Schaefer, 1957). This management principle relies on the idea of an optimal rent and emphasizes losses due to suboptimal exploitation (Barange et al., 2010). This idea will be discussed in detail in Chapter 16.

1.4 Global Governance and Modelling

There is currently a significant move towards global governance of the oceans (Costanza et al., 1998; Ostrom, 2008; Worm et al., 2009) and a key feature of current scientific research into fisheries lies in the definition of fisheries management at the international and transnational levels (Allison, 2001; Ostrom et al., 1999). Every country operates an economic exclusive zone (EEZ) and most fishing is carried out inshore. However, many fish stocks cross boundaries and many fishing fleets operate offshore in shared oceans. This raises the questions: What regulation is necessary? Possible? How can it be implemented in practice? Modeling can play a part in finding answers to these questions.

Current developments are the latest chapter in a story that is now quite old. The idea of the global management of natural resources first became popular several decades ago, when awareness of the limited nature of resources and the anticipated level of their exploitation grew. The most significant event was the publication in 1972 of the Club of Rome report (Meadows et al., 1972). This began a worldwide debate and, in some ways, was the start of political ecology. Today, this debate has changed: it is about planetary boundaries (Galaz et al., 2012), the governance of global supply chains (Boons et al., 2012; Christensen and Maclean, 2011); it is still lively. What is at stake is the potential for (or necessity of) global management tools (Ostrom et al., 1999; Hempel, 1996).

It should be noted that debates about the global management of natural processes rely on models[3]. For example in the debate that followed the publication of the Club of Rome report, the core of the argument was a model of the global economy based on system dynamics. This is still the case today in the debate on global warming. The expression "greenhouse effect" in and of itself, defines a conceptual model and a large part of the debate relies on climate models that illustrate possible futures for our environment, assuming an increase in anthropic greenhouse gases.

Any modeling attempt, by determining endogenous and exogenous variables, and identifying deterministic or non deterministic factors, implicitly consists of setting a scene. From a management perspective this has a marked effect on the definition of actors and their role. It naturally follows that modeling will play a key role in the definition of the future framework for the international governance of marine ecosystems. A few dominant models will define the future paradigms used in the global management of natural resources. There is a need to find alternatives to the system dynamics approach that formed the background to the Club of Rome's analysis. With the benefit of hindsight, this approach is now seen as overly aggregated, too mechanistic, and failing to take into account market feedback (Quirin and Sethi, 1977). Network economics represents such an alternative.

1.5 Interdisciplinarity

This book is aimed at scientists from various disciplines: ecologists, economists, and mathematicians who need an overall view of marine ecosystems and current changes in their management. In order to solve issues that are fundamentally interdisciplinary, it is important to illustrate how problems from one discipline bring about new theoretical developments in another. This implies a deeper relationship than simply the transfer of tools from one discipline to another.

[3]We recommend Breckling et al. (2011) for a discussion of the importance of models in ecological debates, both in theory and practice.

This book will show how to integrate elementary concepts from economics, ecology, and mathematics. It brings together (1) the law of supply and demand and inverse demand functions found in economics, (2) processes of trophic assimilation and somatic maintenance found in ecology, and (3) convex optimization concepts from mathematics. These extremely simplified approaches, brought together in a network framework, make it possible to integrate the various processes necessary to model an exploited natural resource. Oversimplification is sometimes the price that must be paid for effective interdisciplinary research.

Interdisciplinary research is never easy, especially when it involves mathematics[4]. It is not enough to share an intuitive explanation of the results of calculations. We have chosen to present fundamental network economics concepts in mathematical form, including definitions, propositions and proofs. The necessary mathematical background is elementary; it consists of basic linear algebra (matrices and vectors), and basic calculus (continuous functions, derivatives, compact sets). Moreover, mathematical theories are straightforward; we look at specific classes of objects that are realistic (the results can be applied to current ecological or economic questions) but where huge simplifications can be made in definitions and demonstrations. Our work can be seen as an introduction to current optimization theories that aim to obtain similar results for more general classes of mathematical objects.

[4]We are well aware of recent research suggesting that the number of mathematical equations in a paper is inversely related to the number of its citations (Fawcett and Higginson, 2012). Interdisciplinarity can be made effective when scientists from other disciplines understand how mathematicians think.

Chapter 2

An Illustrated Introduction to Networks From the Network Economics Perspective

2.1 Network Economics and Other Network Theories

Network economics is only a small part of research into networks. It deals mainly with the statics and dynamics of networks in which the main characteristics of nodes relate to stocks (capital, materials, people, and biomass) and the main characteristics of links relate to flows (exchanges, migration, and predation). As discussed in the introduction the emphasis on structure, the underlying topology, and the arrangement of entities (in terms of nodes and links) makes it a part of network theory, and not simply an extension of system dynamics.

This chapter briefly reviews how network representations are used in different scientific fields: ecology, economics, physics, social science, biology, and mathematics[1]. For each case, we describe the relationship, if any, with network economics.

[1]Several books about networks have recently been published that are very diverse in origin, perspective, and style; we may mention Wulff et al. (1989); Pascual and Dunne (2005); Boccaletti et al. (2006); Carrington et al. (2005); Mendes et al. (2003); Jackson (2010); Easley and Kleinberg (2010).

2.2 Common Definitions Concerning Networks

Basic definitions. We will use the following definitions that are common to network theories. See figure 2.1.

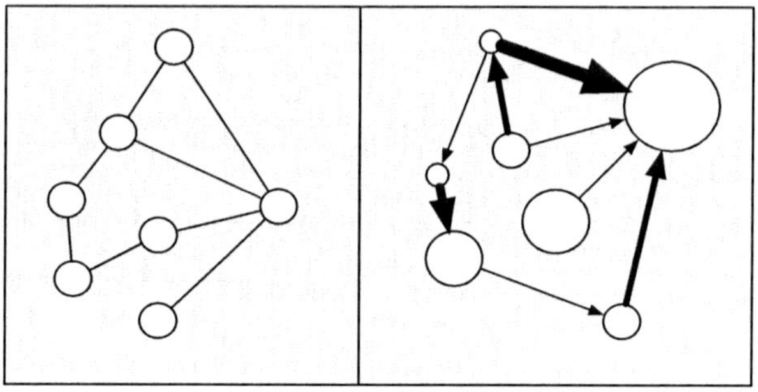

Figure 2.1: Graphical representation of networks. Left: undirected and unweighted network. Right: directed and weighted network.

Definition 2.2.1 *A network is a pair made of a finite set of nodes and a finite set of links. To a link l are associated two nodes a and b; a and b are the extremities of link l.*

Definition 2.2.2 *Two links are said adjacent if they have a common extremity.*

Definition 2.2.3 *A path between two nodes o and e is a set of adjacent links starting from o and ending in e.*

A link l may be assigned a direction or a weight; a node may be assigned a weight. Networks can be directed or weighted. In a directed network, one extremity of a link is its origin; the other is its destination.

Figure 2.2 shows some typical networks with different topologies, sizes, and degrees of connectivity.

Incidence matrix. This is a powerful tool used in the mathematical approach to networks (figure 2.3).

Definition 2.2.4 *The incidence matrix of a non oriented network is a matrix whose lines and columns correspond to nodes and terms are defined as $M_{ij} = 1$ if there exists a link between i and j and $M_{ij} = 0$ elsewhere.*

The powers M^n and the spectrum (the set of its eigenvalues) of the incidence matrix, provide useful indicators of the network structure. For example, let M^n be the matrix M multiplied n times by itself, then M^n_{ij} is the number of path of length n starting at i and ending at j.

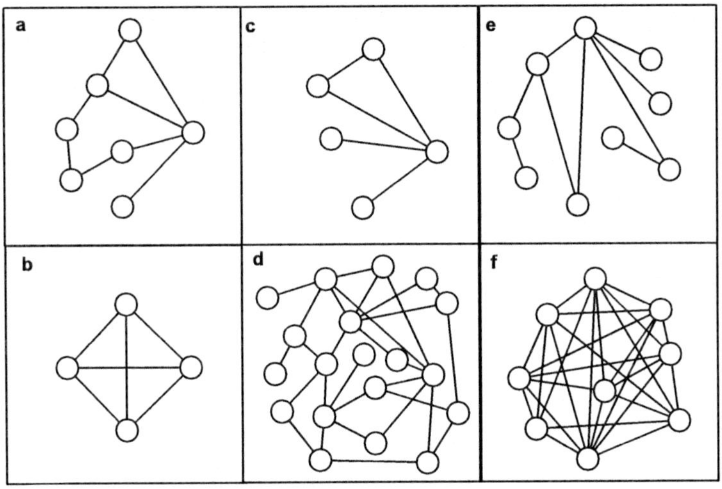

Figure 2.2: Example of networks. Left: networks with irregular (a) and regular (b) topologies. Middle: networks with low (c) and high granularity (d). Right: networks with low (e) and high connectivity (f).

	1	2	3	4	5	6	7	8
1	0	0	0	0	0	0	0	0
2	0	0	0	0	0	0	0	1
3	0	0	0	0	0	0	1	1
4	0	0	0	0	0	1	0	1
5	0	0	0	0	0	0	1	0
6	0	0	0	1	0	0	1	0
7	0	0	1	0	1	1	0	0
8	0	1	1	1	0	0	0	0

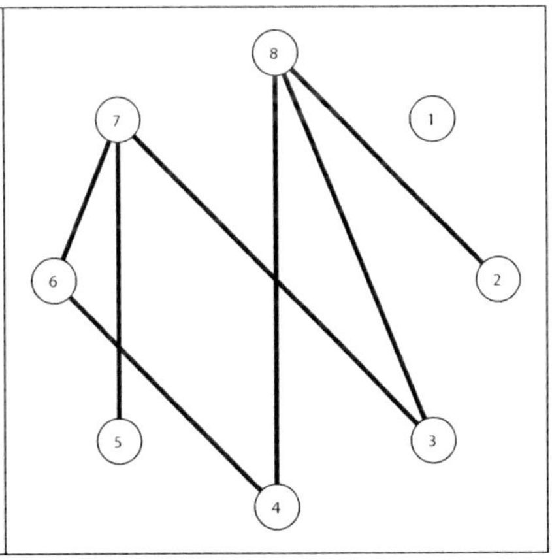

Figure 2.3: Incidence matrix. Left: the incidence matrix. Right: a picture of the related network.

Similarly, a weighted network can be represented by a matrix where the rows and columns correspond to nodes, and values correspond to the weight of the link between two nodes.

Global indices. These relate to the measurements used to characterize the structure of networks. Firstly, there are indices that characterize the whole network. Many relate to network connectivity. For example, let n be the number of nodes, l be the number of links. We may be interested in the number of links per node:

$$I = \frac{l}{n}$$

Or in the connectance of the network:

$$C = \frac{l}{n^2}$$

The Harary (efficiency) index H is also useful. Let s_{ij} be the length of the shortest path between nodes i and j; then H is defined as

$$H = \frac{1}{2}\sum_{i \neq j} \frac{1}{s_{ij}}$$

Characteristic path length is defined as:

$$L = \frac{1}{n(n-1)} \sum_{i \neq j} s_{ij}$$

There are other indices that characterize the overall structure of a weighted network[2]. Ulanowicz (1997) put forward the idea of the ascendancy index to characterize the level of complexity of ecological systems. The ascendancy index U is calculated from the weighted matrix $W = w_{ij}$:

$$U = \sum_{ij} w_{ij} \log \frac{w_{ij} w_{..}}{w_{i.} w_{.j}}$$

Where $w_{i.} = \sum_j w_{ij}$, $w_{.j} = \sum_i w_{ij}$, $w_{..} = \sum_{ij} w_{ij}$.

Fath (2004b) provides another way to approach the global analysis of weighted networks such as trophic structures. The idea is to associate to $W = w_{ij}$, the vector of external outflows $f_i = \sum_{j \neq i} w_{ij}$, the matrix of relative inflows $G = g_{ij}$ where $g_{ij} = w_{ij}/f_i$, the matrix $D = (1 - G)^{-1}$ and to analyze the signs in D in terms of the advantages of belonging to the network (diagonal elements) or being in a mutualistic or antagonistic relationship (other elements).

Local indices. Other indices characterize a particular network component, for example a node, a pair of nodes, or a link. If we define the degree of a node as the number of links starting or ending at it, then the distribution of degrees characterizes many network features (figure 2.4).

It is often useful to identify key nodes. These nodes drastically change the structure of the network if they are removed. This results in centrality indicators, which can be defined in several ways. Some are based on the structure of network paths, for example the shortest or longest path between

[2]We recommand (Latham, 2006) for a long and, however non-exhaustive list of such indices.

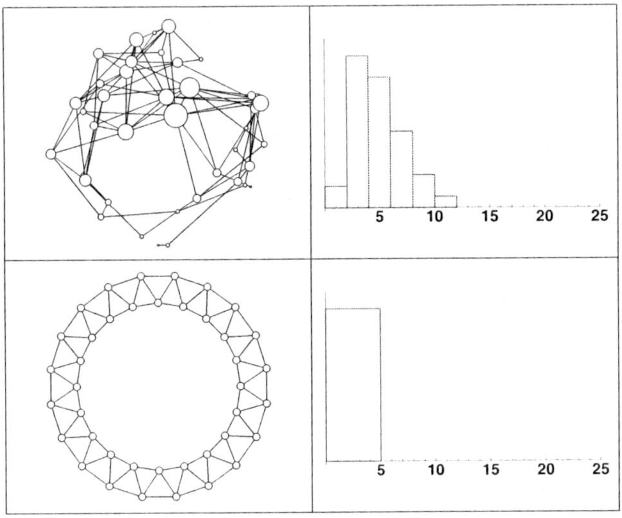

Figure 2.4: Network characteristics: degrees. Left part: a picture of the network. Right part: the distribution of related degrees. Size of nodes is related to their degree. Top: an irregular network: a few number of nodes have a high degree. Bottom: a regular network: almost all nodes have similar degree (low in that case).

two nodes, or the node with the greatest number of adjacent links without node repetition, etc.

Figure 2.5 shows a network and two measures of node centrality. The closeness of a node is the inverse of the sum of its distances to all other nodes (the mean of the length of its shortest path to all other nodes). The betweenness of a node is the ratio of the mean of the shortest paths between pairs of nodes that pass through this node to the mean of all shortest paths between pairs of nodes.

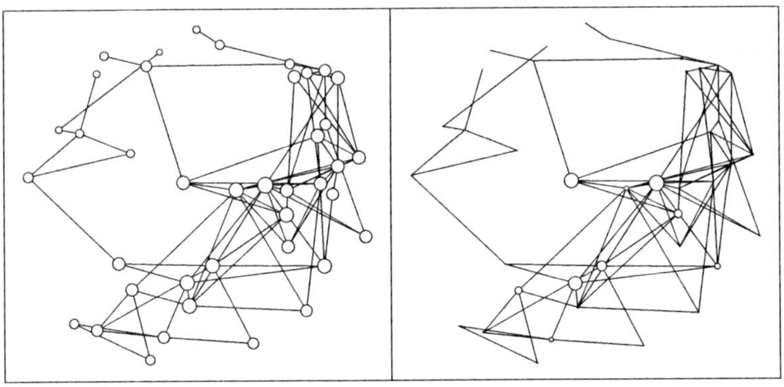

Figure 2.5: Two measures of centrality for the same network. Left: size of nodes proportional to closeness. Right: size of nodes proportional to betweenness.

In weighted networks, there are several other metrics that indicate the importance of a node or a link. Figure 2.6 plots the results of metrics indicating the importance of a link. Further information about such tools can be found in Latham (2006).

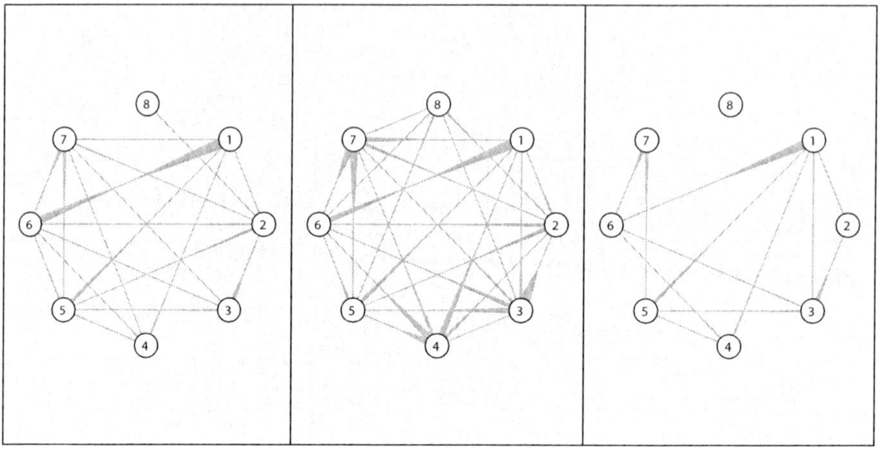

Figure 2.6: Three ways for characterizing the importance of links of the same network. Left: the initial network, in which links are represented according to their weight. Middle: links are represented with a size proportional to their importance as an intermediate link on paths between nodes. Right: links are represented according to their asymmetry.

2.3 Systems Relevant of Network Economics

2.3.1 Ecology

Ecological systems are part of the systems that will be analyzed using network economics as it is described here. Ecological studies are commonly supported research is often supported by a graphical representation of an ecological system as a network[3].

Figure 2.7 shows one of the ecological systems that will serve as the basis for the application of the network economics approach. The figure represents the trophic structure of the coastal ecosystem of Chile (Neira et al., 2004). The flow structure of this system appears to depend on plankton production (affected by climate change), the availability of food (affected by climate change and turbulence), and fishing pressure (affected by economic globalization). The interesting point here is the potential for switches between different trophic chains.

[3]See (May et al., 2006; Milo et al., 2002; Montoya et al., 2006).

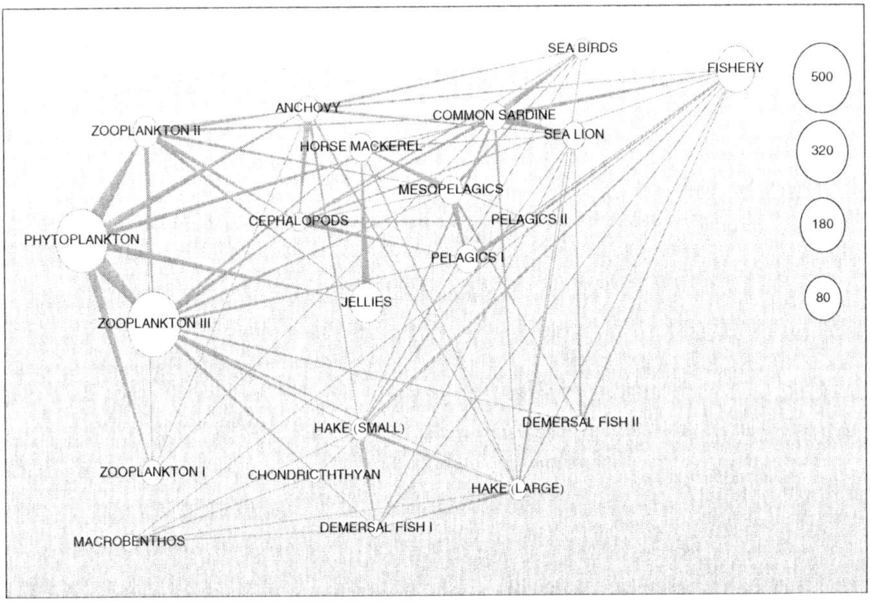

Figure 2.7: The trophic network of the coastal Chilean ecosystem. Circles represent species; their size is proportional to abundance. Arrows represent grazing or predation; thickness is proportional to flow. The horizontal axis corresponds to trophic level; low-level species (phytoplankton) are shown on the left, high-level species (birds, mammals) appear on the right. The vertical axis corresponds to the distribution depths of species. Details about species groups are given in chapter 13.

Ecological theory conventionally distinguishes between two types of network representation: functional and structural (as food web).

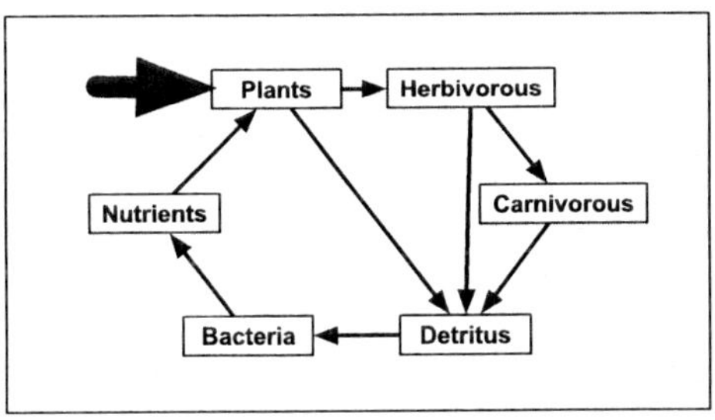

Figure 2.8: The functional point of view on an ecological network. A cycling ecological system. Nodes correspond to functional units. Arrows correspond to energy exchanges between functional units. The large arrow corresponds to the inflow of solar energy.

Figure 2.8 is a functional representation of an ecological system. It focuses on (a) functional units (plants, herbivorous and carnivorous animals, detritus, and nutrients), and (b) energy exchanges between these functional units. This way of representing an ecological system has existed throughout the history of ecological theory, beginning with Lindeman and the introduction of solar energy into a food web (Lindeman, 1942).

A food web (Pimm, 1979) represents the trophic interactions (predation, grazing) between species (see figure 2.9). This second approach has been another constant in ecology, beginning with the pioneering analysis of the food web at Spitsbergen and Bear Island (Summerhayes and Elton, 1923) and continuing with recent work (Allesina et al., 2008; Camacho and Arenas, 2005; Pascual and Dunne, 2006). Much of this research is concerned with the question of biological diversity: identifying keystones (Power et al., 1996; Libralato et al., 2006), bottlenecks (Allesina and Bodini, 2004), or the risk of extinction (Allesina et al., 2006). One important feature is the systematic use of network theory tools to analyze the structure and functioning of ecosystems (Jordan and Scheuring, 2004; Jordan et al., 2005; Krause et al., 2003), or the relationship between fragility and complexity in a food web (Montoya et al., 2003, 2006).

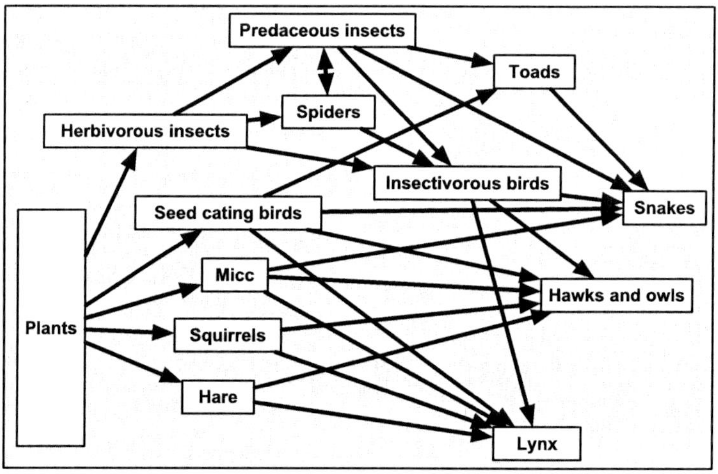

Figure 2.9: The trophic point of view on an ecological network. Network representation of a trophic system. From Daniel Stouffer. Nodes correspond to species. Links correspond to biomass transfers. Emphasis is on the structure of the network.

The functional and structural approaches are clearly different and have led to disputes between their followers (McGill et al., 2006). A network approach can provide formalisms than can reconcile both approaches. An important step in this direction was made by Odum and Odum (2000) who proposed a formalism to represent energetic exchanges in an ecosystem. His work was related to system theory and principles. It made it possible to represent, in a shared framework, flows of both matter and information and it became very popular in the period 1970–1990. It had a particular impact on the debate about the nature of flows (Jorgensen et al., 1995; Suh, 2005; Jorgensen, 2007). The value of the network approach to ecological problems has been also underlined in studies of the resilience of ecosystems (Janssen et al., 2006) and study population dynamics (May et al., 2006).

2.3.2 Economics

Economic systems can also be studied using network economics. The representation of an economic system as a network appeared quite early in economics. In the eighteenth century, Francois Quesnay made one of the first attempts by representing the cycle of exchanges in a simplified economic system (figure 2.10). It should be noted that Quesnay was a physician, and that the economic system he developed was almost certainly conceived through analogy with a living organism.

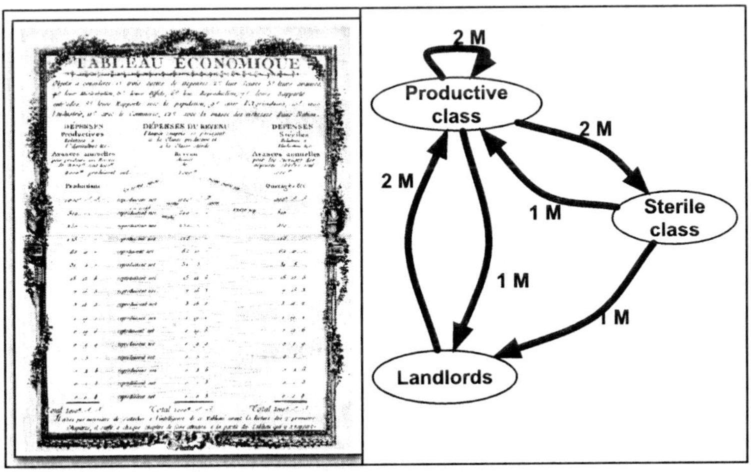

Figure 2.10: Le Tableau economique. Left: the economic cycle in Francois Quesnay's booklet. Right: a network representation. Let us pay attention to the fact that the left part is clearly dynamical while the right part is static.

It has since become common to find network representations of economic systems (Goyal, 2007). For example, figure 2.11 represents financial flows in a national economy. Using the analogy with ecology, this could be said to be a functional approach.

A relevant question is whether there are structural representations of economic systems. An input-output analysis attempts to represent for

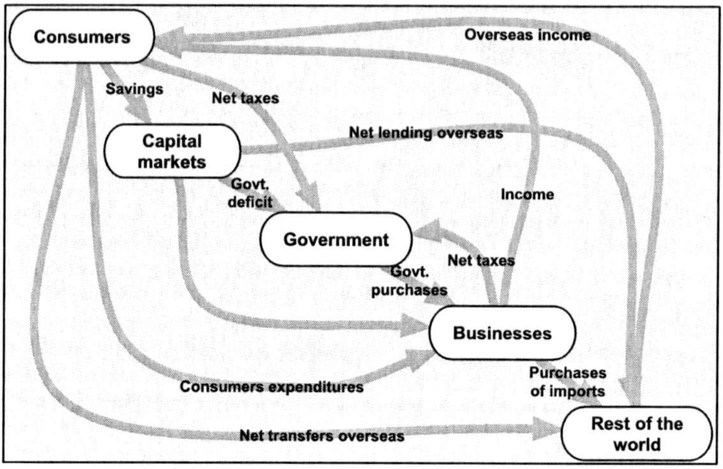

Figure 2.11: A functional point of view on an economic network. Financial flows of national economic system. Nodes are functional units. Such figures can be found in many macroeconomics books.

example, the economic system of a country as an exchange matrix between sectors (Leontief, 1951a). It is straightforward to associate a directed weighted network (figure 2.12) with an input-output matrix, and consequently to relate it to network representations (Abel, 2004; Andrasfai, 1989). This can be seen as a strcutural representation of an economic system. In this case, the focus is on the numerical linear relationships expressed by the matrices. However, unlike ecologists, economists do not usually use network representation of the structural features of an economic system (a notable exception being Brown (2004)). The unwillingness of economists to use a graphical representation of economic networks can be related to the arbitrariness of 2-D representations of general economic systems that are not spatial (although the same could be said for food webs). The structural analysis of economic networks may be considered as a recent research field[4].

Figure 2.13 plots the structure of the trade system that connects the production of small pelagic fish (such as anchovies and sardines) and the markets for fishmeal and fish oil (mainly used for aquaculture). This is a global market and is an example of the kind of system we will analyze with network economics tools.

2.3.3 Interplay between Economics and Ecology

Underlying analogies between the network structure of ecological and economic systems has been a constant feature of the interplay between

[4]See Serrano and Boguná (2003); Wilhite (2006); Blöchl et al. (2010).

	agri	elec	gaz	water	oil	food	transp	advert	CONSO	EXP	TOTAL
agri	2300	0	0	0	30	12000	0	0	7000	1062	22392
elec	500	5000	30	100	700	400	70	120	6000	457	13377
gaz	0	3000	0	0	350	176	110	0	1200	480	5316
water	300	80	0	0	350	200	110	0	2300	284	3624
oil	1400	10	0	400	11000	450	10	0	13000	1241	27511
food	6000	0	0	0	100	11000	0	0	20000	1084	38184
transp	200	200	0	0	1300	2000	500	30	7000	1040	12270
advert	2	60	0	0	30	30	30	2500	3000	300	5952
IMP	11690	5027	5286	3124	13651	11928	11440	3302			
TOTAL	22392	13377	5316	3624	27511	38184	12270	5952			

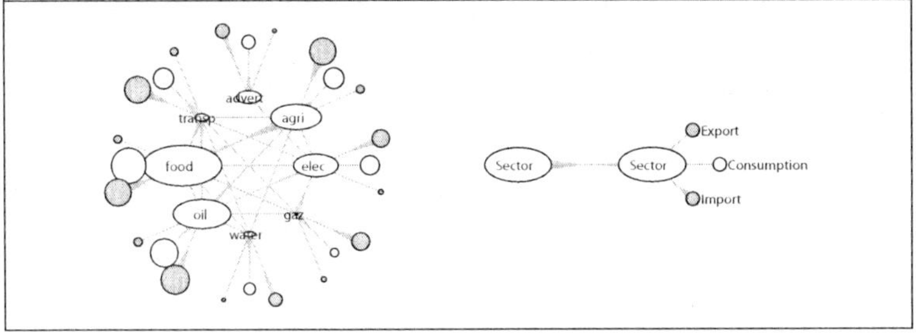

Figure 2.12: Input and output matrix. Top: an exchange matrix where values represent the economic flow from the corresponding row to the corresponding column. Bottom: the network representation, which highlights the relationships between sectors. This is a structural view of an economic system.

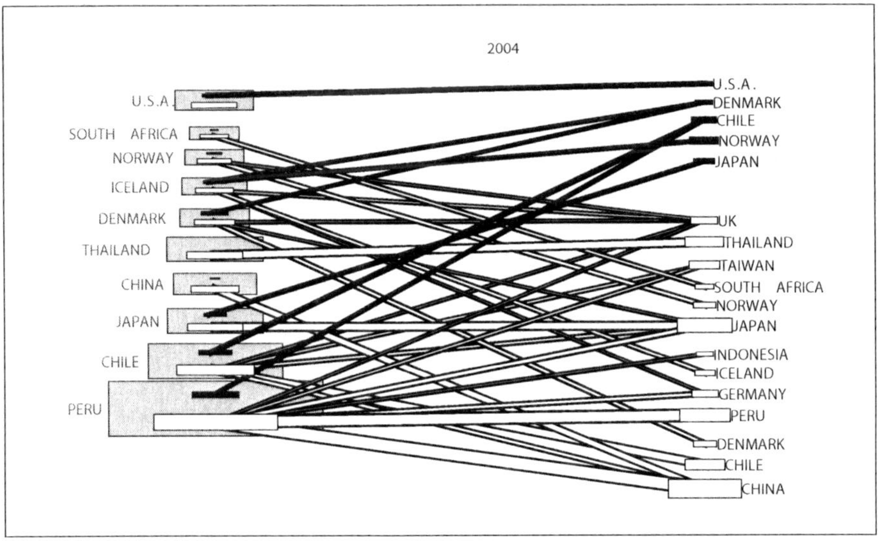

Figure 2.13: The global network linking producers and consumers of fishmeal and fish oil, Left: producing countries; gray rectangles correspond to catches, white rectangles to fishmeal production and gray rectangles fish oil production. Right: consuming countries of fishmeal (bottom) and fish oil (top). Leftright arrows correspond to trade flows. The size of rectangles and the thickness of arrows are proportional to quantities.

economics and ecology. This interplay has been reviewed by Hannon (1973) and has led to new fields of research based on theories of material flows (Suh, 2005), and industrial ecology (Ayres et al., 1996). For example, the study of material and energy flows through industrial systems using principles from ecology provides a wide range of assessment tools.

2.3.4 Electricity

Network economics can be used to analyze electricity networks (Neuhoff et al., 2005). The representation of electric systems in terms of circuits and networks is credited to Gustav Kirchhoff in 1845. This approach makes it possible to formulate universal laws of electricity known as Kirchhoff's laws[5] and allows exhaustive computations for all kinds of electric circuits at all scales. It is important in the context of this book to note that network formalism is currently an efficient tool for analyzing the electricity market (Neuhoff et al., 2005), and tackling the issue of pricing (Bunn, 2004).

We briefly develop the analogy between electric circuits and ecological theories. In a simple electric network such as the one represented in figure 2.14, currents are completely determined by Kirchoff's laws: on nodes, we have the conservation equation: $i_0 = i_1 + i_2$, $i_1 + i_2 = i_3$; between closed loops, we have the complementarity equation: $r_1 i_1 = r_2 i_2$. Solutions of these equations, knowing i_0, r_1, r_2, are:

$$i_1 = \frac{r_2}{r_1 + r_2} i_0$$

$$i_2 = \frac{r_1}{r_1 + r_2} i_0$$

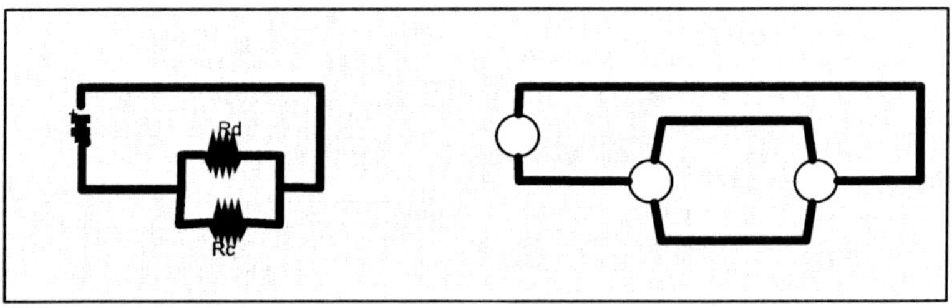

Figure 2.14: Electric networks. A very simple electric circuit (left) and the associated network (right). Nodes correspond to connections; links to connected components.

From these equations, we get the well-known expression of the global resistance: $\frac{1}{r} = \frac{1}{r_1} + \frac{1}{r_2}$. It is straightforward to find an ecological interpretation of this equation; ecological meaning here density dependent.

[5]See for example Chen and Satyanarayana (2007); Gonzalez (2007).

Consider the case where a dispatcher decides where to send i_0 electrons with minimal transportation costs, where these costs are density dependent. It sends i_1 electrons along path 1, i_2 along path 2, and the resulting unitary costs are $r_1 i_1$ and $r_2 i_2$. In this case equilibrium occurs when transportation costs are equal on all paths: that is $r_1 i_1 = r_2 i_2$. Another equivalent possibility is to consider that every electron selects a path depending on transportation costs. The electron then changes its path if the chosen path is more expensive. This is a conventional economic reasoning, starting with an equilibrium principle and ending with a marginal value theorem. This demonstrates the value of mixing disciplinary approaches. Next, we look at the incorporation of reasoning from physics into ecology. For example, algebraic equations can be used to express conservation relationships, taking into account cost considerations. These equations can be solved and an equilibrium state can be determined. This approach avoids considering the "goals" of ecological entities.

2.3.5 Structural Mechanics

Network economics is used in structural mechanics (Christensen, 2008). Let us examine the chain bridge shown in figure 2.15. A state of the system is a set of values $X = (x_a, x_b, \ldots)$ characterizing the position of its elements. The

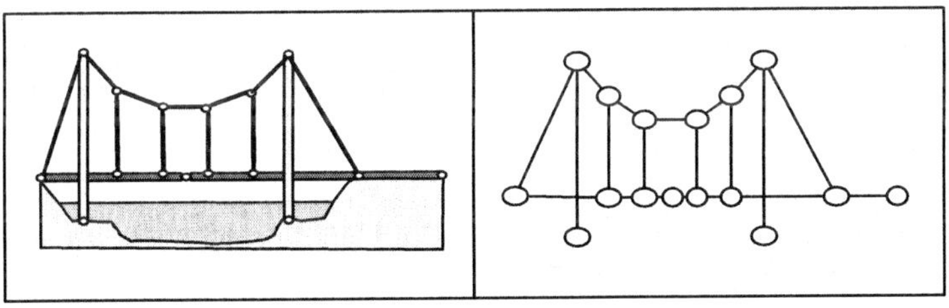

Figure 2.15: A network representation of a chain bridge. Left: a solid structure. Right: the network representation of this structure. Nodes correspond to elements. Links correspond to contact point.

balance of forces in a contact point i, $e_i(X)$ can be computed. The distance between a contact point to the soil $d_i(X)$ can be computed as well. Suppose that X corresponds to an equilibrium configuration of the structure. At a point internal i to the structure, distance to the soil is positive $d_i(X) > 0$ and the balance of forces is zero $e_i(X) = 0$. At a contact point with the soil $d_i(X) = 0$, the balance of forces is positive $e_i(X) > 0$: it is the opposite of the soil's reaction force. Altogether we have a system of equations and inequalities:

$$\begin{aligned} e_i(X) &\geq 0 \\ d_i(X) &\geq 0 \\ e_i(X) = 0 \quad &\text{or} \quad d_i(X) = 0 \end{aligned}$$

This kind of alternatives is called a complementarity equation. Solving it, when it is possible, gives the equilibrium state of the structure.

2.3.6 Transportation

Network economics can be used to analyze transportation systems (Dafermos, 1980a). A road network (figure 2.16) is made of a set of crossings (nodes) and roads (links). There is a cost to taking a road corresponding to tolls, distance, and time. Costs particularly depend on the volume of traffic. The cost of an itinerary is the sum of the costs of the intermediary roads. The seminal work on traffic analysis was carried out by J. G. Wardrop (1952) and concerns the statics of a traffic network. According to Wardrop's principle, drivers determine the equilibrium of the network by concurrently optimizing their travel costs. An economic approach to a road network consists in considering both the conservation equation and the complementarity principle. The conservation equation concerns crossroads; it says that for all crossroads, inflow equals outflow. The complementarity principle concerns itineraries: if there are two itineraries that end at the same point, they will either have same costs and will both be used, or they will have different costs and only the less expensive is used.

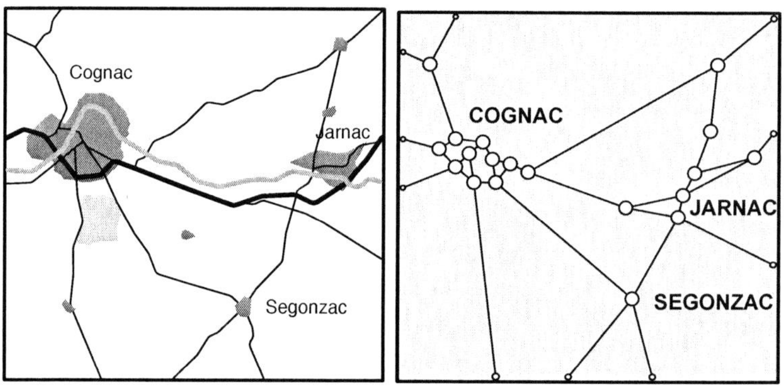

Figure 2.16: A road network. A road map and the associated network structure. Nodes correspond to crossings. Links to roads between crossings.

To illustrate the originality of the network economics approach to traffic studies, we take the example of the Braess's paradox (Braess, 1968), which shows how increasing the number of roads in a network may result in increasing costs for all users.

In the Braess's network (figure 2.17), there are 6 vehicles that must travel from left node A to right node D. Initially, paths are two small busy roads AB, CD and two large roads with a toll AC, BD. If traffic loads are denoted:

q_{AB}, q_{AC}, q_{BD}, q_{CD}, then travelling costs of vehicles are given by:

$$c_{AB}(q_{AB}) = 10\,q_{AB}$$
$$c_{AC}(q_{AC}) = 50 + q_{AC}$$
$$c_{BD}(q_{BD}) = 50 + q_{BD}$$
$$c_{CD}(q_{CD}) = 10\,q_{CD}$$

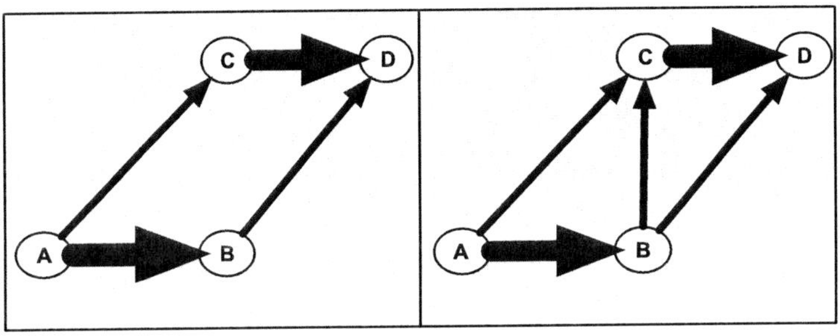

Figure 2.17: The Braess' paradox. Left: the initial road network. Right: the same with a shortcut e between B and C.

There are two itineraries: ABD, ACD. At, equilibrium, there is the same number of vehicles (3) on each itinerary: $X^*_{ABD} = X^*_{ACD} = 3$, $q^*_{AB} = q^*_{AC} = q^*_{BD} = q^*_{CD} = 3$, $C^*_{ABD} = C^*_{ACD} = 83$.

Now, we add a shortcut, a path BC, its traffic load q_{BC}; the travelling costs for a vehicle on this path are given by:

$$c_{BC}(q) = 10 + q_{BC}$$

There are now three itineraries: $ABD, ACD, ABCD$; at equilibrium, we have: $X^*_{ABD} = X^*_{ACD} = X^*_{ABCD} = 2$; $q^*_{AB} = q^*_{CD} = 4, q^*_{AC} = q^*_{BD} = q^*_{BC} = 2$; $C^*_{ABD} = C^*_{ACD} = C^*_{ABCD} = 92$. Compared to previous situation, travelling costs are greater for all vehicles.

The importance of the Braess's paradox must not be underestimated. The fact that the efficiency of a network is not related to its connectivity has been, and is still a real surprise.

2.4 Systems Not Relevant of Network Economics

2.4.1 Social Science

There is a long tradition of using network approaches in social sciences[6]. A frequently analyzed social network concerns the structure of the Florentine

[6] About social networks, we recommend Cook and Whitmeyer, 1992); Freeman (2004); Jackson (2010); Borgatti et al. (2009); Borgatti and Halgin (2011).

society at Renaissance (Padgett and McLean, 2006); it is shown in figure 2.18. The most important field of researches in terms of social networks concerns the development of Internet and of the World Wide Web (Barabsi, 2003; Watts, 2004a).

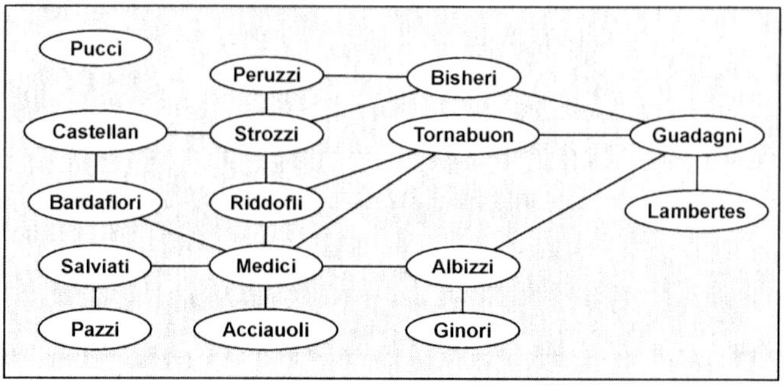

Figure 2.18: The networks of Florentine families. Nodes correspond to families. Links correspond to weddings. Does the central position of the Medici's family explain its historical importance?

An interesting feature of social network theory concerns the weak ties hypothesis (Granovetter, 1973; Borgatti and Halgin, 2011): Does the strength of a social system rely on strong (e.g. hierarchical or familial) ties, or on weak ties such as sharing a common means of transportation? If there are strong ties between A and B, and A and C, what is the nature of the tie between B and C?

An important ongoing question is about small worlds (Watts and Strogatz, 1998; Watts, 2004b; Tang et al., 2010). As figure 2.19 shows, in a network in which there is a small number of highly connected nodes (i.e., they have a high degree) the shortest paths are very short! The most famous example of this is the Erdos's number, which is the length of the shortest path linking a person to Erdos himself as measured by the authorship of mathematical papers. In this network, nodes correspond to mathematicians and there is a link between two mathematicians if they have co-authored a paper. The Erdos's number of a mathematician is the length of the shortest path between Erdos and the mathematician. It appears that, as Erdos has co-authored many papers (many more than other mathematicians) the Erdos's number of most mathematicians is less than five. Of course, this is the distribution of shortest paths that concentrates the interest in small world theories.

To our knowledge, there are no examples of the application of network economics to social networks. We will however note the idea of network externality (the price given to what is gained or lost by being a member of a network) that may provide a link between social networks and network economics (Liebowitz and Margolis, 1994). At the same time, this is our

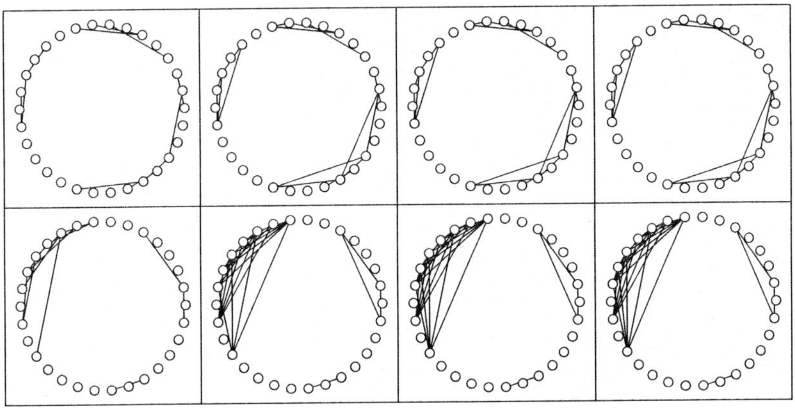

Figure 2.19: Small worlds. Moving from left to right this shows the existence of a path (link) between two nodes with length less than 1, 2, 4, or 8. Top: a network with only short links. Bottom: a network with a small number of long-distance links. When there are a small number of long-distance paths of length 1 (bottom), almost all nodes are connected with a path of length less than 8.

opinion that social network theories should provide tools and ideas that can be useful in network economics.

2.4.2 Biology

There is ongoing, intense research[7] with into metabolic networks (nodes correspond to metabolic and physical processes that determine the properties of a cell; links correspond to the interactions between processes), and into genetic regulatory networks (nodes correspond to the transcription activity, for example the mRNA concentration, of a gene in the network, links correspond to interactions between genes). Figure 2.20 shows a very simple regulatory network (most regulatory processes are represented by networks that are 10–100 times more complicated). To the best of our knowledge, there are no current attempts to model this kind of system using network economics tools.[8]

2.4.3 Mathematics

What we here call a network is commonly called a graph in mathematics[9]. The structure of mathematical graphs is very general and they are used to solve problems in all branches of mathematics.

[7] See for example Junker et al. (2008).

[8] See however Julius et al. (2009).

[9] For interested readers we recommend graph theory textbooks such as those by Harary and Norman (1953); Bollobs (1998); Gross and Yellen (2004).

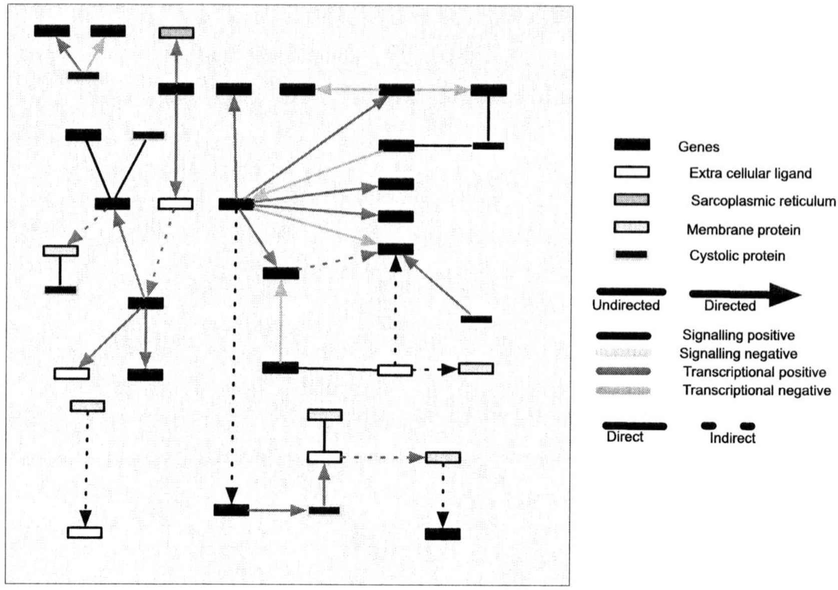

Figure 2.20: A regulatory network. Nodes correspond to types of molecules. Links correspond to an interaction between molecules. Colour and shapes are related to the kind of molecule, to the kind of interaction.

The following anecdote is illustrative and can be considered as the first part of a long story. The mathematician Leonhard Euler initiated the mathematical theory of graphs. He formulated and studied the Koenisberg Bridges problem[10]. In a city with a river, several islands, and several bridges (figure 2.21) does a path exist that makes it possible, starting from one point and arriving at another, to cross all the bridges, once and only once? The enumeration of all possible itineraries proved to be a gigantic task. Eulers methods provided an elegant solution to the problem, whose answer turns out to be no.

Following Euler, the approach was extended and has resulted in solutions to many interesting and difficult problems. For example, for any given network, it is usually interesting to find: (1) the shortest path between two nodes, (2) the best way to subdivide the set of nodes, such that e.g., the number of internal links is high (graph cutting), and (3) to characterize graphs that can be drawn on the 2-D plane without link intersections (planar graphs). These problems appear to be both very general, and very difficult to solve. However, they have applications in many areas of mathematics. Similarly,

[10]See Gribkovskaia et al. (2007).

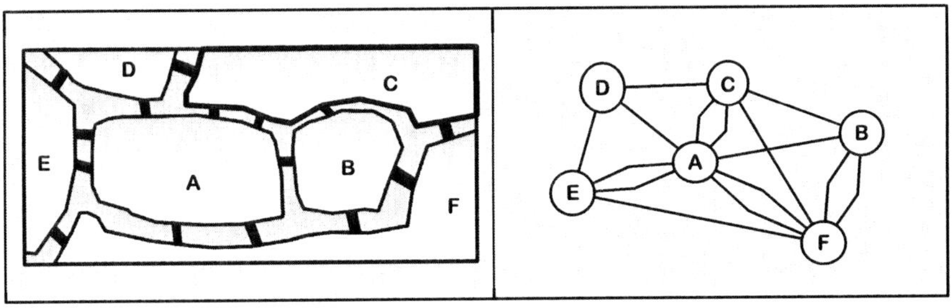

Figure 2.21: The Koenisberg bridges problem. Left: A map of Knigsberg showing the river, bridges and islands. Right: the associated network. Nodes correspond to land and links correspond to bridges.

many combinatory problems are naturally expressed in terms of trees and networks.

There is no known application of network economics to mathematical graphs. However, it is obvious that the mathematical theory of graphs provides tools and ideas that can be useful for network economics.

2.4.4 Statistical Physics

Statistical physics of networks[11] focuses on networks with a very large number of nodes and links (from systems with millions of entities to others of Avogadro's number magnitude). It originates in the following question addressed by Hungarian mathematicians Erdos and Rényi (1960): Assuming a network in which links are randomly distributed between nodes, what can be said about the length of paths? The statistical physics of networks makes it possible to represent systems with a huge number of components, to define phase transition mechanisms, and classes of universality between systems. For example there is a link between the statistical physics of networks and fractal theory, and it has become usual to identify power laws ($N \simeq P^{-\alpha}$) in the description of network structure and dynamics. These distributions are characteristic of fractal systems, i.e., systems with some "auto-similarity": they look like the same at different scales. This makes it possible to identify phase transitions, i.e., states typified by discontinuity. For example, the calculation of a variable X as a function of, for example temperature x, can be written $X(x) \simeq \|x - x_0\|^{-\beta}$; the value of β is related to the phase transition and the topological structure of the network. This makes it possible to reason in terms of universality classes (Barabási et al., 1999).

[11] We recommend Park and Newman (2004); Boccaletti et al. (2006).

The statistical physics of networks represents an important branch of network theory that has little in common with network economics in its current state. It is too early to speculate about how such a highly theoretical approach could be used in conjunction with network economics. However, the questions it addresses are likely to become, in the near future, the subject of joint research.

2.4.5 Evolving Networks

Statistical physics is concerned with how evolve networks[12]. This question is becoming a key question of network theory. A first type of changes concerns the values or weights characterizing links or nodes, not the structure of the network. Then, it appears that quite similar models apply to the propagation of an epidemic, an opinion and to the phase transition in ferromagnetism and disordered matter (see figure 2.22). The epidemic model represents a network in which nodes are individuals and links correspond to contact between people. At the beginning of the epidemic, only one node is affected. At subsequent times, a node is affected if one of its neighbors was affected the previous time. In the ferromagnetism model (such as the Ising model) nodes correspond to atoms and links correspond to interactions. The topology of the network can be regular (Ising, figure 2.22, left) or random (spin glass). Any node has a state: ± 1 corresponding to the value of the spin. States are initially random. At each subsequent time, (1) the state of a node can randomly change according to the temperature, and (2) a node adopts the most prevalent state present in its neighborhood. This leads to an illustrative model of phase transitions in ferromagnetism.

Rewiring, or growing networks processes theories[13] focus on how are created and destructed the links and the nodes of a network (figure 2.23). Rewiring networks have important applications in biology, analysis of Internet (Pennock et al., 2002) and in social sciences (Ahmed and Xing, 2009). The theory of growing networks, for example of the "diffusion limited aggregation processes" (Halsey, 2000) has applications in the physics of crystals, soils (Sander, 2000) and in the biology of development, with an emphasis on resulting fractal structures (Vicsek, 1992) and on discontinuous dynamics.

---

[12]See Mendes et al. (2003); Dorogovtsev and Mendes (2002); Evans and Plato (2008).
[13]See Bornholdt (2005); Ihmels et al. (2005).

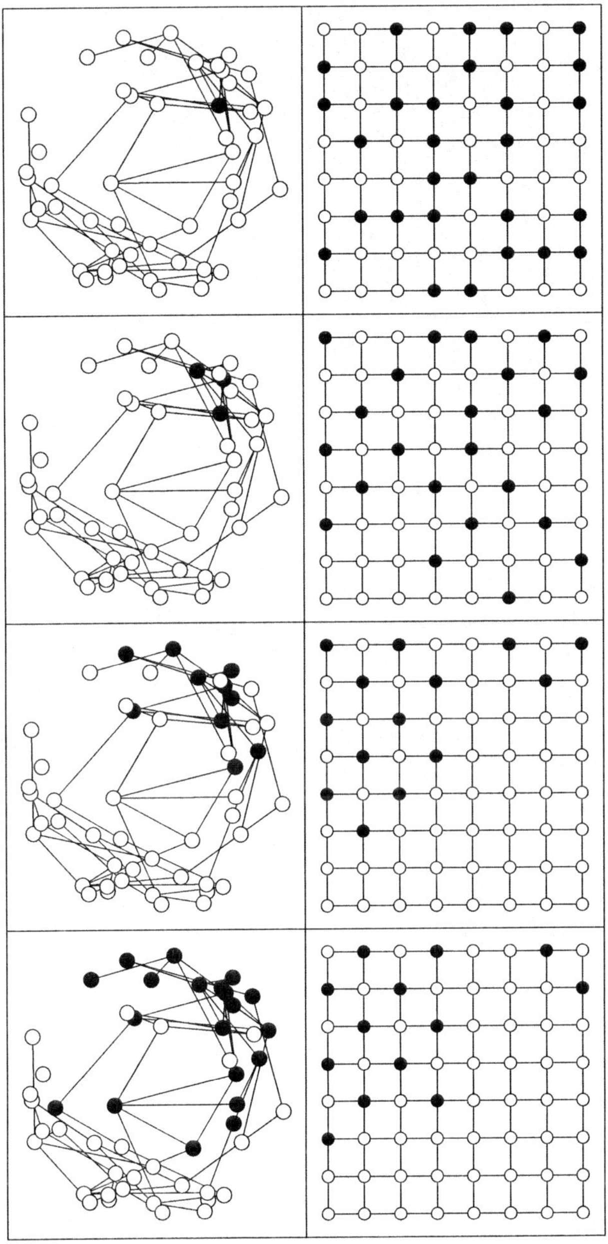

Figure 2.22: Propagation and synchronization on a network at several times steps. Left: propagation of an epidemic through a network; at step 1 (top), a node is infected; then, at following steps, it infects its neighbors. Right: the synchronization of an Ising network; at step 1, a random configuration, then in following steps (2, 5, 10) spins change their orientation accorind to the orientation of their neighbours and to a random effect (related to temperature).

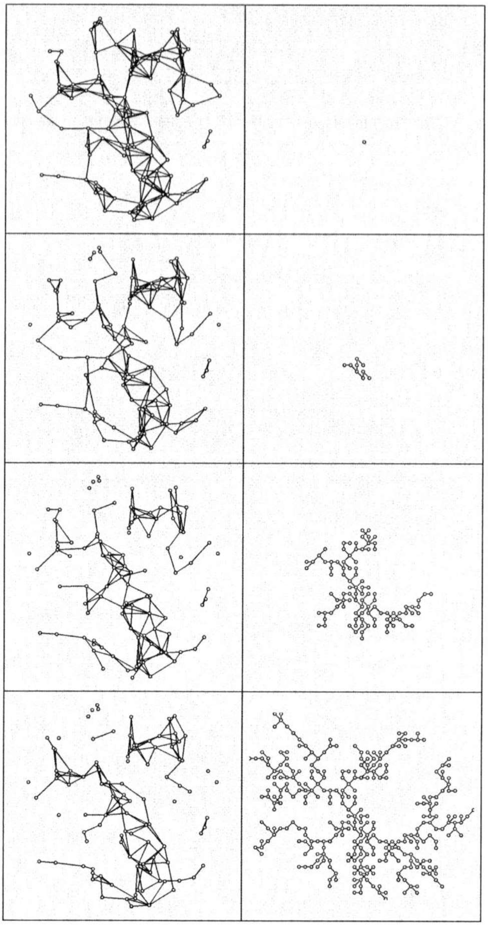

Figure 2.23: Rewiring and growing networks. Left: rewiring. At each time t, a new link is added or removed according to the following procedure: an existing link is destroyed with probability pa and an unconnected pair of nodes is created with a probability pb. Right: diffusion limited aggregation showing steps 1, 51, 151, and 201. At each time step a particle enters the system, and randomly moves. When it moves close to an existing particle, a link is created with this particle and it stops moving.

2.5 About the Diagrammatic Representation of Networks

One of the main features of network approaches lies in the potential to create diagrammatic representations. Networks approaches are part of the techniques used for the visual display of quantitative information[14].

[14]Tufte and Howard (1983) is a major reference; Ragan (2009) provides a very rich illustration of this issue in ecology; see also (Brandes et al., 2006; Brown, 2004; Eco, 2009).

Some types of networks are naturally spatial (in 2-D or 3-D) and their graphical representation is constrained by their structure. This is the case for traffic or electricity networks. For other kinds of network, there is no underlying spatial structure. This is the case in functional ecology where there are no geographical characteristics proper to a system component. In this case, a graphical representation may be misleading. For non-spatial networks, the graphical representation is arbitrary and there is a need to identify networks independently of their 2-D representation (figure 2.24).

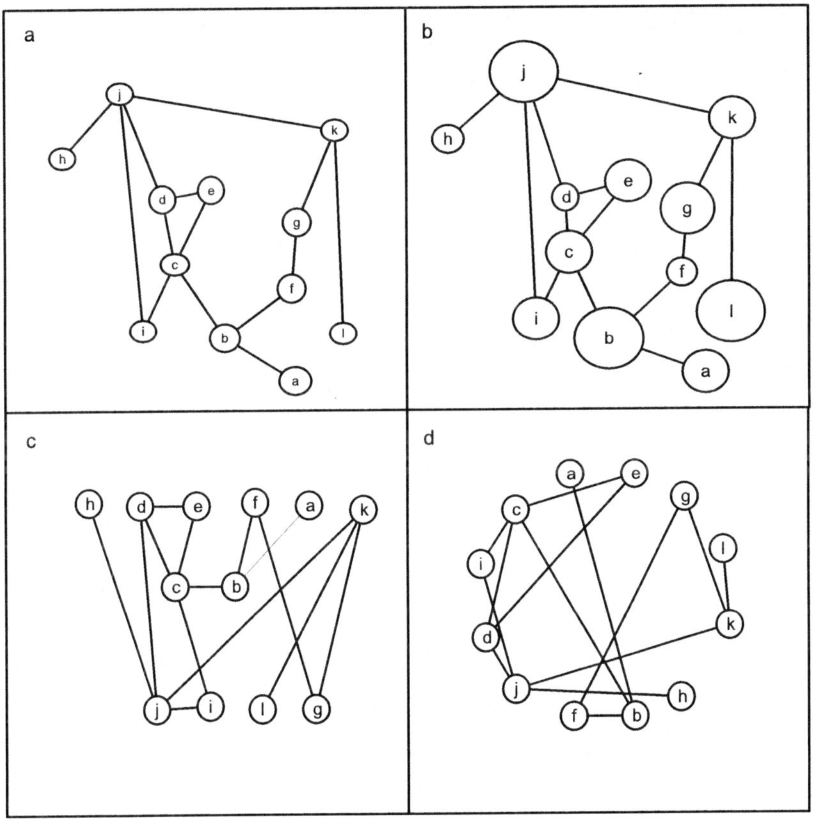

Figure 2.24: Four graphical representations of the same network. Top: he network with different representations of the size of nodes. Bottom: left, a multi-layer representation. Right: a representation of nodes on circles.

A desire for an aesthetically pleasing graphical representation can lead to misunderstandings, or even changes in ways of thinking[15]. We may illustrate what is at stake in the choice of a graphical representation with the history of the representation of living organisms. The point can be illustrated by

[15]One such example is the Power Point cognitive style (Tufte, 2003).

the evolution of the representation of living organisms. In the Middle Ages, the living world was represented by lists and network trees (Ragan, 2009). Figure 2.25 shows the arbitrariness of networks or trees, which reflect a desire to show the world in a particular way. It has been common to express natural laws and to justify their necessity with such representations. It reflects an underlying Platonist position that consists in relating an idea to a form. This point should be kept in mind when developing network approaches.

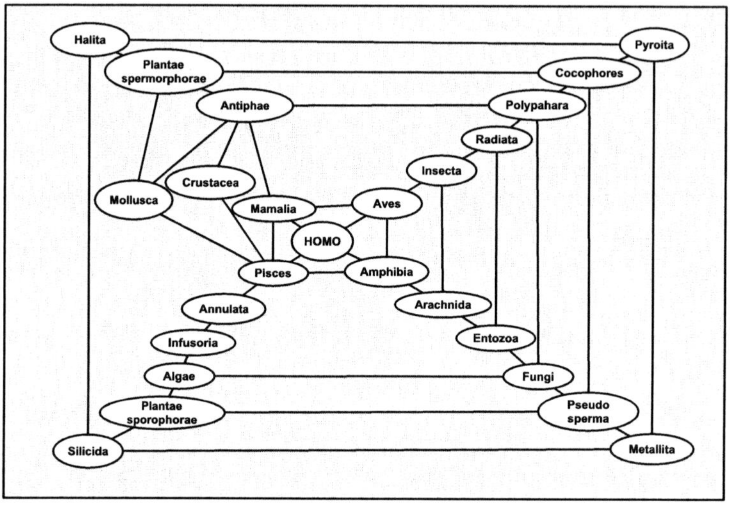

Figure 2.25: Homo at centre. Universal system of nature, from the Primae lineae systematis naturae of Paulus Horaninow (1834). From (Ragan, 2009).

Harcourt (2011) analyzed the destiny of the Quesnay's economic table described earlier. This has puzzled (and it still does) economists (including Karl Marx) who have spent many hours trying to understand it, and provide explanations and mathematical formulations. Some economists[16] have confessed that, despite their effort, the underlying theory was still not clear to them, that the economic flows shown in Quesnays work might be meaningless and that the model only works because it creates an immediate feeling of understanding. For example Harcourt (2011), says: *But what matters is not whether the representation of the system worked; what matters is that a visual representation of the system was created that caught the eye and imagination of a great many of Quesnay's contemporaries.*

[16]See Fox-Genovese and Fox (1976).

Chapter 3

Network Economics, Statics and Dynamics

3.1 Statics and Dynamics of Networks

Network economics can illustrate questions of equilibrium and dynamics in ecology and economics using an example from a totally different field such as structural mechanics. In this example we use an articulated puppet (shown in figure 3.1) to highlight the underlying network structure.

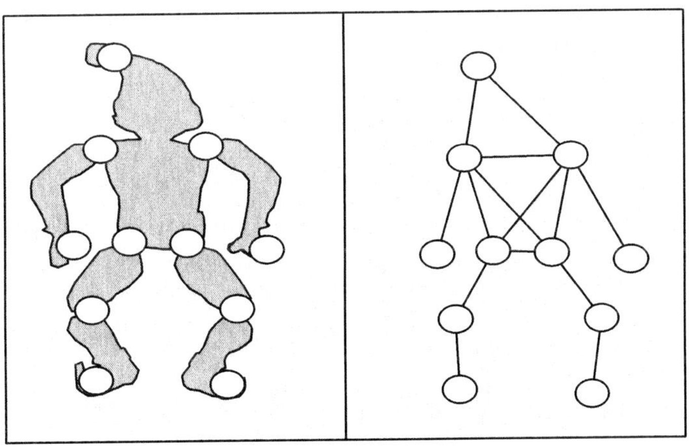

Figure 3.1: An articulated puppet. The underlying network structure of the puppet

3.1.1 Equilibrium Analysis

Suppose that the puppet is unbreakable. We throw it into the air and observe its stationary state when it hits the floor (figure 3.2). As the earlier example of the chain bridge showed, where the puppet is in contact with the floor there is no balance between internal forces, while at other points the internal forces are balanced. Expressing these complementarity relationships makes it possible to compute the equilibrium states of the puppet. As figure 3.2 suggests, there can be many equilibrium states.

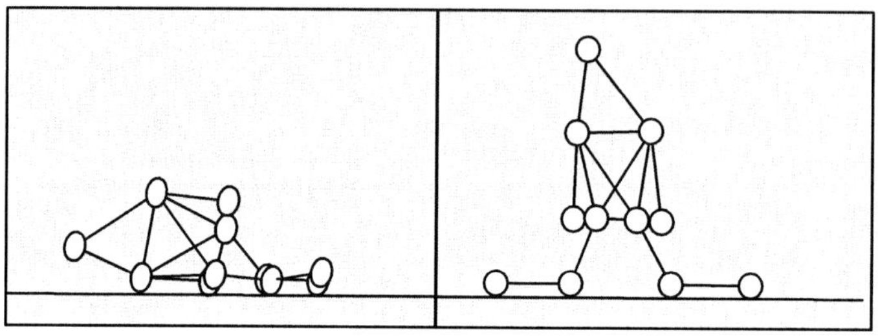

Figure 3.2: Multiple equilibrium states. Different equilibrium states of the puppet lying on the floor.

3.1.2 Dynamical Analysis

Now, suppose that we nail the puppet to a pole outside on a windy day (figure 3.3). Several situations may occur.

(1) If the wind is constant the puppet, once it has stabilized, reaches an equilibrium state. This tends to be the desired state, regardless of wind intensity. Unfortunately it is easy to imagine that in most situations there is no stable equilibrium or indeed any equilibrium at all.

(2) The direction and strength of the wind changes gently. The problem arises of how to relate the movement of the puppet to changes in the wind? This is a problem of smooth dynamics.

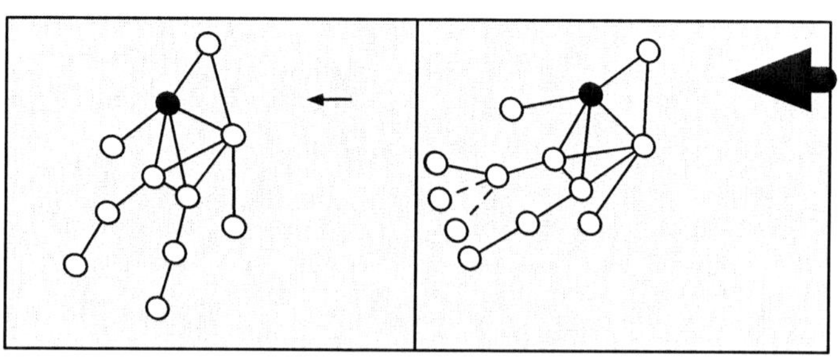

Figure 3.3: Different dynamical regimes. Different states of the puppet under the wind, once nailed to a pole.

(3) The wind is gusty. The same problem arises. This situation usually requires a qualitative description of the dynamics. There is friction in the articulation of the puppet. In this case it is easier to reach a state of equilibrium, although there is a temptation to resort to the friction hypothesis.

(4) Over time the articulation of the puppet changes. As the puppets joints age, local movements become unpredictable.

3.2 Statics and Dynamics of Marine Systems Under the Double Exposure

Marine ecosystems are under pressure from both climate change and economic globalization, and a network representation of this double exposure was shown in figure 1.2. An analogy can be made with the example of the puppet described in the previous section. Both networks have structural constraints that can be expressed as complementarity and conservation equations. The puppet example offers two ways to approach the analysis.

The equilibrium analysis approach. Given specific values for pressure parameters (related to climate change or economic globalization) is there a state of the system such that all conservation and complementarity equations are satisfied? If there is such a state, is it unique? Could such a state be called an equilibrium state? If it is not unique, what is the set of equilibrium states? How sensitive is the set of equilibrium states to pressure parameters?

The dynamical analysis approach. How does the set of equilibrium states change under constant pressure, for example an increase of four degrees centigrade? How does the set of equilibrium states change under slightly changing pressure, for example an increase of 0.2 degrees centigrade per year, or a constant increase in the demand for fish due to population change? How does the set of equilibrium states change in response to sudden changes in demand, for example from an economic boom, or a stock collapse? What is the effect of inertia in investment in the fishing industries? Or unpredictable changes in investment behavior?

3.2.1 Networks: Structure, Dynamics and Equilibrium

Such questions lead to a focus on the structural constraints of the network. Is it rigid? And if so, why? How does it evolve in a context of climate change or globalization? Recall that, in this book, we ask how does evolve an ecological network in a context of climate change or globalization? In the chapters that follow, we use different approaches to give answers to these questions.

A first approach is based on a static analysis: (1) Evaluate the typical characteristics of climate change and economic globalization, (2) Compute the resulting set of equilibrium states, and (3) Perform a sensitivity analysis of the impact of the climate change or economic globalization variable on a characteristic of the set of equilibrium states. We take this approach to the study of the global supply chain for fish.

A second approach is based on a discrete time dynamical analysis: (1) Evaluate the initial state of the network including the characteristics of nodes, links, climate change and economic globalization at time $t = 0$, (2) Compute the resulting network equilibrium; evaluate the new characteristics of nodes and links, (3) Compute how these new characteristics changed from time $t = 0$ to time $t + 1$; update climate change and economic globalization characteristics at time $t + 1$, and (4) Repeat. We take this approach to the study of the global supply chain for tuna.

There is a third approach, which is outside the scope of this book. This is based on a continuous time dynamical analysis: (1) Evaluate the initial state of the network including the characteristics of nodes, links, climate change and economic globalization at time $t = 0$, (2) Compute the differential equation expressing how network characteristics move towards equilibrium at time $t + dt$, and how nodes and links change between time $t = 0$ and time $t + dt$, (3) Update climate change and economic globalization characteristics at time $t + dt$. This approach, known as the projected dynamical systems theory (Nagurney and Zhang, 1996), However it will not be applied in this book, although there is a brief introduction in section 6.6.

Figure 1.2 let imagine some other issues related to network dynamics. The most important relate to changes in network topology, and the creation or destruction of new links or nodes. These are also outside the scope of this book.

3.3 Digression: Statics and Dynamics in Ecology and Economics

This digression is not specific to network economics and can be skipped on a first reading. It deals with the interplay between statics and dynamics in the analysis of systems. Here, we present our own perspective (i.e., that of an interdisciplinary research practitioner). We discuss this issue because network economics suggests a focus on the equilibrium of networks, and afterwards represent dynamics with related tools, such as the projected dynamical systems. Results obtained by network economics approach raise conventional arguments of this topic; and this may be anticipated.

The notion of equilibrium is not easy to fully understand and has been a long-standing subject of discussion in both ecology and economics[1]. The following quote from Paul Samuelson (1943) underlines how old and intense the debate is: *Often in the writings of economists the words "dynamic" and "static" are used as nothing more than synonyms for good and bad, realistic and unrealistic, simple and complex. We damn another man's theory by terming it static, and advertise our own by calling it dynamic. Examples of this are too plentiful to citation.* Have things changed since Samuelson? In theory, yes: various approaches have been developed based on the idea of non-equilibrium, both in eccology with, for example, the intermediate disturbance hypothesis described by Sousa (1979) and in economics with, for example, the disequilibrium foundations of equilibrium found in (Fisher, 1989) or the idea of the real business cycle, etc. In practice, no: the arguments are still vehement (Kaldor, 1972; Blaug, 2002).

Samuelson's comments remain valid and the distinctions he makes later between static and stationary/historical, and dynamic and causal/historical are still pertinent. Therefore, when discussing statics and dynamics it remains particularly important to be unambiguous.

It is common that network economics is used in an interdisciplinary context. Then differences in vocabulary, differences in the use of elementary concepts can become a danger. We will mention in the forthcoming sections (1) the issues raised by the different meanings of the world equilibrium, (2) the different uses of the idea of equilibrium in physical sciences and in ecology or economics, (3) with the example of functional responses in ecology, the consequences of a wish for a mechanistic system, (4) with the example of the IS-LM model, the consequences of a wish for dynamical system.

3.3.1 Equilibrium: Meanings in Ecology and Economics

There are (at least) three possible uses of the word equilibrium in ecology and economics: equilibrium as a balanced state, a stationary state, and a non-stressed state. These uses are not exclusive. It is a surprise to observe that for some economists or some ecologists there no contradiction between these definitions. Let us be more explicit.

Firstly, equilibrium can be related to conservation laws. In ecology, a quantity (size of a population, number of members of a species, amount of energy or biomass) is conserved or transformed in a deterministic way through ecological processes (predation, assimilation, individual growth,

[1] Interested readers can also consult Machlup (1959) for long list of ways to tackle the issue; how it is addressed in ecology is discussed in De Angelis and Waterhouse (1987); how it is addressed in economics is discussed in Frisch (1936); Ingrao and Israel (1990).

spawning, dispersal). Many ecologists have tried to find such conservation laws not only in biological processes, but also in the ecological data itself, for example population data (Ginzburg and Colyvan, 2004). In economics, conservation laws mainly relate to accounting for money, materials, people, etc. All such input-output approaches (Leontief, 1951; Miller and Blair, 1985) are based on a balance between the inflow of an economic sector and its outflow. Therefore, accounting equations are balance equations. Hannon has shown how formal input-output models can be exported from economics to ecology (Hannon, 1973)[2].

Secondly, the idea of equilibrium relates to the mathematical formulation of a dynamic system (finite difference equations, differential equations, and partial differential equations) and the subsequent study of its dynamics, which consists mainly of the description of its steady states, or more generally its attractors or attracting basins (Scheffer and Carpenter, 2003). In this case, equilibrium is a stationary state and there is a tendency to return to it.

Thirdly, the idea of equilibrium relates to equality between forces, such as supply and demand in the marketplace (Walras, 1874; Patinkin et al., 1987). It is the most straightforward example of a complementarity relationship: if supply equals demand, price is positive and when supply exceeds demand there is no price. This idea of market equilibrium can be extended to more complex systems, with many and various agents, and leads to a general equilibrium theory. It is also the basis for game theory (Von Neumann and Morgenstern, 1972; Aumann and Hart, 1994). This idea of equilibrium has been exported from economics to ecology. For example, what happens if trophic flows between two compartments (species) are related to economic behavior? It is the basis for optimal foraging theory[3] (MacArthur and Pianka, 1966), of the marginal value theorem[4] (Charnov, 1976) and the ideal free distribution theory[5] (Fretwell and Lucas, 1970a).

3.3.2 Different Uses of the Idea of Equilibrium in Physical Sciences and in Ecology or Economics

In economics and ecology, equilibrium-based reasoning is very common (Hutchinson, 1953; Cuddington, 2001). In this case, a part of the living world or a set of economic behaviors is observed and identified as an entity. This means that there must be persistence in the observations and a relationship between the time scale of the entity itself and the observation

[2]See also (Polovina, 1984; Christensen and Walters, 2004).

[3]Predators maximize their net energy intake; they make a trade off between the prey they consume and the energy the expend in catching them.

[4]If a predator consumes two prey, the benefits of predation of both are equal.

[5]The distribution of a predator population consuming prey distributed in several patches is the one that minimizes resource competition and maximize fitness.

time scale. A focus is possible. For example the definition of an ecosystem (analogous to the definition of the environment of an organism) implies that there is homeostasis. Although ecosystems cannot be analyzed in terms of evanescent dynamics, they are at the same time complex changing entities[6].

There is a tendency, in ecological studies (De Angelis and Waterhouse, 1987) as well as in economic studies (Frisch, 1936; Ingrao and Israel, 1990), to consider a balance of nature (Pimm, 1991; Cuddington, 2001), a succession towards a climax (Clements, 1936; Turner et al., 1993; Turner, 1989), an economic order (Hayek, 1980), a regular equilibrium growth. All these attitudes are relevant of the same quest for equilibrium.

This quest for equilibrium has to be related to the idea of conservation laws that is a fundamental concept of physics, especially of mechanics. Recall that in this discipline, the progression has been as follows. There has been successively (1) the definition of conservation laws (Kepler), (2) the definition of a relativity principle between measurements of physical quantities by different observers (Galileo), (3) an explanation in terms of forces (Galileo, Newton), (4) a mathematical formulation in terms of differential equations (from Newton to Poincaré). The key point is that this is the movement and its characteristics, as they can be defined according to the relativity principle, that are conserved. At all steps, the idea of equilibrium was present.

Ecologists and economists appear to be looking for what is at the origin of physics: conservation laws. Since these disciplines were established, scientists in these domains have been confronted by a huge problem. They are interested in processes that involve different space and time scales but they have not seen the same success in finding laws that are valid on a large range of scales that has been possible in physics.

In their quest to identify conservation laws economists and ecologists: (1) look for regularities in the quantities obtained from observations (statics), and regularities in the variations of these quantities (dynamics), (2) express such regularities as equalities, (3) define equilibrium as a long time period with no notable changes. This leads to focus on possible explanatory equilibration mechanism. Mathematically, it is simple to express dynamical systems as those where one of the few possible equilibrium states corresponds to the one defined, and in which exist retroaction mechanisms that can be interpreted. Equilibrium is a stable reference state where the balance between internal forces and external forces has been defined. There is the temptation to go further and to interpret the terms of the differential equations as forces. This is not straightforward. It has been noticed how

---

[6] Jax et al. (1998) has suggested a method to deal with a related issue that he calls "The self identity of ecological units" (Grimm et al., 1998).

difficult it was to export the idea of force from physics to economics (Knight, 1934; Mirowski, 1984; Donzelli, 1997) or ecology (Ulanowicz, 1999). An interesting attempt in ecology can be found in the works of Ginsburg and Colyvan (Colyvan and Ginzburg, 2003; Ginzburg and Colyvan, 2004).

3.3.3 The Wish for a Mechanistic System

The following is an analysis, in the best-known ecological model, of the idea of a functional response. This is understood as a functional relationship linking predation flows to prey and predator biomass. It was based on a mathematical formulation (developed by Lottka and Volterra), which was generalized (by Holling amongst others) and became the emblem of mathematical models of ecological systems. This created a new field for researchers (in other words, many scientific papers) and ended in poorly formulated dogma: functional responses. These points are discussed in detail below.

The first step consisted of building a dynamical model of ecological principles. Ecological modeling of dynamical systems is based on the following principles: (1) For species, biomass changes equal inflow minus outflow. (2) Outflow is due to predation by other species plus other losses (mortality, somatic maintenance). (3) Inflow represents assimilated predation on other species. (4) Functional responses relate trophic flows and species biomass.

Mathematically, this is formulated as follows. Let $i, j, k, \ldots$ represent species, b_i the biomass of species i, x_{ij} the trophic flow from species i to species j. Then predation of i on other species is given by $\sum_j x_{ji}$; predation of other species on i is $\sum_j x_{ij}$. Assuming a constant coefficient of assimilation γ_i, inflow of species i is $\gamma_i \sum_j x_{ji}$. Assuming a constant coefficient for somatic maintenance μ_i relating losses of a species i to its biomass b_i, outflow is $\sum_j x_{ij} + \mu_i b_i$. Assuming functional responses $(b_i, b_j) \rightarrow F_{ij}(b_i, b_j)$ that relate flows and biomass, we get a dynamical system:

$$\frac{db_i}{dt} = \gamma_i \sum_j F_{ji}(b_j, b_i) - \sum_j F_{ij}(b_i, b_j) - \mu_i b_i$$

The second step consists of examining the nature of the underlying ecological relationships. The differential equation given above can in many cases be linearized and represented as:

$$\frac{db_i}{dt} = \alpha_i b_i + \sum_j \beta_{ij} b_i b_j$$

Coefficients α_i are positive for autotrophic species, negative for heterotrophic species. Coefficients β_{ji} express the nature of the interaction between species i and j:

$$\beta_{ij} = 0 \text{ and } \beta_{ji} = 0 \rightarrow \text{ indifference}$$
$$\beta_{ij} > 0 \text{ and } \beta_{ji} > 0 \rightarrow \text{ mutualism}$$
$$\beta_{ij} > 0 \text{ and } \beta_{ji} < 0 \rightarrow \text{ predator prey}$$
$$\beta_{ij} < 0 \text{ and } \beta_{ji} > 0 \rightarrow \text{ prey predator}$$
$$\beta_{ij} < 0 \text{ and } \beta_{ji} < 0 \rightarrow \text{ competition}$$

The third step consists of analyzing the mathematical properties of the dynamical system. There are standard tools to characterize the stability of stationary points in such systems (Carpenter, 1994; Scheffer and Carpenter, 2003).

Altogether we have: (1) the expression of a dynamical system, (2) its interpretation in terms of ecological processes, and (3) a way to analyze the resulting dynamics. However, the critical point is that the existence of functional responses has not been proven, there is no empirical evidence of their validity, and there are many ways to mathematically formulate and parameterize them. In fact, there is no way to determine how ecological systems are deterministic. As an example we cite (Koen-Alonso and Yodzis, 2005): *The functional response is a core structural feature of a trophodynamic model. Its precise mathematical form represents the researcher's view of the biological details of that particular predation process and profoundly affects model behaviour. In most cases, however, so little is known of these biological details that there is only weak evidence, if any, that points towards a given specific mathematical formulation. Under these circumstances, it is essential to avoid the temptation to assume some particular form for the functional response and proceed on that basis to get answers that are reassuringly, but meaninglessly, definitive. Rather, we need to explore the range of behaviours that are consistent with what we do know about the system.*

In conclusion, it should not be forgotten that functional responses tend to be assumed to reflect reality; a situation that can lead us far from the actual reality of ecosystems. The noticing of the multiplicity of possible functional responses, of their arbitrariness leads to look for different representations of the prey-predator mechanism, not mechanistic, for example of economic type (this is the assumption of network economics), or not deterministic.

3.3.4 The Wish for a Dynamical System

The following example can be found in any introductory book on macroeconomics. It is about the relationship, at a national level, between interest rates, savings, demand for money and money supply[7]. It starts with a (controversial) formulation (Hicks, 1937) of principles expressed in Keynes's general theory. This was based on (1) the assumption that investment (I), savings (S), demand for money (L) and supply of money (M) all depend on production Y and interest rate i, and (2) on the equalities: investment equals savings and supply for money equals demand for money.

$$I(Y, i) = S(Y, i)$$
$$L(Y, i) = M(Y, i)$$

With these assumptions, it appears that the equations $I = S$ and $L = M$ define two curves in the (Y, i) plane, and that these two curves intersect at a unique point (figure 3.4). The analysis of this static model is in itself very revealing. If some of the most immediate assumptions are relaxed it raises questions about equilibrium states, such as their existence (or not), uniqueness (or not), and their sensitivity (or not) to the parameters used to define functional relationships.

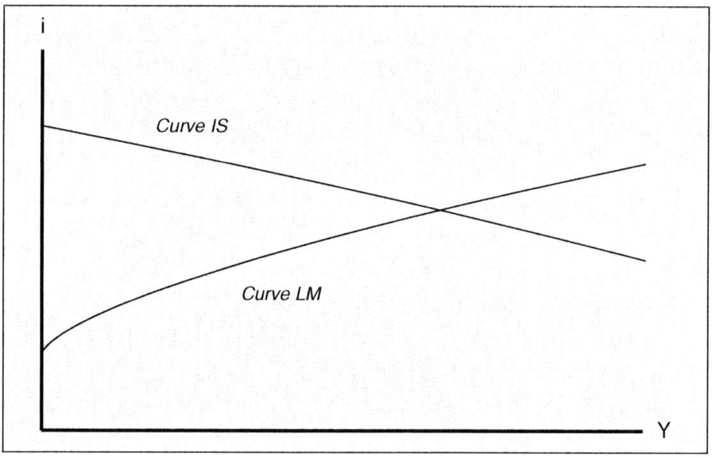

Figure 3.4: The IS-LM model. Diagramm phase

A second step consists of building an associated dynamical model. A mathematical formulation that makes it possible to study relationships in terms of equilibrium and dynamics has been developed by Samuelson (1941)[8]. Assumptions were made as simple as possible: the static model must tend to the equilibrium state. The simplest dynamical system that has these properties is:

[7] see Vroey and Hoover (2004) about the history of the IS-LM model.
[8] See also (Chang and Smyth, 1972).

$$\dot{Y} = a \times (I(Y,i) - S(Y,i))$$
$$\dot{i} = b \times (L(Y,i) - M(Y,i))$$

Coefficients a and b correspond to the adjustment speed. Let $\overline{Y}, \overline{i}$ a stationary state of previous differential equation; we have $(I(Y,i) - S(Y,i)) = (L(Y,i) - M(Y,i)) = 0$; it is easy to show that in order that this state is a stable equilibrium sate of above differential equation, functions I, S, L, M must be such that, all derivatives $\partial(I-S)/\partial Y$, $\partial(L-M)/\partial Y$, $\partial(I-S)/\partial i$, $\partial(L-M)/\partial i$ are non zero and:

$$a\frac{\partial}{\partial Y}(I-S) + b\frac{\partial}{\partial i}(L-M) < 0$$
$$\frac{\partial}{\partial Y}(I-S)\frac{\partial}{\partial i}(L-M) - \frac{\partial}{\partial i}(I-S)\frac{\partial}{\partial Y}(L-M) > 0$$

Here again, the meaning of these equations for economics has resulted in a long list of theoretical debates[9]. But the key point is that the condition about a and b is easy to fulfil on a mathematical point of view, and that there exists a tendency to give these parameters "good" values.

The IS-LM model, both in its static and its dynamic version, has been the core of macro economic politics in the 1960–1970 period until it triggered a deep criticism even if then there occur periodic reconsideration (King, 1993). Now in the reasons to criticize the IS-LM model, we underline the following ones. (1) The static model is wrong; relationships such as $I(Y,i)$, $S(Y,i)$, $L(Y,i)$, $M(Y,i)$ are either not stable, or they do not exist, or if they do exist they are only statistical correlations and not causal relationships, etc. (2) Assumptions resulting in a stable equilibrium were too optimistic, there has been a tendency to neglect situations of unstable equilibrium; and (3) Dynamical equations are based on a physical time, but the time of equilibration in an economic system has nothing to do with physical time (Garrison, 2000). What are realistic speed adjustment coefficients? What is the nature of the dynamics of economic processes?

We favor the second and third explanations, i.e., the hurry to build the dynamical system and the tendency to be satisfied with retroaction forces once they result in identified stationary states.

3.3.5 Concluding Remark

We have discussed the equilibrium versus dynamic issue in economics and ecology and underlined the point that the quest for equilibrium may be dangerous. We must be aware of this issue and systematically question: What is the nature of the functional relationships that form the basis for

[9] See (Chang and Smyth, 1972).

dynamical systems? Do they really exist? Or are they defined simply to give meaning to mathematical equations? Has the possibility of using different methods for timing events (other than the physical calendar) been considered? How can confusion between the timing of a computation (steps, the timing of an iterative algorithm), and timing of the process itself (Winfree, 2001) be avoided?

Part II

A Gentle Introduction to the Mathematics of Equilibrium and Optimization

Chapter 4

Introduction

In this part we show how to represent an economic or ecological problem with a network structure, and how to compute and analyze the associated equilibrium. This consists of solving mathematical equations occurring on both nodes and links. These local equations are often, but not always, related to an individual optimization. Network economics can be seen as a way to represent, through the network, the coordination of optimization behaviors. Chapter 5 presents the background to mathematical optimization in such a way that the main tools of network economics, namely the variational inequality problem and the complementarity problem appear as immediate generalizations of mathematical optimization. These tools will be presented in chapter 6.

4.1 Warning

This introduction to convex optimization is as short and as illustrated as possible. We restrict definitions to the simplest cases: functions are twice continuously differentiable; convex sets are compact polyhedrons. Fortunately, this is sufficient for almost all of the results that are presented in this book and means that mathematical proofs can be given for the most important theorems. The mathematical theory of convex optimization makes it possible to go much further using more general definitions: set-valued sets, semi-continuous functions, etc[1].

[1] Reference books are Rockafellar (1990); Avriel et al. (1988); Borwein and Lewis (2006); Boyd and Vandenberghe (2004); Hadjisavvas et al. (2005); Lee et al. (2005).

The results presented here are elementary and are presented in a mathematical style. Definitions are precise, results appear as theorems, and proofs of theorems are provided. This presentation of simplified versions of mathematical theories in a mathematical style (definitions, theorems, and proofs) is intended to help to develop a fruitful, reciprocal dialog between mathematicians and ecologists and economists.

Mathematicians are often surprised to see that theories and concepts that they consider difficult are spontaneously used by ecologists or economists. This is particularly the case for differential, partial derivative or stochastic equations and seems to be due to the availability of powerful simulation software (Stella, Vinsim, Matlab). Similarly, we hope that optimization theory is not too complicated and can be understood when presented to non-mathematicians in a mathematical style. That is the main purpose of this part.

4.2 Complementarity

The principal mathematical idea developed in this book is very, very simple; it concerns complementarity. Mathematically, it is defined as follows.

Definition 4.2.1 *There is complementarity between two numbers a and b if $a \geq 0$, $b \geq 0$, $ab = 0$. Then we put $a \perp b$.*

Complementarity is a highly non-linear relationship (figure 4.1).

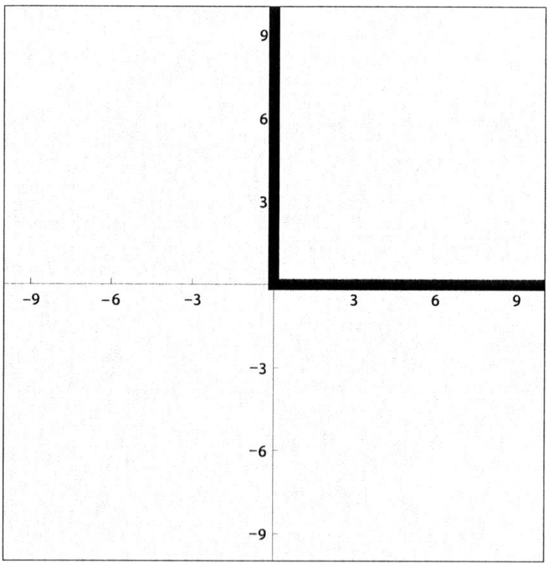

Figure 4.1: Complementarity. Solutions of the complementarity problem are shown by the thick lines.

Definition 4.2.2 *There is complementarity between two vectors* $A = (a_i)$ *and* $B = (b_i)$ *if, for all* i*, we have* $a_i \geq 0$, $b_i \geq 0$, $a_i b_i = 0$. *Then we put* $A \perp B$.

4.3 Notations

Notations for vectors and matrices, notation for optimization problems are as follows:

$\mathbb{R}$	Real numbers
$\mathbb{R}_+$	Non negative real numbers
x, y, z	Lowercase letter: scalar
α, β, γ	Greek letter: scalar parameters
X, Y, Z	Uppercase letter: Vectors $X = (x_1, \ldots, x_n)$
$\mathbf{A}, \mathbf{B}, \mathbf{C}$	Uppercase bold letter: Matrices
$X \cdot Y$	Scalar product: $X \cdot Y = \sum_i x_i y_i$
$\|X\|$	Norm: $\|X\| = \sqrt{\sum_i x_i^2}$
$f(x)$	Scalar function $f : \mathbb{R} \to \mathbb{R}$
$f(X)$	Scalar function $f : \mathbb{R}^n \to \mathbb{R}$
$F(X)$	Vector function $F : \mathbb{R}^n \to \mathbb{R}^m$
$\mathbf{A} \cdot X, X \cdot \mathbf{A}$	Multiplication of a matrix and a vector
$\mathbf{A} \cdot \mathbf{B}$	Multiplication of two matrices
${}^T\mathbf{A} \cdot \mathbf{B}$	Transpose of a matrix
$\mathcal{MIN}(K, f)$	Minimization problem with convex set $K \subset \mathbb{R}^n$ and function $f : K \to \mathbb{R}$
$\mathcal{MIN}(\mathbf{A}, B, f)$	Minimization problem with convex set $K = \{X \in \mathbb{R}^n \mid \mathbf{A} \cdot X \geq B\}$ and linear function $f : K \to \mathbb{R}$
$\mathcal{VI}(\mathcal{F}, \mathcal{K})$	Variational inequality problem with convex set $K \subset \mathbb{R}^n$ and function $F : K \to \mathbb{R}^n$
$\mathcal{VI}(\mathbf{A}, B, F)$	Variational inequality problem with convex set $K = \{X \in \mathbb{R}^n \mid \mathbf{A} \cdot X \geq B\}$ and function $F : K \to \mathbb{R}^n$
$\mathcal{AVI}(\mathbf{A}, B, \mathbf{M}, N)$	Affine variational inequality problem with convex set $K = \{X \in \mathbb{R}^n \mid \mathbf{A} \cdot X \geq B\}$ and function $F : K \to \mathbb{R}^n$, $F(X) = \mathbf{M} \cdot X + N$
$\mathcal{LCP}(\mathbf{M}, N)$	Linear complementarity problem with linear function $F : \mathbb{R}^n \to \mathbb{R}^n$, $F(X) = \mathbf{M} \cdot X + N$
$\mathcal{CP}(\mathbf{M}, N)$	Complementarity problem with function $F : \mathbb{R}^n \to \mathbb{R}^n$
$\mathcal{NEP}$	Nash equilibrium problem
$\mathcal{GNEP}$	Generalized Nash equilibrium problem

We mark the end of a demonstration with symbol $\square$.

Chapter 5

Convexity and Optimization

5.1 Convex Sets

5.1.1 Convex Sets

A large part of mathematical optimization theory relies on the idea of convex sets (figure 5.1) and convex functions.

Definition 5.1.1 *A subset $K \subset \mathbb{R}^n$ is convex if for all pairs of points $X \in K, Y \in K$ the line segment joining X and Y is included in K: for all $\alpha \in [0,1]$, $\alpha X + (1-\alpha)Y \in K$.*

5.1.2 Examples

Let us give some basic examples of convex sets (see figure 5.2). A line L in the plane $\mathbb{R}^2$ has an equation $ax + by = c$; it splits the whole plane into half planes: $H = \{(x,y) \mid ax + by \geq c\}$ and $L = \{(x,y) \mid ax + by \leq c\}$. Both half planes are convex subsets of $\mathbb{R}^2$ (figure 5.2, top, left). Let us remark that the vector A with components (a,b) is orthogonal to the line L. We generalize this definition in arbitrary dimensions. An affine hyper space H, orthogonal to a vector $A = (a_1, \ldots, a_n)$ is defined by an equation

$$\{(x_1, \ldots, x_n) \mid a_1 x_1 + \cdots + a_n x_n = c\}$$

Definition 5.1.2 *A half space in $\mathbb{R}^n$ is defined as*

$$H = \{(x_1, \ldots, x_n) \mid a_1 x_1 + \cdots + a_n x_n \geq c\}$$

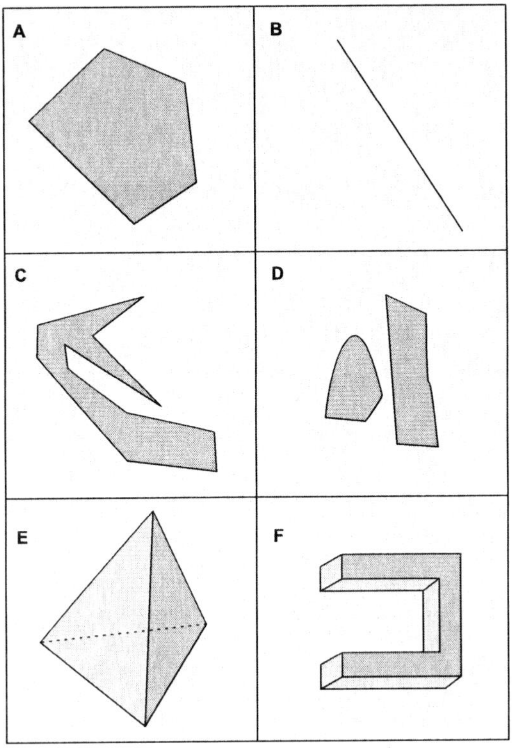

Figure 5.1: Convexity. Top: examples of convex sets in the plane $\mathbb{R}^2$: (a) a polyhedron; (b) a line. Middle: examples of non-convex sets in the plane: (c) set with a "concavity", (d) set with several disjoint parts. Bottom: (e) a convex set in $\mathbb{R}^3$, (f) a non-convex set in 3d.

Definition 5.1.3 *A convex polyhedron is an intersection of half spaces (figure 5.2, top, right). It is defined by a finite number of linear equations:*

$$H = \{(x_1, \ldots, x_n) \mid a_1^k x_1 + \cdots + a_n^k x_n \geq c^k\}$$

Definition 5.1.4 *A cone $K \subset \mathbb{R}^n$ is a convex subset that for all $X \in K$, all $\lambda \geq 0$, $\lambda X \in K$ (figure 5.2, bottom, left).*

Definition 5.1.5 *The generators of a cone (figure 5.2, bottom, right) are a set of vectors $X_1, \ldots, X_p$ such that for every $X \in K$ there exists coefficients $\lambda_i \geq 0$ such that:*

$$X = \lambda_1 X_1 + \cdots + \lambda_p X_p = \sum_i \lambda_i X_i$$

Example: vectors $(1, 0)$, $(0, 1)$ are generators of the cone $\mathbb{R}_+^2$.

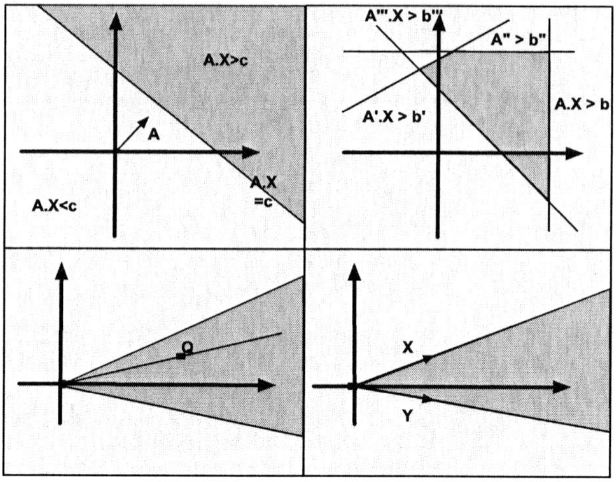

Figure 5.2: Convex sets. Top, left: a half-space in a 2d-space defined by $A \cdot X > c$; vector A is normal (orthogonal) to line $A \cdot X = c$. Top, right: a polyhedron in the plane as the intersection of half-spaces $A \cdot X > c$, $A' \cdot X > c'$, $A'' \cdot X > c''$, $A''' \cdot X > c'''$. Bottom, left: a cone in the plane: it contains the whole line through a point Q as soon it contains Q. Bottom, right: its generators X and Y.

5.1.3 Main Theorems

First important theorem is the projection theorem (figure 5.3, left).

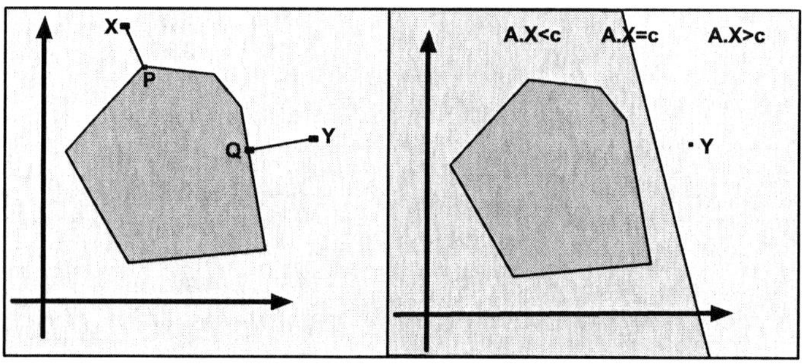

Figure 5.3: Projection on a convex set. Left: The projection of external points on a convex set in the plane. Right: A line separating a point Y and a convex polyhedron in the plane.

Theorem 5.1.1 *Let K be a closed convex set. For all X, there exists a unique point $P \in K$ minimizing the distance with K.*

Proof. Let us prove firstly the existence. Let Q a point in K. Let $r > d(X, Q)$. Let $B(X, r)$ the ball centred in X with radius r. For a given X, the function $Y \to d(X, Y)$ is a continuous functions. It reaches its minimum on the non-empty compact set $K \cap B(X, r)$. Let us prove the unicity. Suppose that two distinct points P_1 and P_2 minimize the distance of X to K. Then middle point $P = (P_1 + P_2)$ is such that $d(P, X) < d(P_1, X) = d(P_2, X)$ because the length of the height of an isosceles triangle is less than the length of adjacent sides. □

Definition 5.1.6 *The projection of X on K is the point $P \in K$ minimizing the distance with K. We note it $P_K(X)$*

We have the following characterization theorem:

Theorem 5.1.2 *P is the projection of X on K if and only if for all $Y \in K$, $(P - X) \cdot (P - Y) \leq 0$.*

Proof. If $X \in K$ then $P = X$ and the theorem is obvious. Suppose that $X \notin K$. Let $Y \in K$. Then $\|X - Y\|^2 > 0$, $\|X - P\|^2 > 0$. Suppose that $(P-X) \cdot (P-Y) > 0$. Let $\alpha \in [0, 1]$. Let $Q_\alpha = (1-\alpha)P + \alpha Y$. Then $Q_\alpha \in K$ and it easy to show that, if $\|X - Q_\alpha\|^2 = N$, then:

$$\begin{aligned} N &= \|X - P\|^2 - 2\alpha(P - X) \cdot (P - Y) \\ &\quad + \alpha^2(\|X - Y\|^2 - \|X - P\|^2 - 2(P - X) \cdot (P - Y)) \\ &= A\alpha^2 - 2B\alpha + \|X - P\|^2 \end{aligned}$$

With coefficients $A = \|X - Y\|^2 - \|X - P\|^2 - 2(P - X) \cdot (P - Y)$ and $B = (P - X) \cdot (P - Y)$. Recall that $B > 0$. If $A \leq 0$, we obviously have $\|X - Q_\alpha\|^2 < \|X - P\|^2$ for all $\alpha \in [0, 1]$. This is a contradiction with the definition of P. If $A > 0$, we choose $\alpha = Min(1/2, B/A)$, and we get also $\|X - Q_\alpha\|^2 < \|X - P\|^2$. This is still a contradiction. □

According to previous theorem, there is an obtuse angle between (1) the line joining X to its projection on K and (2) the line joining X to any point $Y \in K$.

Theorem 5.1.3 *The projection function P_K is contracting; we have $\|P_K(X) - P_K(Y)\|^2 \leq \|X - Y\|^2$.*

Proof. Let $X \neq Y$, $P = P_K(X)$ and $Q = P_K(Y)$. According to previous theorem, we have: $(P - X) \cdot (P - Q) \leq 0$ and $(Q - Y) \cdot (Q - P) \leq 0$. We sum these two inequalities. We get: $(P - X + Y - Q) \cdot (Q - P) \geq 0$ and then:

$$\begin{aligned} \|X - Y\|^2 &= \|P - X + Y - Q + Q - P\|^2 \\ &= \|P - X + Y - Q\|^2 + \|Q - P\|^2 \\ &\quad + 2(P - X + Y - Q) \cdot (Q - P) \\ &\geq \|Q - P\|^2 \end{aligned}$$

□

The second important theorem is the separation theorem (figure 5.3, right).

Theorem 5.1.4 *Let $K \subset \mathbb{R}^n$ be a polyhedron. Let $X \notin K$. There exists an hyper plane which separates them, i.e., a pair $(A \in \mathbb{R}^n, c \in \mathbb{R})$ such that $A \cdot X < c$ and $A \cdot Y > c$ for all $Y \in K$.*

Proof. Let $P = P_K(X)$ the projection of X on K. Let $A = X - P$. Let $c = (X - P) \cdot (P + X)/2$. We have

$$\begin{aligned} A \cdot X - c &= (X - P) \cdot X - (X - P) \cdot (X + P))/2 \\ &= (X - P) \cdot (X - P)/2 \geq 0 \end{aligned}$$

For $Y \in K$, we have, according to theorem 5.1.2, $(P - X) \cdot (P - Y) \leq 0$; thus $A \cdot P \geq A \cdot Y$. Then

$$\begin{aligned} A \cdot Y - c &\leq (X - P) \cdot P - (X - P) \cdot (P + X)/2 \\ &= (X - P)(P - (P + X)/2 \\ &= -(X - P) \cdot (X - P)/2 \leq 0 \end{aligned}$$

□

5.1.4 Normal Cone

The normal cone of a polyhedron K at a point P is related to the points whose projection on K is exactly P.

Definition 5.1.7 *Let $K \subset \mathbb{R}^n$ be a polyhedron. Let $P \in K$. The normal cone to K at P, $N_K(P)$ is the set of $Z \in \mathbb{R}^n$ such that $P_K(P + Z) = P$.*

Theorem 5.1.5 *Let K be a polyhedron, $P \in K$. Then $N_K(P)$ is the set of $Z \in \mathbb{R}^n$ such that for all points $Y \in K$, we have: $Z \cdot (Y - P) \leq 0$.*

Proof. This is an immediate consequence of theorem 5.1.2. Let Z such that $P_K(P + Z) = P$; then for all $Y \in K$, $(P - (P + Z)) \cdot (P - Y) \leq 0$; whence for all $Y \in K$, $Z \cdot (Y - P) \leq 0$. Reciprocally, let Z such that for all Y in K, $Z \cdot (Y - P) \leq 0$, then $(P - (P + Z) \cdot (P - Y) \leq 0$. Thus P is the projection on K of $P + Z$. □

In some cases the normal cone to a polyhedron is easy to identify (see figure 5.4): if P is interior to K, then $N_K(P) = \{0\}$; if P is on the boundary of K and if the boundary is "regular" in P, then $N_K(P)$ is the half line "normal" to K; if P is on the boundary of K and if the boundary is "irregular" in P, then $N_K(P)$ is a convex cone.

A third important theorem of convex sets theory is about the characterization of normal cones[1]. We have:

[1]This theorem is also known as the Farkas-Minkowski lemma.

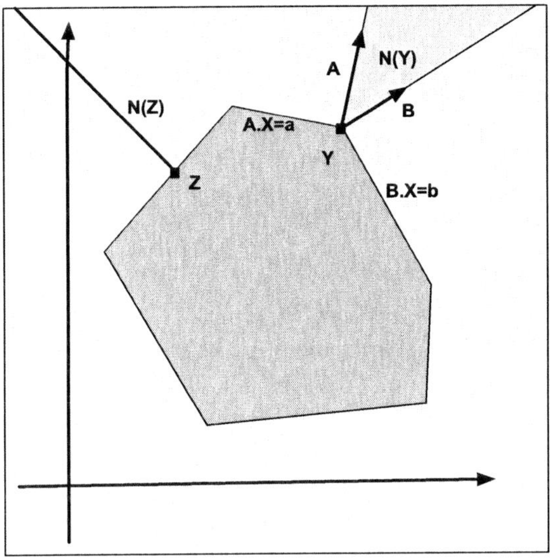

Figure 5.4: Normal cone. Normal cones to a polyhedron in several points. The point Z is on the hyper plane $C \cdot X = c$; $N_K(Z)$ is the cone generated by C. The point Y is at the intersection of the hyper planes $A \cdot X = a$ and $B \cdot X = b$; $N_K(Y)$ is the cone generated by A and B.

Theorem 5.1.6 *Let K a polyhedron defined by linear constraints: $K = \{X \mid X \cdot A_i \geq b_i$. Let $P \in K$. Let I_0 the set of indices such that $A_i \cdot P - b_i = 0$. Then $N_K(P)$ is the cone with generators $-A_i$ for $i \in I_0$*

Proof. We have to prove that $Y \in N_K(P)$ is equivalent to $Y = -\sum_i \alpha_i A_i$, with, for all $\alpha_i \geq 0$ and either $\alpha_i = 0$ or $A_i \cdot P - b_i = 0$.

According to theorem 5.1.5, we know that the normal cone $N_K(P)$ is the set of $V \in \mathbb{R}^n$ such that for all $Y \in K$, we have $V \cdot (Y - P) \leq 0$.

Let $H_K(P)$ be the cone with generators $-A_i$ for $i \in I_0$; we have $H_K(P) = \{Y \mid Y = -\sum_{i \in I_0} \alpha_i A_i$ with $\alpha_i \geq 0\}$.

Firstly, let us prove that $H_K(P) \subset N_K(P)$. Let V a vector in $H_K(P)$; we have $V = -\sum_{i \in I_0} \alpha_i A_i$. Then, we have:

$$\begin{aligned} V \cdot (Y - P) &= -\sum_{i \in I_0} \alpha_i A_i (Y - P) \\ &= -\sum_{i \in I_0} \alpha_i Y \cdot A_i + \sum_{i \in I_0} \alpha_i P \cdot A_i \end{aligned}$$

If $i \in I_0$, then, by definition $P \cdot A_i = b_i$. As $Y \in K$ then, by definition, $Y \cdot A_i \geq b_i$. Thus, for all $Y \in K$, $V \cdot (Y - P) \leq 0$ and $V \in N_K(P)$.

Now, let us prove that $N_K(P) \subset H_K(P)$. Suppose that there exists $Z \in N_K(P)$ such that $Z \notin H_K(P)$ and let us show that this is a contradiction.

As $H_K(P)$ is a convex set, there exists an hyper plane that separates Z and $H_K(P)$: there exists $U \in \mathbb{R}^n$ and $c \in \mathbb{R}$ such that $U \cdot Z > c$ and $U \cdot Y < c$ for all $Y \in H_K(P)$.

As $H_K(P)$ is a cone, we have: $-\alpha A_i \in H_K(P)$ for all $i \in I_0$ and all $\alpha \geq 0$. Thus $U \cdot (-\alpha A_i) < c$ for all $i \in I_0$ and all $\alpha \geq 0$. This is possible only if (1) $-U \cdot A_i \leq 0$ for $i \in I_0$ and (2) $c \geq 0$.

Let $Y = P - \beta U$. For $i \in I_0$, $A_i \cdot P = b_i$ and $A_i \cdot U \geq 0$; whence $A_i \cdot Y \geq b_i$. For $i \notin I_0$, $A_i \cdot P > b_i$; $A_i \cdot U \geq 0$, $A_i \cdot Y = A_i P - \beta A_i U \geq b_i$, for $0 < \beta < Min_i((A_i P - b_i)/A_i \cdot U)$.

We deduce that $Y \in K$. As $Z \in N_K(P)$, we have: $Z \cdot (Y - P) \leq 0$. On an other hand: $Z \cdot (Y - P) = \beta Z \cdot U \geq \beta c > 0$. This is a contradiction. □

5.2 Calculus

In this part, we briefly summarize the most important definitions and principles of calculus: the definitions of gradients, vector fields and Taylor's formula. A classical reference on this subject is Lang (1986). As everywhere in this book we consider only continuously twice differentiable functions.

5.2.1 Gradients

The gradient of a 1-D function $f : \mathbb{R} \to \mathbb{R}$ is its derivative: $\nabla f(x) = f'(x)$ (figure 5.5). ∇f is a function $\mathbb{R} \to \mathbb{R}$.

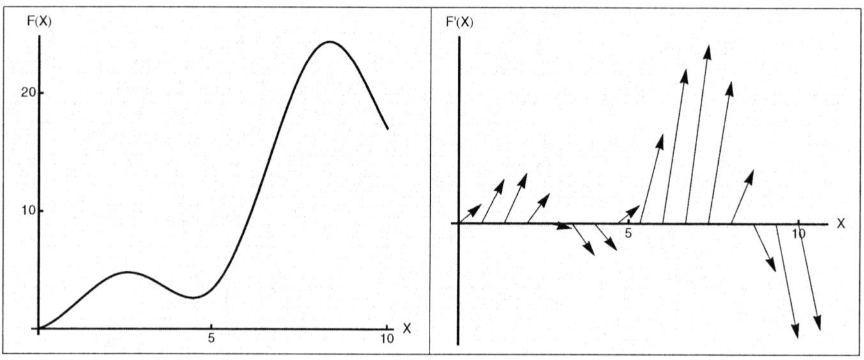

Figure 5.5: Gradients in dimension 1. A 1-D function. Its gradient plotted as a vector field.

In this section, we make an intensive use of plots of 2-D functions to illustrate further definitions and results: either as a 3-D plot, either as a density plot (figures. 5.6 and 5.7).

The gradient of a 2-D function $f : \mathbb{R}^2 \to \mathbb{R}$ is the function $\nabla f : \mathbb{R}^2 \to \mathbb{R}^2$ defined by:

$$\nabla f(X) = Df(x, y) \quad = (\frac{\partial f(x,y)}{\partial x}, \frac{\partial f(x,y)}{\partial y})$$

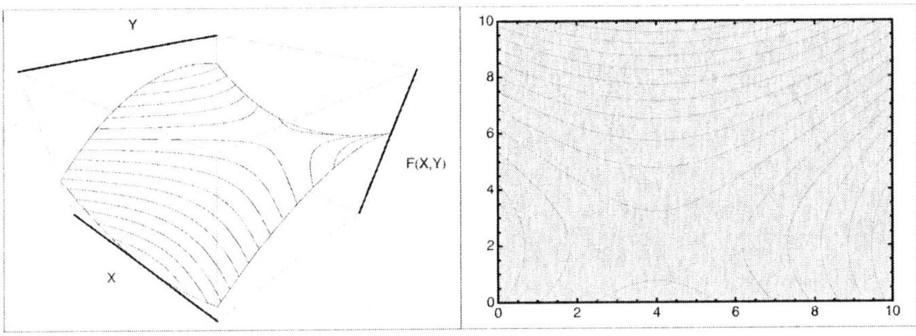

Figure 5.6: Gradients in dimension 2 (a). A two dimensional function in 3-D. Its density plot.

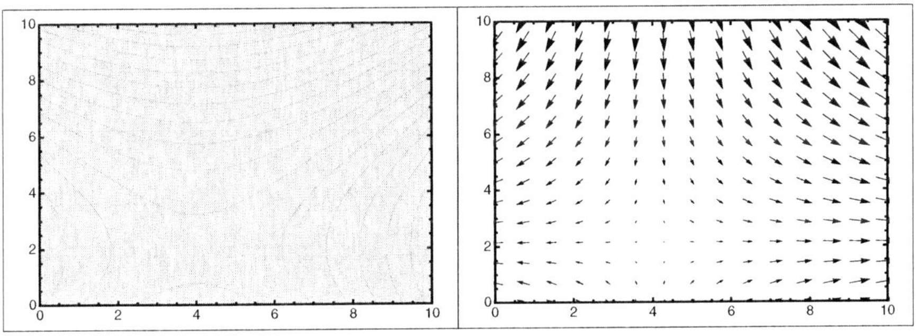

Figure 5.7: Gradients in dimension 2 (b). The density plot of a two dimensional function. Its gradient plotted as a vector field.

Definition 5.2.1 *The gradient of a function* $f : \mathbb{R}^n \to \mathbb{R}$ *is the function* $Df : \mathbb{R}^n \to \mathbb{R}^n$ *defined by:*

$$Df(X) = Df(x_1, \dots, x_n)$$

$$= (\frac{\partial f(x_1, \dots, x_n)}{\partial x_1}, \dots, \frac{\partial f(x_1, \dots, x_n)}{\partial x_n})$$

Taylor's formula is as follows. At order 1, in 1-D, we have:

$$f(x + h) = f(x) + h\frac{\partial f(x)}{\partial x} + hg(h) \text{ with } \lim_{h \to 0} g(h) = 0$$

At order 1, in 2-D, we have:

$$f(x+h, y+k) = f(x,y) + \frac{\partial f(x,y)}{\partial x}h + \frac{\partial f(x,y)}{\partial y}k$$
$$+ \|(h,k)\|g(h,k)$$
$$\text{with} \lim_{\|(h,k)\|\to 0} g(h,k) = 0$$

At order 1, in n-D, we have

Theorem 5.2.1 *Let f be a function $H \subset \mathbb{R}^n \to \mathbb{R}$. Let $\nabla f(X)$ its gradient at X. Then*

$$f(X+H) = f(X) + \nabla f(X) \cdot H + g(H)\|H\|$$
$$\textit{with} \lim_{\|H\|\to 0} g(H) = 0$$

Proof is a simple reformulation of the definition of a derivative in 1-D. It is a little bit more complicated in n-D and can be found in any introduction to calculus. □

5.2.2 Second Order

Now, we define second order derivatives.

Definition 5.2.2 *To a function $f : \mathbb{R}^n \to \mathbb{R}$, we associate its second order derivative:* $\mathbf{D^2f} : \mathbb{R}^n \to \mathbb{R}^{n\times n}$:

$$\nabla^2 \mathbf{f}(X) = \nabla^2 \mathbf{f}(x_1, \ldots, x_n)$$
$$= \begin{pmatrix} \frac{\partial^2 f(x_1,\ldots,x_n)}{\partial x_1 \partial x_1} & \ldots & \frac{\partial^2 f(x_1,\ldots,x_n)}{\partial x_1 \partial x_n}) \\ \ldots & \ldots & \ldots \\ \frac{\partial^2 f(x_1,\ldots,x_n)}{\partial x_n \partial x_1} & \ldots & \frac{\partial^2 f(x_1,\ldots,x_n)}{\partial x_n \partial x_n}) \end{pmatrix}$$

Taylor's formula at order 2 is given in next theorem.

Theorem 5.2.2 *Let f be a function $\mathbb{R}^n \to \mathbb{R}$. Let $\nabla f(X)$ its gradient at X. Let $\nabla^2 \mathbf{f}(X)$ its second order derivative at X. Then:*

$$f(X+H) = f(X) + \nabla f(X) \cdot H + \frac{1}{2} H \cdot \nabla^2 \mathbf{f}(X) \cdot H$$
$$+ g(H)\|H\|^2$$
$$\textit{with} \lim_{\|H\|\to 0} g(H) = 0$$

That is, using components notation instead of vector notation:

$$\begin{aligned} f(x_1+h_1,\dots,x_n+h_n) = {} & f(x_1,\dots,x_n) \\ & + \sum_i \frac{\partial f(x_1,\dots,x_n)}{\partial x_i} h_i \\ & + \frac{1}{2}\sum_{ij} \frac{\partial f(x_1,\dots,x_n)}{\partial x_i \partial x_j} h_i h_j \\ & + g(h_1,\dots,h_n)\|(h_1,\dots,h_n)\|^2 \\ & \text{with} \lim_{\|(h_1,\dots,h_n)\|\to 0} g(h_1,\dots,h_n) = 0 \end{aligned}$$

5.2.3 Derivative of a Function $\mathbb{R}^n \to \mathbb{R}^m$

Generalizations of gradients for multi dimensional functions: $F : \mathbb{R}^n \to \mathbb{R}^m$ are immediate to define.

Definition 5.2.3 *The derivative of a function $\mathbb{R}^n \to \mathbb{R}^m$, given by $F(X) = (f_1(X),\dots,f_m(X))$, at a point X is a matrix:*

$$\mathbf{JF}(X) = \begin{pmatrix} \frac{\partial f_1(x_1,\dots,x_n)}{\partial x_1} & \dots & \frac{\partial f_1(x_1,\dots,x_n)}{\partial x_n} \\ \dots & \dots & \dots \\ \frac{\partial f_m(x_1,\dots,x_n)}{\partial x_1} & \dots & \frac{\partial f_m(x_1,\dots,x_n)}{\partial x_n} \end{pmatrix}$$

5.3 Convex Functions

5.3.1 Convex Functions in Dimension 1

We still consider only continuous, twice continuously differentiable functions.

Definition 5.3.1 *A function $f : \mathbb{R}^n \to \mathbb{R}$ is said convex if for all $\alpha \in [0,1]$, we have:*

$$f(\alpha x + (1-\alpha)y) \quad \leq \quad \alpha f(x) + (1-\alpha)f(y)$$

A function $f : \mathbb{R} \to \mathbb{R}$ is said strictly convex if for all $\alpha \in [0,1]$, we have:

$$f(\alpha x + (1-\alpha)y) \quad < \quad \alpha f(x) + (1-\alpha)f(y)$$

Differences between convex and non-convex functions are depicted in figure 5.8.

Theorem 5.3.1 *Let $f : \mathbb{R} \to \mathbb{R}$; then f is a convex function if and only if $f''(x) \geq 0$.*

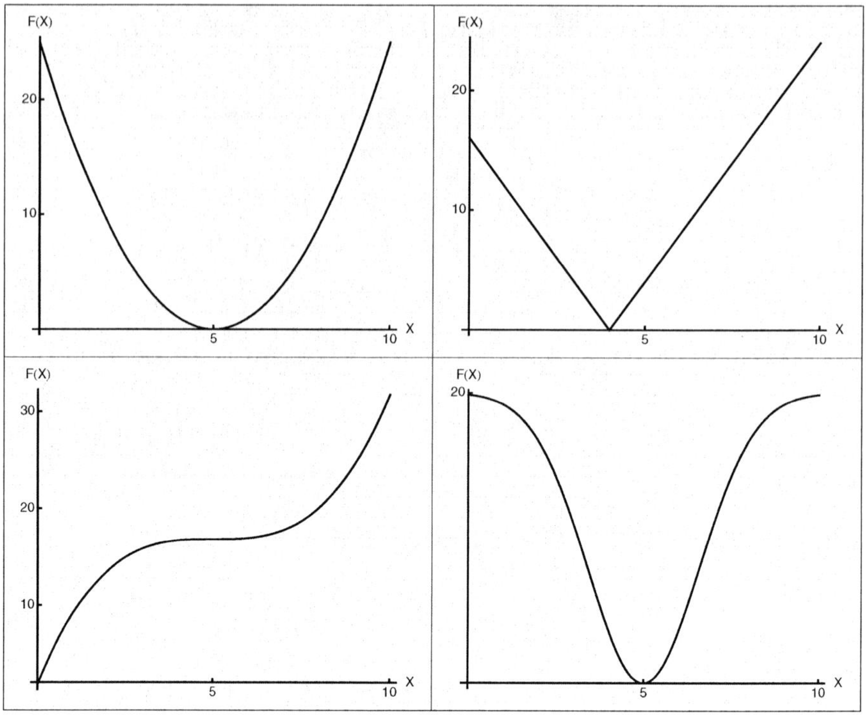

Figure 5.8: Convex and non convex functions. Top: convex functions. Bottom: non-convex functions.

Proof. Suppose that $f''(x) \geq 0$ for all x. Then f' is non-decreasing. Let $x \leq y$ and $z = \alpha x + (1 - \alpha y)$. We have:

$$f(z) - f(x) = \int_x^z f'(t)dt \leq f'(z)(z - x)$$
$$f(y) - f(z) = \int_z^y f'(t)dt \geq f'(z)(y - z)$$

Thus

$$f(z) \leq f(x) + f'(z)\lambda(y - x)$$
$$f(z) \leq f(y) - f'(z)(1 - \lambda)(y - x)$$

We multiply the first inequality by $(1 - \alpha)$, the second by α and add the inequalities. We get:

$$f(\alpha x + (1 - \alpha y)) \leq (1 - \alpha)f(x) + \alpha f(y)$$

Thus, by definition, f is convex.

Conversely: suppose that $f''(x) < 0$. By continuity, there exists $y > x$ such $f''(z) < 0$ for $z \in [x, y]$. Then, we use exactly the same argument as before and get:

$$f(\alpha x + (1 - \alpha y)) > (1 - \alpha)f(x) + \alpha f(y)$$

Which means that f is not convex. □

Theorem 5.3.2 *Let $f : \mathbb{R} \to \mathbb{R}$; then f is a strictly convex function if and only if $f''(x) > 0$.*

Proof is similar to previous one, changing inequalities to strict inequalities.

5.3.2 Convex Functions in Higher Dimensions

Definition 5.3.2 *$f : \mathbb{R}^n \to \mathbb{R}$ is said convex if for all $\alpha \in [0, 1]$, we have:*

$$f(\alpha X + (1 - \alpha)Y) \leq \alpha f(X) + (1 - \alpha)f(Y)$$

It is said strictly convex if for all $\alpha \in [0, 1]$, we have:

$$f(\alpha X + (1 - \alpha)Y) < \alpha f(X) + (1 - \alpha)f(Y)$$

Theorem 5.3.3 *Let $f : \mathbb{R}^n \to \mathbb{R}$; then f is a convex function if and only if is second order derivative $\nabla^2 \mathbf{f}(X)$ is a positive matrix for all X.*

Proof. We have to prove that, for all vectors V, $V \cdot \nabla^2 \mathbf{f}(X) \cdot V \geq 0$. We define the function $u(x) = f(X + xV)$. Then, u is a convex function:

$$\begin{aligned} u(\alpha x + (1 - \alpha y)) &= f(\alpha(X + xV) + (1 - \alpha)(X + yV)) \\ &\leq \alpha f(X + xV) + (1 - \alpha)f(X + yV) \\ &= \alpha u(x) + (1 - \alpha)u(y) \end{aligned}$$

Thus $u''(x) \geq 0$. And, in particular $u''(0) \geq 0$. But

$$\begin{aligned} u''(0) &= \lim_{h \to 0} \frac{u(h) - 2u(0) + u(-h)}{h^2} \\ &= \lim_{h \to 0} \frac{f(X + hV) - 2f(X) + f(X - hV)}{h^2} \\ &= V \cdot \nabla^2 \mathbf{f}(X) \cdot V \end{aligned}$$

□

Theorem 5.3.4 *Let $f : \mathbb{R}^n \to \mathbb{R}$; then f is a strictly convex function if and only if is second order derivative $\nabla^2 \mathbf{f}(X)$ is a strictly positive matrix for all X.*

Proof. Similar to previous one. □

5.4 Non Constrained Optimization

The goal of mathematical optimization lies in finding the minimum or maximum value of a given function and where it occurs. Recall that we have deliberately restricted our presentation to: (1) continuously differentiable functions; and (2) optimization on polyhedrons. Almost all the results we present here are valid for general convex subsets, but in more complicated forms, with more complicated proofs[2].

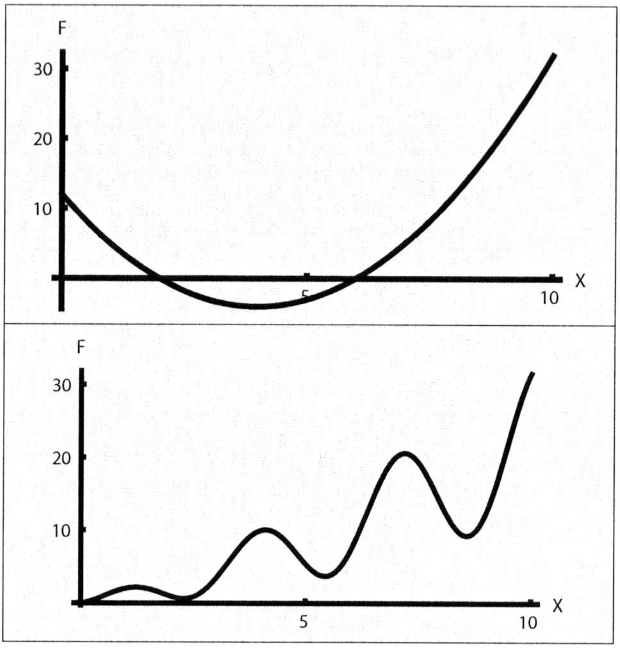

Figure 5.9: Optimization of a 1 dimensional function. How do we find the values of x where $f(x)$ is globally minimal? locally minimal? locally maximal? What happen in these points in terms of derivatives (slope), in terms of second order derivatives?

Let us give several definitions. Let f be a function from $\mathbb{R}^n$ to $\mathbb{R}$.

Definition 5.4.1 *A point $X^* \in \mathbb{R}^n$ is a global minimum of f is for all $X \in \mathbb{R}^n$, $f(X^*) \leq f(X)$.*

Definition 5.4.2 *A point $X^* \in \mathbb{R}^n$ is a local minimum of f if there is a neighbourhood of X^* in which $f(X^*) \leq f(X)$.*

5.4.1 Non Constrained Optimization in Dimension 1

In dimension 1, for a function $f : \mathbb{R} \to \mathbb{R}$, (see figure 5.9), we have:

[2]Modern and detailed presentations of optimization theory can be found in Boyd and Vandenberghe (2004) or Borwein and Lewis (2006).

Theorem 5.4.1 *Let x is be local minimum of function $f : \mathbb{R} \to \mathbb{R}$. Then $f'(x) = 0$.*

Proof. We use Taylor's formula at order 1.

$$f(x+h) = f(x) + f'(x)h + g(h)h \quad \text{with} \quad \lim_{h \to 0} g(h) = 0$$

Suppose that x is a local minimum, then $f(x) \leq f(x+h)$ for a small enough value of h. We have then: $f'(x)h + g(h)h \geq 0$. We choose $h = af'(x)$. For a small enough, we have:

$$a(f'(x))^2 + af'(x)g(af'(x)) \geq 0$$

$$af'(x)(f'(x)) + g(af'(x)) \geq 0$$

We divide previous expression by $af'(x)$, with a having the opposite sign to the one of $f'(x)$; we get $f'(x) + g(af'(x)) \leq 0$; we let a tend to 0, then as $g(af'(x))$ tends to 0, we get $f'(x) \leq 0$. This is a contradiction. □

Theorem 5.4.2 *Let x be a local minimum of function $f : \mathbb{R} \to \mathbb{R}$. Then $f'(x) = 0$ and $f''(x) \geq 0$. Reciprocally if $f'(x) = 0$ and $f''(x) > 0$, then x is a local minimum of function f.*

If x is a local minimum of function f, then $f(x+h) \geq f(x)$ for h small enough. We use Taylor's formula at order 2, together with the fact that $f'(x) = 0$. We have:

$$\frac{1}{2}h^2 f''(x) + h^2 g(h) \geq 0 \quad \text{with} \quad \lim_{h \to 0} g(h) = 0$$

With the same argument as in the proof of theorem 5.4.1, we show that $f''(x) \geq 0$. □

5.4.2 Non Constrained Optimization in Higher Dimensions

There is a relationship between optima of a n-D function (figure 5.10) and its gradients or higher order derivatives. At a minimum point, there is an increase in all directions: slopes must be positive in all directions; this is possible only if they are zero. More precisely, we have:

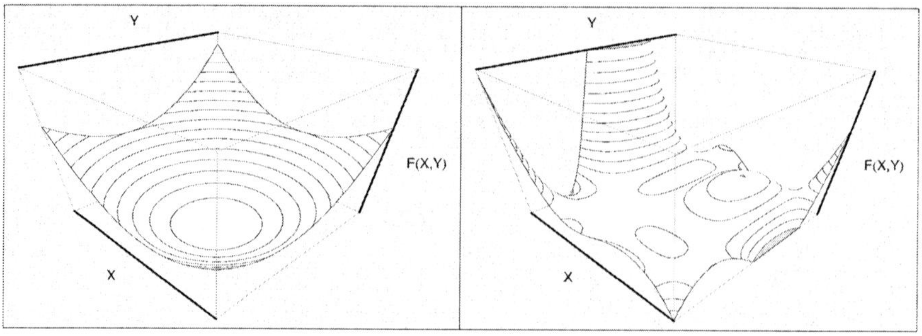

Figure 5.10: 3 dimensional plots of a 2 dimensional function. An optimization problem consists in finding the values of (x, y) where $f(x, y)$ is globally minimal, locally minimal?

Theorem 5.4.3 *Let X be a local minimum of a function $f : \mathbb{R}^n \to \mathbb{R}$. Then* $\nabla f(X) = 0$.

Proof. If X is a minimum point for F, According to Taylor's formula 5.2.1, we get:

$$\nabla f(X) \cdot H + g(H)\|(H)\| \geq 0 \text{ with } \lim_{\|(H)\| \to 0} g(H) = 0$$

With $H = a\nabla f(X)$, we get:

$$a(\nabla f(X) \cdot \nabla f(X)) + \|a\nabla f(X)\| g(a(\nabla f(X))) \geq 0$$

That is

$$a\|\nabla f(X)\|^2 + |a|\|\nabla f(X)\| g(a\nabla f(X)) \geq 0$$

Suppose $\|\nabla f(X)\| > 0$. With $a < 0$, we divide by $a\|\nabla f(X)\|$:

$$\|\nabla f(X)\| - ag(a(\nabla f(X))) \leq 0$$

We let a tend toward 0, we get $\|\nabla f(X)\| \leq 0$. This is a contradiction. And $\|\nabla f(X)\| = 0$ □

Theorem 5.4.4 *Let X is be local minimum of function $f : \mathbb{R}^n \to \mathbb{R}$. Then $\nabla f(X) = 0$ and $\nabla^2 \mathbf{f}(X)$ is a non-negative matrix, i.e., $H \cdot \nabla^2 \mathbf{f}(X) \cdot H \geq 0$ for all $H \in \mathbb{R}^n$.*

Proof. Let X be a local minimum of F. Let H a direction. We use Taylor's formula at order 2, with aH, together with the fact that $\nabla f(X) = 0$.

$$f(X + aH) = f(X) + a^2 \frac{1}{2} H \cdot \nabla^2 \mathbf{f}(X) \cdot H + \|aH\|^2 g(aH)$$

with $\lim_{a\to 0}\|g(aH)\| = 0$. Thus, dividing by a^2:

$$H \cdot \nabla^2 \mathbf{f}(X) \cdot H + 2\|H\|^2 g(aH) \geq 0$$

with $\lim_{a\to 0}\|g(aH)\| = 0$. This implies: $H \cdot \nabla^2 \mathbf{f}(X) \cdot H \geq 0$. As it is true for all vectors H, it means that $\nabla^2 \mathbf{f}$ is a non-negative matrix. □

Theorem 5.4.5 *Suppose that $f : \mathbb{R}^n \to \mathbb{R}$ and $X \in \mathbb{R}^n$ is such that $\nabla f(X) = 0$ and $\nabla^2 \mathbf{f}(X)$ is a positive matrix, i.e., $H \cdot \nabla^2 \mathbf{f}(X) \cdot H > 0$ for all $H \in \mathbb{R}^n$, then X is a strict local minimum of f.*

Proof. Still an immediate use of Taylor's formula at order 2. □

5.4.3 Minimization of Convex Functions

Theorem 5.4.6 *Let $f : \mathbb{R} \to \mathbb{R}$ a convex function: if f has a local minimum, this minimum is also a global minimum.*

Proof. Suppose that f has two local minima $x < y$; then $f'(x) = f'(y) = 0$ and as f' is increasing, $f'(z) = 0$ for $z \in [x, y]$. This means that f is constant on $[x, y]$ and the local minima have the same values. □

Theorem 5.4.7 *Let $f : \mathbb{R} \to \mathbb{R}$ a strictly convex function: if f has a local minimum, this minimum is an unique global minimum.*

Proof. Suppose that f has two global minima $x < y$; then $f(x) = f(y)$ according to previous theorem. But

$$f(\frac{x+y}{2}) \quad < \quad \frac{f(x)+f(y)}{2} = f(x)$$

□

Theorem 5.4.8 *Let $f : \mathbb{R}^n \to \mathbb{R}$ a convex function: if f has a local minimum, this minimum is also a global minimum.*

Proof. Suppose that f has two local minima X, Y; we define $g : \mathbb{R} \to \mathbb{R}$ defined by $g(t) = f(tX + (1-t)Y)$; it is obviously convex and has two minima in $t = 0$ and $t = 1$; we apply theorem 5.4.6 to g and get $g(0) = g(1)$ i.e., $f(X) = f(Y)$. □

Theorem 5.4.9 *Let $f : \mathbb{R}^n \to \mathbb{R}$ a strictly convex function: if f has a local minimum, this minimum is an unique global minimum.* □

5.5 Constrained Optimization

Now we are interested in finding the minimum of a function defined on a polyhedron. Several possible situations in 2-D are shown in figure 5.11.

Definition 5.5.1 *Given a polyhedron $K \subset \mathbb{R}^n$ and a function $f : K \to \mathbb{R}$, a solution of the constrained minimization problem $\mathcal{MIN}(f, K)$ is a point $X^* \in K$ such that, for all $X \in K$, we have $f(X^*) \leq f(X)$.*

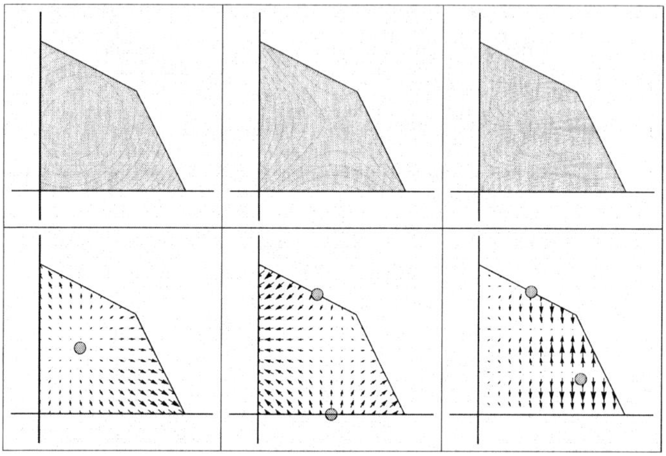

Figure 5.11: Finding the minimum of a function defined on a polyhedron. Different possibilities. Right: the function has one minimum value in the interior of the polyhedron. Middle: the function has minimal values on the boundaries. Left: the function has a complicated set of minimal values. Top: the function. Bottom: its gradient.

Theorem 5.5.1 *Let K a convex polyhedron. If X^* is a solution of the constrained program $\mathcal{MIN}(f, K)$ then $-\nabla f(X^*) \in N_K(X^*)$*

This theorem is illustrated in figure 5.12.

Proof. Let us recall the normal cone characterization theorem (5.1.5): $N_K(X)$ is the set of $Y \in \mathbb{R}^n$ such that for all $Z \in K$, we have: $(X - Z) \cdot Y \leq 0$. Now, let us express the Taylor's formula (Definition 5.2.1);

$$f(X + H) = f(X) + \nabla f(X) \cdot H + g(H)\|H\|$$

with $\lim_{\|H\| \to 0} g(H) = 0$.

Let us prove that, if $X^* \in K$ is a local minimum, then for all H such that $X^* + H \in K$, then $H \cdot \nabla f(X^*) \geq 0$. Let H be such that $X^* + H \in K$; then, K being convex, $X^* + aH \in K$ for $a \leq 1$. For a small enough, we have $f(X^* + aH) \geq f(X^*)$, i.e., $a\nabla f(X^*) \cdot H + g(aH)\|aH\| > 0$; we divide

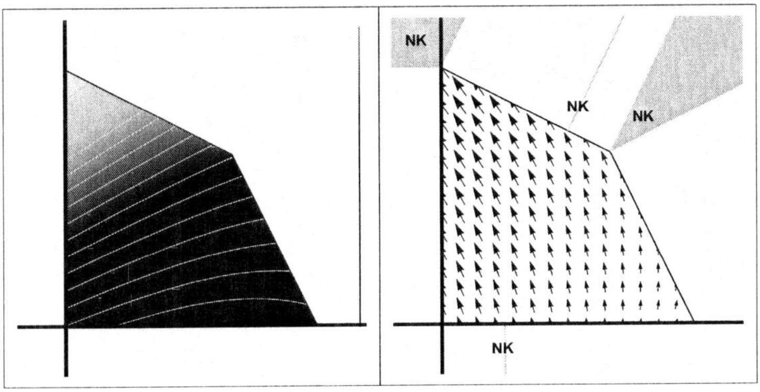

Figure 5.12: Constrained programming and normal cones. Minimum of the function occur at points in which the gradient of the function is opposite to the normal cone.

this inequality by a and let a tend towards 0. We get the expected result: $H \cdot \nabla f(X^*) \geq 0$.

Finally, we write $H \cdot \nabla f(X^*) = (X^* + H - X^*) \cdot \nabla f(X^*)$. We conclude that, for all $Z \in K$, $(Z - X^*) \cdot \nabla f(X^*) \geq 0$. This is equivalent to say that: $-\nabla f(X^*) \in N_K(X^*)$. □

5.6 Karush-Kuhn-Tucker Theorem

This is undoubtedly the most important result of the theory of constrained optimization. We give it in the simplified form allowed by our assumptions: (1) the continuity and differentiability of the function to optimize, and (2) the polyhedral structure of the constrained set.

Recall the definition 4.2.1 of the complementarity relationship between two numbers a and b: $a \perp b$ if $a \geq 0, b \geq 0, ab = 0$.

5.6.1 Karush-Kuhn-Tucker Theorem: Direct Form

Theorem 5.6.1 *Let K be a convex polyhedron defined by constraints $\sum_i a_{ki} x_i \geq b_k$. Let f be a function $K \to \mathbb{R}$. Then, if $X^* \in K$ is a local minimum of the function f, there exists coefficients $\lambda_k^* \geq 0$ such that*

$$0 = -\frac{\partial f}{\partial x_i}(X^*) + \sum_k \lambda_k^* a_{ki}$$

$$\lambda_k^* \perp (\sum_i a_{ki} x_i^* - b_k)$$

Proof. According to theorem 5.5.1, we have $-\nabla f(X^*) \in N_K(X^*)$. We use the normal cone characterization theorem 5.1.6: let

$K_0 = \{k \mid \sum_i a_{ki}x^*i - c_k = 0\}$ then $N_K(X^*)$ is generated by $-A^k$ for $k \in K_0$. Thus $-\nabla f(X^*) = -\sum_{k\in K_0} \lambda_k^* A^k$ for some $\lambda_k^* \geq 0$. We put $\lambda_k^* = 0$ for $k \notin K_0$ and we get the theorem. □

Coefficients λ_k^* are termed multipliers. Equivalent formulations of the Karush-Kuhn-Tucker theorem are:

Theorem 5.6.2 *Let K be a convex polyhedron defined by constraints $x_i \geq 0$ and $\sum_i a_{ki}x_i \geq b_k$. Let f be a function $K \to \mathbb{R}$. Then, if $X^* \in K$ is a local minimum of the function f, there exists λ_k^* such that*

$$\lambda_k^* \perp (\sum_i a_{ki}x_i^* - b_k)$$

$$x_i^* \perp (\frac{\partial f}{\partial x_i}(X^*) - \sum_k \lambda_k^* a_{ki})$$

Proof. According to previous theorem 5.6.1, there exists λ_k^* and ν_i^* such that

$$0 = -\frac{\partial f}{\partial x_i}(X^*) + \sum_k \lambda_k^* a_{ki} + \nu_i^*$$

$$\lambda_k^* \perp (\sum_i a_{ki}x_i^* - b_k)$$

$$\nu_i^* \perp x_i^*$$

We have:

$$\nu_i^* = \frac{\partial f}{\partial x_i}(X^*) - \sum_k \lambda_k^* a_{ki}$$

And the theorem is proved. □

Theorem 5.6.3 *Let K be a convex polyhedron defined by constraints $x_i \geq 0$ and $\sum_i a_{ki}x_i \geq b_k$. Let f be a function $K \to \mathbb{R}$. Then, if $X^* \in K$ is a local maximum of the function f, there exists λ_k^* such that*

$$\lambda_k^* \perp (\sum_i a_{ki}x_i^* - b_k)$$

$$x_i^* \perp (-\frac{\partial f}{\partial x_i}(X^*) - \sum_k \lambda_k^* a_{ki})$$

Proof is immediate. □

5.6.2 Karush-Kuhn-Tucker: Inverse Form

Theorem 5.6.4 *Let K be a convex polyhedron defined by constraints $\sum_i a_{ki}x_i \geq b_k$. Let f be a convex function $K \to \mathbb{R}$. Let $X^* \in K$ be such that* *there exists coefficients* $\lambda_k^* \geq 0$ *verifying*

$$0 = -\frac{\partial f(X^*)}{\partial x_i} + \sum_k \lambda_k^* a_{ki}$$

$$\lambda_k^* \perp (\sum_i a_{ki} x_i^* - b_k)$$

Then X^* *is a minimum of the function* f *on* K

Proof. We write Taylor's formula at order two and use the positivity of $\nabla^2 \mathbf{f}(X^*)$:

$$f(X^* + \alpha H) = f(X^*) + \alpha \nabla f(X^*) \cdot H$$
$$+ \alpha^2 \frac{1}{2} H \cdot \nabla^2 \mathbf{f}(X^*) \cdot H + \|\alpha H\|^2 g(\alpha H)$$
$$\textit{with } \lim_{\alpha \to 0} \|g(\alpha H)\| = 0$$

As $\nabla^2 \mathbf{f}(X^*)$ is a positive matrix, we have

$$f(X^* + \alpha H) \geq f(X^*) + \alpha \nabla f(X^*) \cdot H + \|\alpha H\|^2 g(\alpha H)$$
$$\textit{with } \lim_{\alpha \to 0} \|g(\alpha H)\| = 0$$

As in previous demonstrations, it is enough to prove that $\alpha \nabla f(X^*) \cdot H \geq 0$ for small enough admissible values of α, i.e., such that $X^* + \alpha H \in K$. By hypothesis, we have $\nabla f(X^*) = \sum_k \lambda_k^* a_{ki}$. Thus:

$$\nabla f(X^*) \cdot H = \sum_i (\sum_k \lambda_k^* a_{ki}) h_i$$

Let K_0 the set of indices k such as $\sum_i a_{ki} x_i^* - b_k > 0$. We have $\lambda_k^* = 0$ for $k \in K_0$. Thus:

$$\alpha \nabla f(X^*) \cdot H = \alpha \sum_i \sum_{k \notin K_0} \lambda_k^* a_{ki} h_i$$
$$= \sum_{k \notin K_0} \lambda_k^* \sum_i a_{ki} \alpha h_i$$
$$= \sum_{k \notin K_0} \lambda_k^* \sum_i a_{ki} (x_i^* + \alpha h_i - x_i^*)$$
$$= \sum_{k \notin K_0} \lambda_k^* (\sum_i a_{ki} (x_i^* + \alpha h_i) - b_k)$$

But $X^* + \alpha H \in K$. Thus, by defiinition of K, α is such $\sum_i a_{ki}(x_i^* + \alpha h_i) \geq b_k$. Thus alpha $\nabla F(X*)$. $H \geq 0$ and X^* is a local minimum. But we have seen with theorem 5.4.6 that a local minimum of a convex function was also a global minimum. Whence the result. □

Theorem 5.6.5 *Let K be a convex polyhedron defined by constraints $\sum_i a_{ki}x_i \geq b_k$. Let f be a strictly convex function $K \to \mathbb{R}$. Let $X^* \in K$ be such that there exists coefficients $\lambda_k^* \geq 0$ such that*

$$0 = -\frac{\partial f(X^*)}{\partial x_i} + \sum_k \lambda_k^* a_{ki}$$

$$\lambda_k^* \perp (\sum_i a_{ki}x_i^* - b_k)$$

Then X^* is the unique minimum of the function f on K

Proof. The principle is similar to previous one. □

5.6.3 Application 1

The Karush-Kuhn-Tucker theorem transforms an optimization problem into an algebraic one. The resulting algebraic problem can be complicated. We illustrate this with the computation of the solution of two constrained minimization problems. There are simpler ways to reach the solution, but this example is illustrative of the algorithmic difficulty of the complementarity problem we will have to deal with later.

The problem consist in finding the minimum of $f(x, y) = 3x^2 - 5xy + y^2 - 2x + 5y$ subject to $x \geq 0$, $y \geq 0$, $x + 2y \leq 1$, $2x + y \leq 1$. We have

$$\nabla f(x, y) = (6x - 5y - 2, -5x + 2y + 5)$$

Karush-Kuhn-Tucker conditions are

$$\begin{aligned} x &\perp 6x - 5y - 2 + \lambda_1 \\ y &\perp -5x + 2y + 5 + \lambda_2 \\ \lambda_1 &\perp 1 - x - 2y \\ \lambda_2 &\perp 1 - 2x - y \end{aligned}$$

We examine all cases.

It is obvious that $x = 0, y = 0$ is not a solution.

Are there solutions in the interior of the polyhedron? we would have $x > 0$, $y > 0$, $1 - x - 2y > 0$, $1 - 2x - y > 0$, thus $\lambda_1 = \lambda_2 = 0$; thus a linear system $6x - 5y - 2 = 0, -5x + 2y + 5 = 0$, whose solutions are $x = 21/13$, $y = 20/13$, they are not such that $x + 2y < 1$.

Are there solutions such that $x = 0$, $y > 0$, $1 - x - 2y > 0$, $1 - 2x - y > 0$? we would have we would have $\lambda_1 = \lambda_2 = 0$ and $Min(y, 2y + 5) = 0$, which is impossible.

Are there solutions such that $x > 0$, $y = 0$, $1 - x - 2y > 0$, $1 - 2x - y > 0$; we would have $\lambda_1 = \lambda_2 = 0$ and $6x - 2 = 0$; that is $x = 1/3$. We easily check that $x = 1/3$, $y = 0$, $\lambda_1 = 0$, $\lambda_2 = 0$ is a solution of the Karush-Kuhn-Tucker conditions.

Are there solutions such that $x > 0$, $y > 0$, $1-x-2y = 0$ and $-2x-y > -1$? we would have $\lambda_1 = 0$; eliminating y and λ_2, we get: $(9 - 17x)/5 \geq 0$, $(13x - 21)/5 \perp 7 - 16x)/5$. That implies $x \leq 9/17$, $x \geq 21/13$, $x \leq 16/7$, which is impossible.

Are there solutions such that $x > 0$, $y > 0$, $1-x-2y > 0$ and $1-2x-y = 0$? we would have: $\lambda_2 = 0$; we get; eliminating x and λ_1, we get $(20-13y)/5+\lambda_1 = 0$, $\lambda_1 \perp (-12y)/5 - 2y) = 0$, $-(5y + 9)/5 \geq 0$. Which is impossible.

Are there solutions such that $x > 0$, $y > 0$, $1 - x - 2y = 0$ and $1 - 2x - y = 0$? we would have $x = y = 1/3$. Then $6(1/3) - 5(1/3) - 2 + \lambda_1 = 0$, $-5(1/3) + 2(1/3) + 5 + \lambda_2 = 0$ and resulting value of λ_2 is negative.

Thus $x = 1/3$, $y = 0$ is the only solution of the Karush-Kuhn-Tucker conditions. This is the sought minimum.

5.6.4 Application 2

The problem consists in finding the minimum of $f(x, y) = 6x^2 - 6y^2 + xy - 3x + 3y$ subject to $x \geq 0$, $y \geq 0$, $x + 2y \leq 1$, $2x + y \leq 1$.
We have

$$\nabla f(x, y) = (12x + y - 3, x - 12y + 3)$$

Karush-Kuhn-Tucker conditions are

$$\begin{aligned} x &\perp 12x + y - 3 + \lambda_1 \\ y &\perp x - 12y + 3 + \lambda_2 \\ \lambda_1 &\perp 1 - x - 2y \\ \lambda_2 &\perp 1 - 2x - y) \end{aligned}$$

Exactly as before, we examine all cases and we find that $(x = 33/145, y = 39/145)$, $(x = 1/4, y = 0)$, $(x = 39/145, y = 33/145)$ are the solutions of the Karush-Kuhn-Tucker conditions. They are the sought local minima.

5.6.5 Karush-Kuhn-Tucker Conditions and Computing

Previous examples showed how to compute the solution of a constrained optimization program by enumerating all cases defined by the Karush-Kuhn-Tucker equations. For problems of a high dimensionality, this approach becomes intractable. We then use iterated algorithms that converge to a solution when applied to a restrictive set of functions. Such algorithms are presented in next chapters.

5.6.6 The Whole Karush-Kuhn-Tucker Theorem

The importance of the Karush-Kuhn-Tucker theorem comes from it simplicity and its generality; it is valid for more general situations, especially optimization not only on a polyhedron but also on a convex set.

Theorem 5.6.6 *Let $f(x_1, \ldots, x_n) \to f(x_1, \ldots, x_n)$ be a continuous function on a convex set K defined by inequalities:*

$$x_i \geq 0, g_k(x_1, \ldots, x_n) \geq 0$$

If $(x_1^*, \ldots, x_n^*)$ is a local minimum, there exist coefficients λ_k^* such that, in complementarity terms:

$$x_i^* \perp -\frac{\partial f}{\partial x_i}(x_1^*, \ldots, x_n^*) - \sum_k \lambda_k^* \frac{\partial g_k}{\partial x_i}(x_1^*, \ldots, x_n^*)$$
$$\lambda_k^* \perp g_k(x_1^*, \ldots, x_n^*)$$

Proof of this theorem and of its inverse can be found in Boyd and Vandenberghe (2004). □

5.6.7 Karush-Kuhn-Tucker Conditions, Multipliers and Prices

Karush-Kuhn-Tucker multipliers are closely related to prices. Suppose (1) that we know how prices p depend on the quantities x that are produced in an economic system; $x \to p(x)$, (2) that our production must satisfy a constraint $g(x) \geq 0$, (3) that we want to maximize our revenue $xp(x)$. Then, the generalization of the Karush-Kuhn-Tucker conditions can be written as:

$$x \perp (-p(x) + xp'(x)) - \lambda g'(x)$$
$$\lambda \perp g(x)$$

When there is an effective solution (i.e., such that $x > 0$, $p(x) > 0$), we get: $xp'(x) - \lambda g'(x) < 0$. So, we are interested in the signs of $p'(x)$ and $g'(x)$. Usually, one considers that $p'(x) < 0$ (prices decrease with the quantity that is sold). If $g'(x) < 0$ then the constraint is related to a maximal threshold. If $g'(x) > 0$ then the constraint is related to a minimal threshold. The first case is the most common. Then $\lambda g'(x)$ can be see as a quantity that is added to prices when constraint is such that $g(x) = 0$. This is the reason why λ is referred to as a shadow price.

5.7 Vector Fields

5.7.1 Definitions

Let us begin with the definition of a vector field in 2-D, which allows illustrative plots.

Definition 5.7.1 *A vector field in 2-D is a function* $V : H \subset \mathbb{R}^2 \to \mathbb{R}^2$ *defined as* $(x, y) \to V(x, y) = (h(x, y), k(x, y))$.

A vector field associates a vector to each point of the plane. A gradient is obviously a vector field. In a n-D space, we have the following definition.

Definition 5.7.2 *A vector field is a function from an open subset of* $\mathbb{R}^n$ *to* $\mathbb{R}^n$. *We have:* $F : H \subset \mathbb{R}^n \to \mathbb{R}^n$. *In term of vector coordinates, we have:*

$$X = (x_1, \ldots, x_n) \to F(X) = (f_1(x_1, \ldots, x_n), \ldots, f_n(x_1, \ldots, x_n))$$

In n-D, as in 2-D, a gradient is obviously a vector field.

5.7.2 Vector Fields and Gradients

Not every vector field is a gradient. One of the most important calculus theorems is the following.

Theorem 5.7.1 *Let* V *be a function* $\mathbb{R}^2 \to \mathbb{R}^2$ *defined by* $V(x, y) = (h(x, y), k(x, y))$. *Then it is the gradient of a function* $F : \mathbb{R}^2 \to \mathbb{R}$ *if only if:*

$$\frac{\partial h(x, y)}{\partial y} = \frac{\partial k(x, y)}{\partial x}$$

Proof. If $V = \nabla f$ is a gradient. Then by definition of $\nabla^2 \mathbf{f}$, it is clear that

$$\frac{\partial h(x, y)}{\partial y} = \frac{\partial k(x, y)}{\partial x} = \frac{\partial^2 f(x, y)}{\partial x \partial y}$$

Now, suppose that this equation is satisfied. We fix a starting point $X_0 = (x_0, y_0) \in H$. Let $X = (x, y) \in H$. We define $f(X)$ defined as

$$f(x, y) = \int_{x_0}^{x} h(x, y_0)dx + \int_{y_0}^{y} k(x, y)dy$$

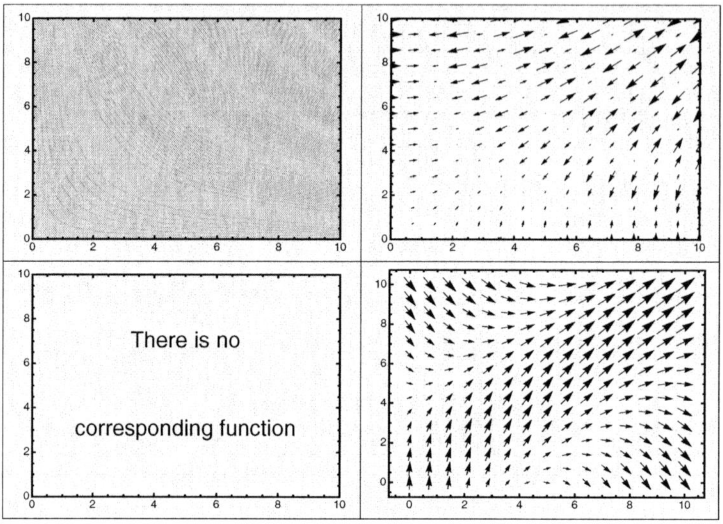

Figure 5.13: Gradients and vector fields. Top: a vector field (right) which is a gradient: $y\cos(xy), x\cos(xy)$ is the gradient of a function $\sin(xy)$ (left). Bottom: a vector field (right) which is not a gradient: $(\sin((x+y)/16), \cos((x-y)/3))$. There are no evident difference between both situations.

Then, we have

$$\begin{aligned}
\frac{\partial f(x,y)}{\partial x} &= h(x,y_0) + \int_{y_0}^{y} \frac{\partial k(x,y)}{\partial x} dy \\
&= h(x,y_0) + \int_{y_0}^{y} \frac{\partial h(x,y)}{\partial k} dy \\
&= h(x,y_0) + (h(x,y) - h(x,y_0)) \\
&= h(x,y) \\
\frac{\partial f(x,y)}{\partial y} &= k(x,y)
\end{aligned}$$

□

Remark. The function $f(x,y)$ is computed as an integral along a path $(x_0,y_0),(x_0,y),(x,y)$. It can be shown that we would have obtained the same solution with any other path from (x_0,y_0) to (x,y). One important issue, in the general case, i.e., when the vector field is not defined everywhere but on a subset $K \subset \mathbb{R}^2$, comes from that, we have then to define a path included in K and that the result of integration may depend on this path: the result of the integral on two paths starting from (x_0,y_0) and ending in (x,y) may differ when it is not possible to continuously transform one of these paths into the other one. The analysis of this situation gives important theorems of calculus (Lang, 1986).

In n-D, we have:

Theorem 5.7.2 *The vector field* $V : \mathbb{R}^n \to \mathbb{R}^n$ *defined by* $V(x_1, \ldots, x_n) = v_1(x_1, \ldots, x_n), \ldots, v_n(x_1, \ldots, x_n)$ *is the gradient of a function* F *if only if*

$$\frac{\partial v_i(x,y)}{\partial x_j} = \frac{\partial v_j(x,y)}{\partial x_i}$$

This condition is obviously equivalent to the symmetry of the Jacobian matrix

$$JF(X) = \begin{pmatrix} \frac{\partial v_1(x_1,\ldots,x_n)}{\partial x_1} & \ldots & \frac{\partial v_1(x_1,\ldots,x_n)}{\partial x_n} \\ \ldots & \ldots & \ldots \\ \frac{\partial v_n(x_1,\ldots,x_n)}{\partial x_1} & \ldots & \frac{\partial v_n(x_1,\ldots,x_n)}{\partial x_n} \end{pmatrix}$$

Here, we must pay attention that condition of theorem 5.7.2 is very restrictive: in n-D, for $n > 2$, most of the vector fields are not gradients.

5.7.3 Vector Fields and Differential Equations

There is a strong link between vector fields and differential equations (see figure 5.14).

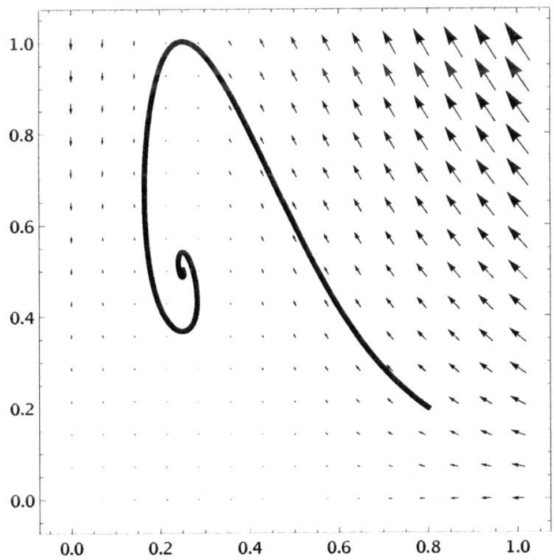

Figure 5.14: Vector field and differential equation. Trajectory.

Definition 5.7.3 *The differential equation associated to a vector field* $V : \mathbb{R}^n \to \mathbb{R}^n$ *is expressed as:*

$$X'(t) = F(X(t))$$
$$X(0) = X_0$$

In figure 5.14, we have depicted, in 2-D,a vector field and the trajectory of the related differential equation. One observes that the trajectory ends at a given point. One suspect that this is the case for all possible trajectories, whatever is their starting point. This can be proven by the mathematical analysis of the differential equation.

5.8 Digression: Fixed Point Theorems

In this section, we sketch a proof of the Bouwer's and Kakutani's theorem that are the most important theorems about convex sets[3]. Mathematically, this section is a little bit more difficult than the other sections of this book.

5.8.1 Brouwer' Theorem

Let $K \subset \mathbb{R}^n$ a compact, convex set and $F : K \to K$ a continuous function (figure 5.15 left).

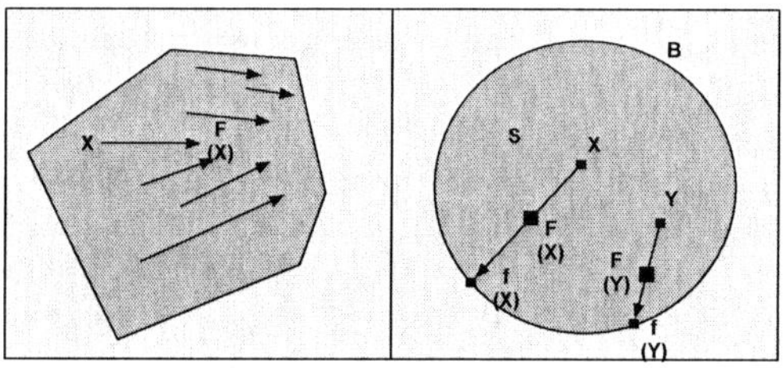

Figure 5.15: Endomorphisms of convex sets and retractions. Left: An application from a convex set of the plane to itself. Right: A function f of the unit ball B to itself that has no fixed point and the corresponding retraction from B^n to sphere S^{n-1}.

Theorem 5.8.1 *F has a fixed point: there exists $X \in K$ such that $F(X) = X$.*

We start the proof of theorem 5.8.1 by the special case of the unit balls. After that, we will prove that it is true for compact convex subsets of $\mathbb{R}^n$ (theorem 5.8.5). Let $B^n = \{X, \|X\| \leq 1\}$ be the unit ball. Let $S = \{X \mid \|X\| = 1\}$ be the unit sphere, the boundary of B^n. Let $B^n - S^{n-1}$ be the interior of B^n. Proof is based on the following definition.

Definition 5.8.1 *A retraction is a continuous function: $G : B^n \to S^{n-1}$ such that $G(X) = X$ if $X \in S$.*

[3]Historical references are Brouwer (1912); Kakutani (1941). Modern presentations can be found in Todd (1976); Border (1989); Istratescu (2001). The proof we give can be found in Rogers (1980).

The outline of the proof of theorem 5.8.1 for the unit ball B^n proceeds as follows. Firstly, we prove the theorem when F is a continuously differentiable function. To do that we will prove that Brouwer's theorem for B^n is equivalent to the non existence of a retraction (theorem 5.8.2) and that a retraction cannot exist (theorem 5.8.3). Then we will prove that the theorem is true for continuous functions (theorem 5.8.4).

Theorem 5.8.2 *Brouwer's theorem for the unit ball B^n and a continuously differentiable function is true if and only if there exists no retraction.*

Only if part of the proof. Suppose that there exists a retraction $F : B^n \to S^{n-1} \subset B^n$. We consider $G : B^n \to B^n$ defined by $G(X) = -F(X)$. Obviously, G is a continuous function without fixed point. Thus Brouwer's theorem is false.

If part of the proof. Suppose now that Brouwer's theorem is false. Let $F : B^n \to B^n$ be continuous and without no fixed point. Then we consider (figure 5.15 right) the function, which to $X \in B^n$ associates the point $Y \in S^{n-1}$ such that X, $F(X)$ and Y are on the same line (in that order). This is a retraction. □

Theorem 5.8.3 *A continuously differentiable function can't be a retraction.*

Proof. We suppose to be known the following results:

- If a polynomial $P(x)$ is identically zero on a non empty open interval $[a, b]$, it is identically zero on $\mathbb{R}$.
- If A and B are matrices $det(A + xB)$ is a polynomial in x.
- A continuous function from a compact $K \subset \mathbb{R}^n$ to $\mathbb{R}$ is the uniform limit of a sequence of differentiable functions from K to $\mathbb{R}$.
- A compact subset of $\mathbb{R}^n$ is homeomorph to a some unit ball.
- If G is a differentiable function from $K \subset \mathbb{R}^n$ to $\mathbb{R}^m$, with derivative JG, then

$$\|G(X) - G(Y)\| \leq Max_{Z \in K} \|JG(Z)\| \|X - Y\|$$

- If G is a diffeomorphism between two open subsets A and B of $\mathbb{R}^n$, with derivative JG, we have

$$\int_A Det(JG(X))dX = \int_B dY$$

Suppose that there exists a retraction G. We put $G_t(X) = (1 - t)$ $X +$ $tG(X)$. Clearly G_t is a function from B^n to B^n. Let c be the maximal value of the Jacobian of F for $X \in B^n$: $c = Sup_{X \in B^n} \|JG(X)\|$. Let $\alpha = 1/(1 + c)$. G_t is characterized in the following lemmas 5.8.1, 5.8.2, 5.8.3, 5.8.4.

Lemma 5.8.1 *If $t < \alpha$, G_t is injective.*

Proof. We have:

$$G_t(X) - G_t(Y) = (1-t)(X-Y) + t(G(X) - G(Y))$$

As $\|G(X) - G(Y)\| \leq c\|X - Y\|$, we deduce

$$\|G_t(X) - G_t(Y)\| \geq |1-t|\|X-Y\| - ct\|X-Y\|$$

We conclude that G_t is injective for $t < 1/(1+c)$. □

Lemma 5.8.2 *If $X \in B^n - S^{n-1}$, then $G_t(X) \in B^n - S^{n-1}$.*

Proof. Let $\|X\| < 1$. We have:

$$\begin{aligned}\|G_t(X)\| &= \|(1-t)X + tG(X)\| \\ &\leq (1-t)\|X\| + t\|G(X)\| \\ &= (1-t)\|X\| + t < 1\end{aligned}$$

□

Lemma 5.8.3 *If $t < \alpha$, G_t is a local diffeomorphism: any point X as a open neighbourhood V such that G_t is a diffeomorphism for U to $G_t(U)$.*

Proof. We have

$$\begin{aligned}\|JG_t(X)\| &= \|(1-t)Id + tJG(X)\| \\ &\geq (1-t)\|Id\| - t\|JG(X)\| \\ &\geq (1-t) - ct\|X\| \geq (1-t) - ct\end{aligned}$$

Thus $\|JG_t(X)\| > 0$ for $t < 1/(1+c)$. This means that G_t is a local diffeomorphism. □

Lemma 5.8.4 *If $t < \alpha$, then G_t is surjective.*

Proof. Let $Y \in B^n$. If $Y \in S$ then $G_t(Y) = Y$ and $Y \in G_t(S)$. Suppose now that $Y \notin G_t(B^n)$. Let $V \in B^n - S^{n-1}$ and $X = G_t(V)$. According to lemma 5.8.3 $X \in B^n - S^{n-1}$ There exists Z on the line between X and Y, which is on the boundary of $G_t(B^n - S^{n-1})$. As X and Y are in $B^n - S^{n-1}$, $Z \in B^n - S^{n-1}$.

According to lemma 5.8.1 G_t is a local diffeomorphism and thus the image of an open set is an open set: $G_t(B^n - S^{n-1})$ is open and Z which is on its boundary is not in $G_t(B^n - S^{n-1})$.

We consider a sequence $X_n \in B^n - S^{n-1}$ such that $Z = \lim_n G_t(X_n)$. We extract from X_n a subsequence converging towards a point U. It is clear that $Z = G_t(U)$.

We have altogether $Z \in G_t(B^n)$ and $Z \notin G_t(B^n - S^{n-1})$. Thus $Z \in G_t(S^{n-1})$: there exists $W \in S$ such that $Z = G_t(W)$; we have $G(W) = W$; but $G_t(W) = (1-t)W + tG(W) = (1-t)Wt + tW = W$. Thus $Z \in S$. This is a contradiction and lemma 5.8.4 is proven. □

End of proof of theorem 5.8.3. We put:

$$P(t) = \int_{\overset{\circ}{B^n}} Det(JG_t(X))dX$$

Then, we have:

- $P(t)$ is a polynomial in t; this is due to the fact that $G_t(W) = W + t(G(W) - W)$ and that the determinant is a polynomial;
- According to previous lemmas, if $t < \alpha$, G_t is a differentiable bijection of $B^n - S^{n-1}$, we may change variables: $X \to Y = G_t(X)$. We have:

$$P(t) = \int_{B^n} dY = Volume(B^n)$$

- We have $G_1 = G$ and G is a retraction; thus for all $X \in B$, $G_1(X) \in S$ and $\|G_1(X)\|^2 = 1$. The gradient $\nabla\|G_1(X)\|^2$ of the constant function $X \to \|G_1(X)\|^2$ is zero. But we have

$$\nabla\|G_1(X)\|^2 = 2JG_1(X) \cdot G1(X)$$

We have proven that for all $X \in B$, for all H in $\mathbb{R}^n$, we have: $(JG_1(X) \cdot H) \cdot G_1(X) = 0$; this is possible, only if $JG_1(X)$ is not a bijection; thus $Det(JG_1(X)) = 0$; as this is valid for all X, we have: $P(1) = 0$.
- Two polynomials which coincide on a open set coincide everywhere, thus: $Volume(B^n) = P(0) = P(1) = 0$.

This is a contradiction. □

Theorem 5.8.4 *Brouwer's theorem. Part 2. Let $F : B^n \to B^n$ a continuous function. Then F has a fixed point.*

Proof. F as the uniform limit of continuously differentiable functions F_n. Each one has a fixed point X_n. We consider X an accumulation point in X_n. Then it is obvious that X is a fixed point of F.

Theorem 5.8.5 *Brouwer's theorem. Part 3. Let K a compact, convex set and $F : K \to K$ a continuous function. Then F has a fixed point.*

If K is a convex set, there is a homeomorphism H between K and an unit sphere B. Then $H^{-1} \circ F \circ H : B \to B \to$ has a fixed point X. $H(X)$ is a fixed point of F.

5.8.2 Kakutanis' theorem

Definition 5.8.2 *A multi application $F : U \mapsto V$ is a function that associate to any $X \in U$ a subset $F(X) \subset V$.*

Definition 5.8.3 *A multi application F is closed if its graph $\{(X,Y) \mid X \in U, Y \in F(X)\}$ is a closed subset of $U \times V$.*

Definition 5.8.4 *A fixed point of a multi application from X to itself, $F : U \mapsto U$ is an element $X \in U$ such that $X \in F(X)$.*

Let us express the important Kakutani's theorem that will be used in the next parts:

Theorem 5.8.6 *Let K a compact convex subset of $\mathbb{R}^n$. Let F a multi application $F : K \mapsto K$ which is closed and such that for all X, $F(X)$ is a non empty convex subset of K. Then F has a fixed point.*

Proof. Let $\epsilon > 0$. And $X_1, \ldots X_n$ such that $K \subset \bigsqcup B(X_i, \epsilon)$. Let G_i an a partition of unity associated to balls $B(X_i, \epsilon)$. We have $\sum_i G_i(X) = 1$ for all $X \in K$ and $G_i(X) = 0$ if $X \notin B(X_i, \epsilon)$. We choose $Y_i \in F(X_i)$. We put $g(X) = \sum_i G_i(X) Y_i$. It is a continuous application from K to K. It has a fixed point Z. In this way, to $\epsilon = 1/n$, we associate Z_n. Then it remains to prove that an accumulation point of the sequence Z_n is a fixed point of the multi application F. If not, as F is closed and $Z \notin F(Z)$ we would have $d(Z, F(Z)) \geq \mu > 0$. Then $U = \{Y \in K \mid d(Y, F(Z)) < d(Z, F(Z))\}$ is an open subset such that $F(Z) \in U$ and $d(Z, U) > 0$. As F is a closed application, there exists ν such that $d(X, Z) \leq \nu$ implies $F(X) \subset U$. Let n such that $1/n + d(Z, Z_n) \leq \nu$. In the computation of $g(Z_n) = \sum_i G_i(Z_n) Y_i$, we have: if $d(X_i, Z_n) > 1/n$ then $G_i(Z_n) = 0$, if $d(X_i, Z_n) \leq 1/n$ then $Y_i \in U$. We conclude that $g(Z_n) \in U$. This is a contradiction. □

Let us prove an important theorem due to John von Neumann.

Theorem 5.8.7 *Let K and H convex compacts. Let U and V two closed subsets of $K \times H$ such that for all $X \in K$, $U(X) = \{Y \mid (X,Y) \in U\}$ is a closed convex and non empty subset of H, for all $Y \in H$, $V(Y) = \{X \mid (X,Y) \in V\}$ is a closed convex and non empty subset of K. Then $U \bigcap V$ is not empty.*

Proof. We consider the multi application $F : K \times H \mapsto K \times H$, $(X,Y) \rightarrow V(Y) \times U(X)$. It is closed. We apply Kakutani's theorem. □

Chapter 6

Variational Inequality and Complementarity

6.1 Variational Inequality Problems

We have shown (1) that vector fields were generalizations of gradients (definition 5.7.2), and (2) that optimization of a function f on a convex set K was related to a condition on the gradient of f and the normal cones $N_K(X)$ in points X of K (theorem 5.5.1): a solution X^* is such that:

$$-\nabla f(X^*) \in N_K(X^*)$$

It therefore follows that the definition of a variational inequality is:

Definition 6.1.1 *Let $K \subset \mathbb{R}^n$ a convex polyhedron and $F : K \to \mathbb{R}^n$ a vector field. The variational inequality problem $\mathcal{VI}(F, K)$ consists in finding $X^* \in K$ such that $-F(X^*) \in N_K(X^*)$.*

According to the characterization theorem of normal cones (Theorem 5.1.6), the definition of a variational inequality problem is equivalent to:

Definition 6.1.2 *Let $K \subset \mathbb{R}^n$ a convex polyhedron and $F : K \to \mathbb{R}^n$ a vector field. The variational inequality problem $\mathcal{VI}(F, K)$ consists in finding $X^* \in K$ such that for all $X \in K$, we have $F(X^*) \cdot (X - X^*) \geq 0$.*

Thus the variational inequality problem can be seen as a generalization of convex programming. In the forthcoming sections we draw upon the work of Nagurney (1993)[1].

[1]Other suggested readings are Dafermos (1980b); Konnov (2001); Facchinei and Pang (2003).

6.1.1 Characterization of Solutions

If X^* is in the interior of K, it is easy to show that it is a solution of the variational inequality if and only if $F(X^*) = 0$: one may find ϵ such that $X \in K$ for all X such that $\|X^* - X\| \leq \epsilon$; we have thus $F(X^*) \cdot H \geq 0$ for all H such that $\|H\| \leq \epsilon$; thus $F(X^*) \cdot H \geq 0$ for all H; this implies $F(X^*) = 0$. Other solutions corresponding to non zero values of $F(X^*)$ are on the boundary of K. Figure 6.1 represents some examples of solutions of variational inequalities.

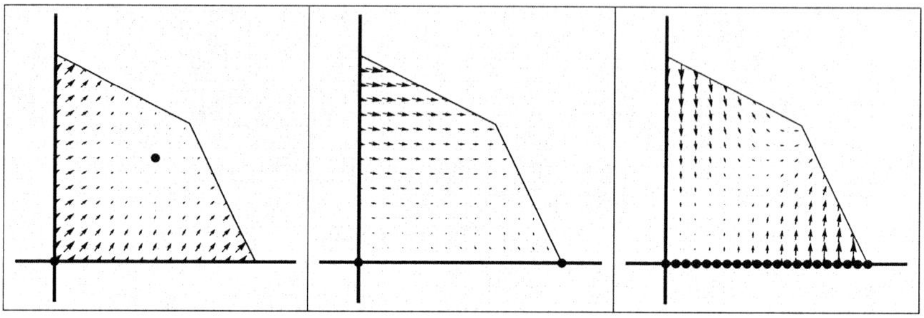

Figure 6.1: Solutions of a variational inequality problem. Left: there is a interior solution and a solution on the boundary. Middle: there are several solutions on the boundary. Right: there is a continuous set of solutions on the boundary.

6.1.2 Example

To illustrate the computational issue addressed by variational inequalities, let us attempt to solve a simple problem. Let the constrained set $K \subset \mathbb{R}^2$ be the set of (x, y) such that $x \geq 0$, $x \geq 0$, $x + 2y \leq 1$, $2x + y \leq 1$, and the vector field is $(x, y) \to F(x, y)$ with $F(x, y) = (x + 5y - 3, y - x + 2)$. As in computations given at the end of section 5.6, we examine the possible cases for a point in K, being in the interior, being on a face, being on an edge; and even with a very small example, there are many alternatives. We successively examine all these alternatives:

- an interior point
- a point (x, y) with $x = 0, 0 < y < 1/2$
- a point (x, y) with $y = 1/2, x = 0$
- a point (x, y) with $0 < x < 1/3, x + 2y = 1$
- a point (x, y) with $0 < y < 1/3, 2x + y = 1$
- $(0, 0)$
- $(0, 1/2)$
- $(1/2, 0)$
- $(1/3, 1/3)$

It is immediate to show that only one of these points is a solution of the variational inequality problem. The solution is unique; it is $(1/2, 0)$. It is easy to see how quickly the number of possibilities increases with the dimension of the problem. Fortunately, there are dedicated algorithms available to compute the solution for special cases of variational inequality problems.

6.1.3 Equivalence Theorems

Definition 6.1.3 *Let $K \subset \mathbb{R}^n$ be a convex polyhedron. Let f be a function $K \to \mathbb{R}$. We note $\mathcal{MIN}(K, f)$ the problem of finding $X^* \in K$ which minimizes $f(X)$ subject to $X \in K$.*

We have:

Theorem 6.1.1 *If X^* is solution of $\mathcal{MIN}(K, f)$ then it is a solution of the variational inequality problem $\mathcal{VI}(K, \nabla f)$.*

Proof. This is an obvious consequence or results of the preceding section: if X^* is solution of $\mathcal{MIN}(K, f)$, then we have: $-\nabla f(X^*) \in N_K(X^*)$: it is a solution of the variational inequality $\mathcal{VI}(C, \nabla f)$. □

Theorem 6.1.2 *Let f be a convex function $K \to \mathbb{R}$. Then any solution of the minimization problem $\mathcal{MIN}(K, f)$ is a solution of the variational inequality problem $\mathcal{VI}(K, \nabla f)$ and reciprocally.*

Proof. Knowing previous theorem, it is enough to prove that a solution of $\mathcal{VI}(K, \nabla f)$ is a solution of $\mathcal{MIN}(K, f)$. Then the proof is immediate using theorems 5.4.4 and 5.4.6. □

Theorem 6.1.3 *Let $K \subset \mathbb{R}^n$ be a convex polyhedron and $F : K \to \mathbb{R}^n$ a vector field. If F is such that the matrix*

$$\mathbf{JF}(X) = \begin{pmatrix} \dfrac{\partial f_1(x_1, \ldots, x_n)}{\partial x_1} & \cdots & \dfrac{\partial f_1(x_1, \ldots, x_n)}{\partial x_n} \\ \cdots & & \cdots\cdots \\ \dfrac{\partial f_n(x_1, \ldots, x_n)}{\partial x_1} & \cdots & \dfrac{\partial f_n(x_1, \ldots, x_n)}{\partial x_n} \end{pmatrix}$$

is symmetric, then a solution of $\mathcal{MIN}(K, f)$, where the function f is such that $F = \nabla f$, is a solution of the variational inequality problem $\mathcal{VI}(F, K)$.

Proof. According to theorem 5.7.2 of preceding section, we know that, due to the symmetry of matrix M, there exists a function f such that $F = \nabla f$. Then a solution of $\mathcal{MIN}(K, f)$ is a solution of $\mathcal{VI}(K, \nabla f)$ that is a solution of $\mathcal{VI}(K, F)$. □

The following theorem is important, both to prove existence theorems and for the construction of algorithms.

Theorem 6.1.4 *Let K be a closed, convex set. Let P_K denote the projection on K. Then X^* is a solution of the variational inequality problem $\mathcal{VI}(F,K)$ if and only if it is a fixed point of the application $X \to P_K(X - \gamma F(X))$; that is if $X^* = P_K(X^* - \gamma F(X^*))$.*

Proof. Let X^* a solution of the variational inequality. For all $X \in K$, we have $F(X^*) \cdot (X - X^*) \geq 0$. We multiply both terms by γ and add $X^* \cdot (X - X^*)$, we get $X^* \cdot (X - X^*) \geq (X^* - \gamma F(X^*)) \cdot (X - X^*)$ which, according to theorem 5.1.2 is a characterization of the projection of $X^* - \gamma F(X^*)$ on K.

If $X^* = P_K(X^* - \gamma F(X^*))$, then according to theorem 5.1.2, we have: $(X^* - (X^* - \gamma F(X^*)) \cdot (X^* - X) \leq$ for all $X \in K$. This is exactly $F(X^*) \cdot (X - X^*) \geq 0$ for all $X \in K$. Thus X^* is a solution of the variational inequality. □

6.1.4 Existence Theorems

Theorem 6.1.5 *If $K \subset \mathbb{R}^n$ is a compact polyhedron, if $F : K \to \mathbb{R}^n$ is continuous function, then $\mathcal{VI}(F,K)$ has a solution.*

Proof. It is sufficient to apply Brouwer's theorem (theorem 5.8.1) to the continuous function $X \to P_K(X - \gamma F(X))$. □

We have previously obtained results about the existence of a solution for a minimization problem $\mathcal{MIN}(f,K)$ with a convex function f. Existence theorems for variational inequalities $\mathcal{VI}(F,K)$ involve assumptions on function F that are generalizations to vector fields, i.e., functions $F : K \subset \mathbb{R}^n \to \mathbb{R}^n$, of the assumption of convexity for a function $f : K \subset \mathbb{R}^n \to \mathbb{R}$.

Definition 6.1.4 *F is monotone on K if $(X - Y) \cdot (F(X) - F(Y)) \geq 0$ for all $X, Y \in C, X \neq Y$.*

It is immediate that a linear function $F(X) = A \cdot X + B$ is monotone if and only if A, that is $\nabla F(X)$, is a non negative matrix. More generally, it can be proven that, if a differentiable function $F : K \subset \mathbb{R}^n \to \mathbb{R}^n$ is monotone, then $\nabla F(X)$ is a non negative.

Definition 6.1.5 *F is strictly monotone on K if $(X - Y) \cdot (F(X) - F(Y)) > 0$ for all $X, Y \in C, X \neq Y$.*

Definition 6.1.6 *F is strongly monotone on K if there exists $c > 0$ such that $(X - Y) \cdot (F(X) - F(Y)) > c\|X - Y\|^2$ for all $X, Y \in K, X \neq Y$.*

Definition 6.1.7 *F is pseudo monotone if for all X, Y, $(Y - X) \cdot F(X) \geq 0$ implies $(Y - X) \cdot F(Y) \geq 0$.*

Theorem 6.1.6 *If F monotone then it is pseudo monotone.*

Proof. Let F monotone. Then $(X - Y) \cdot (F(X) - F(Y)) \geq 0$ for all $X, Y \in C, X \neq Y$. Let $X, Y \in K$ such that $(Y - X) \cdot F(X) \geq 0$, thus $(Y - X) \cdot F(Y) = (Y - X) \cdot (F(Y) - F(X) + F(X)) = (Y - X) \cdot (F(Y) - F(X)) + (Y - X) \cdot F(X)) \geq 0$. □

Theorem 6.1.7 *If F pseudo monotone then the solution set of $\mathcal{VI}(F, K)$ coincides with the set of X^* such that for all Y, $(Y - X^*) \cdot F(Y) \geq 0$.*

Proof. Let X^* in the solution set of $\mathcal{VI}(F, K)$. We have for all Y, $(Y - X^*) \cdot F(X^*) \geq 0$, thus, as F is pseudo monotone, for all Y, $(Y - X^*) \cdot F(Y) \geq 0$.

Reciprocally, let $X^* \in K$ such that for all Y, $(Y - X^*) \cdot F(Y) \geq 0$. Let $\lambda \in [0, 1]$ and $X_\lambda = \lambda Y + (1 - \lambda X)$; X_λ is in K. Then, we have $0 \leq (X_\lambda - X) \cdot F(X_\lambda) = \lambda(Y - X^*) \cdot F(X_\lambda)$; we divide by λ; we get $0 \leq (Y - X^*) \cdot F(X^*)$; we let λ tend towards 0. We get $0 \leq (Y - X^*) \cdot F(X_\lambda)$. Thus X^* is a solution of $\mathcal{VI}(F, K)$. □

Theorem 6.1.8 *If F is pseudo monotone on K, the solution set of $\mathcal{VI}(F, K)$ is a (possibly empty) closed convex subset of K.*

Proof. According to previous theorem, the solution set of $\mathcal{VI}(F, K)$ is the intersection of the closed convex subsets $H_Y = \{X \in K \mid (X - Y) \cdot F(Y) \geq 0$. □

Theorem 6.1.9 *If F strongly monotone on K, then $\mathcal{VI}(F, K)$ has a solution.*

Proof. If K is bounded, it is compact and we apply previous theorem. We may suppose that K is not bounded. Let $X_0 \in K$. Then

$$\begin{aligned}(X - X_0) \cdot F(X) &= (X - X_0) \cdot (F(X) - F(X_0)) \\ &\quad + (X - X_0) \cdot F(X_0) \\ &\geq \tau \|X - X_0\|^2 + (X - X_0) \cdot F(X_0)\end{aligned}$$

Thus $(X - X_0) \cdot F(X) \to +\infty$ when $\|X\| \to +\infty$. There exists r such that for $\|X\| \geq r$, $(X \quad X_0) \cdot F(X) > 0$. Let us consider the problem $\mathcal{VI}(F, K \cap B_r)$; according to theorem 6.1.5, $K \cap B_r$ being compact, this problem has at least a solution: X_r. We have $(X - X_r) \cdot F(X_r) \geq 0$ for all $X \in K$ and such that $\|X\| \leq r$.

Suppose that $\|X_r\| = r$; then (1) $(X_r - X_0) \cdot F(X_r) > 0$ and (2) $(X_0 - X_r) \cdot F(X_r) \geq 0$. This is a contradiction.

Thus $\|X_r\| < r$. Let $X \in K$. There exists ϵ such that $X_r + \epsilon(X - X_r) \in K \bigcap B_r$. Thus $(X_r - (X_r + \epsilon(X - X_r)) \cdot F(X_r)) \geq 0$. We divide by ϵ and get $(X - X_r) \cdot F(X_r) \geq 0$. This means that X_r is a solution of $\mathcal{VI}(F, K)$. □

Theorem 6.1.10 *If F strictly monotone on K then $\mathcal{VI}(F, K)$ has an unique solution.*

Proof. Let X^* and Y^*, two different solutions of $\mathcal{VI}(F, K)$; then, as F is pseudo monotone, we have altogether $(X^* - Y^*) \cdot F(Y^*) \geq 0$, $(Y^* - X^*) \cdot F(X^*) \geq 0$; by summing these expressions, we get $(X^* - Y^*) \cdot (F(Y^*) - F(X^*)) \geq 0$ or $(X^* - Y^*) \cdot (F(X^*) - F(Y^*)) \leq 0$. Which contradicts the fact the F is strictly monotone. □

6.1.5 Projection Algorithm for Variational Inequalities

Recall theorem 6.1.4: X^* is a solution of the variational inequality problem $\mathcal{VI}(F, K)$ if and only if it is a fixed point of the application $X \to P_K(X - \gamma F(X))$; i.e., if: $X^* = P_K(X^* - \gamma F(X^*))$.

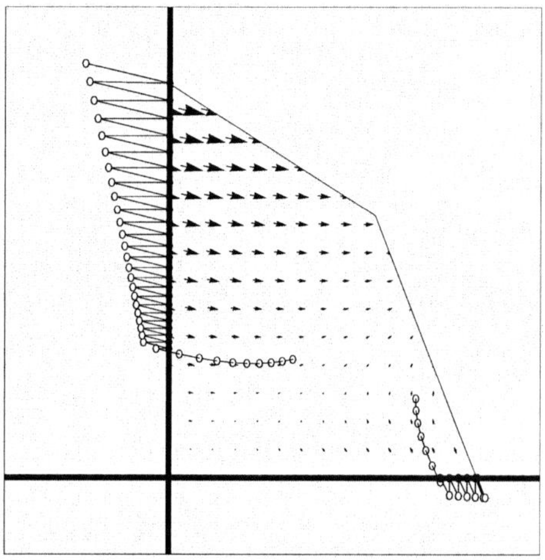

Figure 6.2: Projection algorithm. Start from an interior point. Follows the vector field. When the boundary is reached, project the direction of the vector field then on the boundary. Iterate until stability.

The principle of the projection algorithms consist in finding a fixed point of $X \to P_K(X - \gamma F(X))$. It is given in its simplest form in algorithm 6.1.

1. Step 0:
   - Choose $X_0 \in K$
   - Put $k = 0$
2. Step 1:
   - $X_{k+1} = P_K(X_k - \gamma F(X_k))$
   - $k \leftarrow k + 1$
3. Step 2: if $\|X_{k+1} - X_k\| < \varepsilon$ return X_k, else go to step 1

Algorithm 6.1: Projection algorithm

Figure 6.2 illustrates the principles of the projection algorithm.

Theorem 6.1.11 *If F is strictly monotone, then the projection algorithm converges to the unique solution of the variational inequality problem.*

Proof. It is enough to show that $X \to G(X) = P_K(X - \gamma F(X))$ is a strictly contracting function. By definition of strict monotonicity, we have:

$$(X - Y) \cdot (F(X) - F(Y)) > c\|X - Y\|^2$$

But, according to the Schwarz inequality,

$$(X - Y) \cdot (F(X) - F(Y)) \leq \|X - Y\|\|F(X) - F(Y)\|$$

We deduce:

$$\|F(X) - F(Y)\|^2 > c^2\|X - Y\|^2$$

According to theorem 5.1.3, P_K is contraction:

$$\|P_K(X) - P_K(X)\| \leq \|P_K(X) - P_K(X)\|$$

Thus:

$$\begin{aligned} \|G(X) - G(Y)\|^2 &= \|P_K(X - \gamma F(X)) - P_K(Y - \gamma F(Y))\|^2 \\ &\leq \|X - Y - \gamma(F(X) - F(Y))\|^2 \\ &= \|X - Y\|^2 + \gamma^2\|F(X) - F(Y)\|^2 \\ &\quad - 2\gamma(X - Y) \cdot (F(X) - F(Y)) \\ &\leq (1 - 2\gamma c + \gamma^2 d^2)\|X - Y\|^2 \end{aligned}$$

It is enough to show that $(1 - 2\gamma c + \gamma^2 d^2) < 1$ for an appropriate value of γ. We take $\gamma = c/4d^2$. And it works. □

1. Step 0:
   - Choose $X_0 \in K$
   - Put $k = 0$
2. Step 1:
   - $\widetilde{X_k} = P_K(X_k - \gamma F(X_k))$
   - $X_{k+1} = P_K(X_k - \gamma F(\widetilde{X_k}))$
   - $k \leftarrow k + 1$
3. Step 2: if $\|X_{k+1} - X_k\| < \varepsilon$ return X_k, else go to step 1

Algorithm 6.2: Modified projection algorithm

For other classes of functions, there are many results about the convergence of the projection algorithm and its variants. Main variant is the modified projection algorithm (Solodov and Tseng, 1996), It is given in algorithm 6.2. It is more efficient and works in more general situations, for example for monotone functions (see figure 6.3).

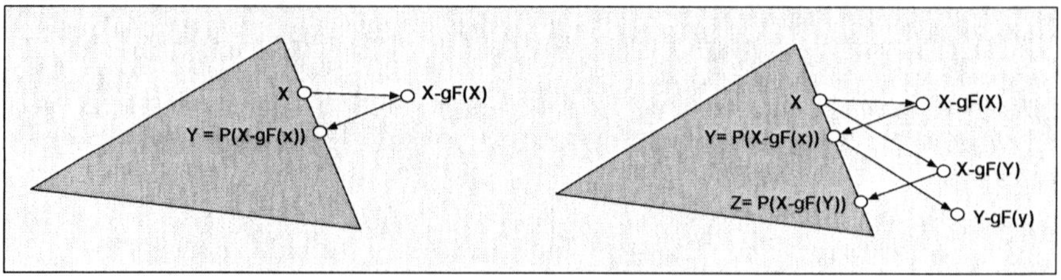

Figure 6.3: Projection algorithms. Left: initial projection algorithm. Right: modified projection algorithm.

6.2 Complementarity Problems

6.2.1 Complementarity Slackness

We have already seen complementarity relationships several times, for example in the formulation of the Karush-Kuhn-Tucker theorem. Linear and non-linear theories based on complementarity provide a general, efficient and surprisingly simple framework for these situations[2].

Recall the normal cone characterization theorem 5.1.6. Let K a polyhedron defined by: $K = \{X \mid A_i \cdot X \geq c_i\}$. Let $P \in K$. Let I_0 the set of indices such that $A_i \cdot P = c_i$. Then $N_K(P)$ is the cone with vertex P and generators $-A_i$ for $i \in I_0$. With previous notations, we may rephrase it, in terms of complementarity, as:

Theorem 6.2.1 *Let K a polyhedron defined by $\mathbf{A} \cdot X \geq B$. Let $X \in K$. Then $Z \in N_K(X)$ if and only if there exists Y such that $Z = -{}^{\mathbf{T}}\mathbf{A} \cdot Y$ and $Y \perp (\mathbf{A} \cdot X - B)$.*

Proof: we consider the vector Y of coefficients of Z in terms of $-A_i$ given by theorem 5.1.6. □

A particular case of the above theorem is:

Theorem 6.2.2 *Let K a polyhedron defined by $\mathbf{A} \cdot X \geq B$ and $X \geq 0$. Let $X \in K$. Then $Z \in N_K(X)$ if and only if there exists Y such that $Y \perp (\mathbf{A} \cdot X - B)$ and $X \perp (-Z - {}^{\mathbf{T}}\mathbf{A} \cdot Y)$*

[2]Introductions to complementarity problems can be found in Cottle et al. (2009), and Facchinei and Pang (2003).

Proof. We put

$$\mathbf{M} = \begin{pmatrix} a_{11} & \dots & a_{1n} \\ \dots & \dots & \dots \\ a_{p1} & \dots & a_{pn} \\ 1 & \dots & 1 \end{pmatrix}, \quad N = \begin{pmatrix} b_1 \\ \dots \\ b_p \\ 0 \end{pmatrix}$$

In a more compact form, we have $\mathbf{M} = (\mathbf{A}, \mathbb{1})$ and $N = (B, \mathbb{0})$. Constraints $\mathbf{A} \cdot X \gg B$ and $X \gg 0$ become $\mathbf{M} \cdot X \gg N$. Thus there exists $V = (Y, U)$ such that $Z = -{}^{\mathbf{T}}\mathbf{M} \cdot V$ and $V \perp (\mathbf{M} \cdot X - N)$. Equality $Z = -{}^{\mathbf{T}}\mathbf{M} \cdot V$ gives $Z = -{}^{\mathbf{T}}\mathbf{A} \cdot Y - U$, i.e., $U = -{}^{\mathbf{T}}\mathbf{A} \cdot Y - Z$. Complementarity slackness $V \perp (\mathbf{M} \cdot X - N)$ gives $Y \perp (\mathbf{A} \cdot X - B)$ and $U \perp X$; i.e., $X \perp (-Z - {}^{\mathbf{T}}\mathbf{A} \cdot Y)$. □

6.2.2 Linear Complementarity Problems

Definition 6.2.1 *Given a $n \times n$ matrix $\mathbf{M}$ and a n-vector N, the linear complementarity problem denoted $\mathcal{LCP}(\mathbf{M}, N)$ consists in finding X such that $X \perp (\mathbf{M} \cdot X + N)$.*

Examples. It is immediate that

- if $\mathbf{M} = \begin{pmatrix} -1 & 0 \\ 0 & -1 \end{pmatrix}$, $N = \begin{pmatrix} -1 \\ -1 \end{pmatrix}$, then $\mathcal{LCP}(\mathbf{M}, N)$ has no solution.
- if $\mathbf{M} = \begin{pmatrix} -1 & 0 \\ 0 & -1 \end{pmatrix}$, $N = \begin{pmatrix} 1 \\ 1 \end{pmatrix}$, then $\mathcal{LCP}(\mathbf{M}, N)$ has a solution $(0, 0)$.
- if $\mathbf{M} = \begin{pmatrix} 1 & 0 \\ 0 & 1 \end{pmatrix}$, $N = \begin{pmatrix} a \\ b \end{pmatrix}$, then $\mathcal{LCP}(\mathbf{M}, N)$ has always a solution $(Max(0, -a), Max(0, -b))$.

Computation 1. Let F be defined by: $F(x, y) = (3x + y - 4, x + y + 5)$. We look for a solution of linear complementarity problem associated to F. Let us examine successively the solutions of type $(x > 0, y > 0)$, $(x > 0, 0)$, $(0, y > 0)$, $(0, 0)$. A solution $(x > 0, y > 0)$ is such that $3x + y - 4 = 0$, $x + y + 5 = 0$, i.e., $x = 9/2, y = -19/2$; it does not work. A solution $(0, y > 0)$, is such that $y - 4 \geq 0$, $y + 5 = 0$, i.e., $x = 0, y = -5$; it does not work. A solution $(x > 0, 0)$, is such that $3x - 4 \geq 0$, $x + 5 = 0$, i.e., $x = -5, y = 0$; it does not work. Obviously a solution $x = 0, y = 0$ does not work as well. The problem has no solution.

Computation 2. Let F be defined by: $F(x, y) = (3x + y - 4, x + y - 5)$. We find, with the same mean, two solutions $x = 0, y = 5$ and $x = 4/3, y = 0$.

6.2.3 Linear Programming and Complementarity

Let K be defined by constraints $\mathbf{A} \cdot X \geq B$ and $X \geq 0$. Suppose that $f(X) = C \cdot X$.

Theorem 6.2.3 *A solution of the linear program MIN(f, K) is a solution of the linear complementarity problem $\mathcal{LCP}(\mathbf{M}, N)$, where:*

$$\mathbf{M} = \begin{pmatrix} 0 & -^{\mathbf{T}}\mathbf{A} \\ \mathbf{A} & 0 \end{pmatrix}, \; N = \begin{pmatrix} C \\ -B \end{pmatrix}$$

Proof: a solution X^* of the minimization problem MIN(F, K) is such that $-\nabla f(X^*) = -C \in N_K(X^*)$. Thus, according to previous theorem, there exists Y^* such that $Y^* \perp (\mathbf{A} \cdot X^* - B)$ and $X^* \perp (C - A^{*T} \cdot Y)$. If we put $Z = \begin{pmatrix} X \\ Y \end{pmatrix}$, it is obvious that we have: $Z^* \perp (\mathbf{M} \cdot Z^* + N)$. □

Therefore for all i, $x_i \geq 0$, $(MX + N)_i \geq 0$ and $x_i = 0$ or $(MX + N)_i = 0$.

6.2.4 Quadratic Programming and Linear Complementarity

Let K be defined by constraints $\mathbf{A} \cdot X \geq B$ and $X \geq 0$. Suppose that $f(X) = \frac{1}{2} X \cdot \mathbf{Q} \cdot X + C \cdot X$.

Theorem 6.2.4 *A solution of the quadratic program $\mathcal{MIN}(f, K)$ is a solution of the linear complementarity problem $\mathcal{LCP}(M, N)$, where:*

$$\mathbf{M} = \begin{pmatrix} Q & -^{\mathbf{T}}\mathbf{A} \\ \mathbf{A} & 0 \end{pmatrix}, \; N = \begin{pmatrix} C \\ -B \end{pmatrix}$$

Proof. Similar to previous one, noticing that $\nabla f(X) = Q \cdot X + C$. □

6.2.5 Affine Variational Inequality and Linear Complementarity

Theorem 6.2.5 *With the same notations, solving the affine variational inequality $\mathcal{AVI}(\mathbf{Q}, C, \mathbb{R}^n_+)$ is equivalent to solving the linear complementarity problem $\mathcal{LCP}(\mathbf{M}, N)$.*

Proof is an immediate consequence of definitions. □

Theorem 6.2.6 *Let K be defined by constraints $\mathbf{A} \cdot X \geq B$ and $X \geq 0$. Let $F(X) = \mathbf{Q} \cdot X + C$. Solving the affine variational inequality $\mathcal{AVI}(\mathbf{Q}, C, K)$ is equivalent to solving the linear complementarity problem $\mathcal{LCP}(M, N)$, where:*

$$M = \begin{pmatrix} Q & -^{T}A \\ A & 0 \end{pmatrix}, \; N = \begin{pmatrix} C \\ -B \end{pmatrix}$$

Proof. We write $-(Q \cdot X^* + C) \in N_K(X^*)$ and the proof is similar to previous ones. □

6.2.6 Existence and Unicity Theorems for the Linear Complementarity Problems

Existence and unicity theorems are related to several types of matrices.

Definition 6.2.2 *The complementarity problem $\mathcal{LCP}(\mathbf{M}, Q)$ is said feasible if there exists X such that $X \geq 0$ and $\mathbf{M} \cdot X + Q \geq 0$*

Definition 6.2.3 *A matrix $\mathbf{M}$ is said a PD matrix (positive definite) if, for all $X \neq 0$, we have $X \cdot \mathbf{M} \cdot X > 0$, a PSD matrix (positive semi definite) if, for all X, we have $X \cdot \mathbf{M} \cdot X \geq 0$.*

Theorem 6.2.7 *Let $\mathbf{M}$ be a positive semi definite matrix. Then, if $\mathcal{LCP}(\mathbf{M}, Q)$ is feasible then it has a solution.*

Proof. Let $K = \{Z \mid Z \geq 0, \mathbf{M} \cdot Z + Q \geq 0\}$ the set of feasible states. By hypothesis K is non empty. In a first step, we prove that putting $f(Z) = Z \cdot (\mathbf{M} \cdot Z + Q)$, the quadratic program $\mathcal{MIN}(f, K)$ has a solution Z^*. In a second step, we prove that Z^* is a solution of $\mathcal{LCP}(\mathbf{M}, Q)$.

First step. We remark that $f(Z) \geq 0$ for all $Z \in K$. We remark also that $f(Z) = Z.(\mathbf{S} \cdot Z + Q)$ with symmetric, positive semi-definite matrix $\mathbf{S} = (\mathbf{M} + {}^{\mathbf{T}}\mathbf{M})/2$. We may consider a basis of $\mathbb{R}^n$ in which S is diagonal. If in this basis, with $Z = (z_1, \ldots, z_n)$, $Q = (q_1, \ldots, q_n)$, $S = Diag(s_1, \ldots, s_n)$ $(s_i \geq 0)$, we have:

$$f(Z) = \sum_i s_i z_i^2 + q_i z_i$$

We may thus reason component by component. Let K_i the intersection of K and the axis corresponding to component i. If $s_i > 0$, the convex quadratic function f reaches its minimum value on the convex set K_i. If $s_i = 0$, the linear function f that is bounded below on K, thus on K_i, reaches its minimum value on K_i. For both cases, it is enough to consider that K_i is an interval; we have $K_i = [u_i, v_i]$ or $[-\infty, v_i]$ or $[v_i, \infty]$.

Second step. Let Z^* be the solution of $\mathcal{MIN}(f, K)$. According to the Karush-Kuhn-Tucker theorem, there are multipliers Λ^* such that

$$Z^* \perp \mathbf{S} \cdot X^* + Q - {}^{\mathbf{T}}\mathbf{M} \cdot \Lambda^*$$
$$\Lambda^* \perp \mathbf{M} \cdot Z^* + Q$$

As $Z^* \geq 0$ and $\mathbf{M} \cdot Z^* + Q \geq 0$, we have $Z^* \cdot (\mathbf{M} \cdot Z^* + Q) \geq 0$. On the other hand, from previous equations, we have:

$$\begin{aligned}
Z^* \cdot (\mathbf{M} \cdot Z^* + Q) &= (Z^* - \Lambda^*) \cdot (\mathbf{M} \cdot Z^* + Q) \\
&\quad \text{because } \Lambda^* \cdot (\mathbf{M} \cdot Z^* + Q) = 0 \\
&= (Z^* - \Lambda^*) \cdot (\mathbf{M} \cdot Z^* + Q) \\
&\quad + (Z^* - \Lambda^*) \cdot ({}^{\mathbf{T}}\mathbf{M} \cdot (Z^* - \Lambda^*) - {}^{\mathbf{T}}\mathbf{M} \cdot (Z^* - \Lambda^*)) \\
&\leq (Z^* - \Lambda^*) \cdot (\mathbf{M} \cdot Z^* + Q + ({}^{\mathbf{T}}\mathbf{M} \cdot (Z^* - \Lambda^*)) \\
&\quad \text{because } {}^{\mathbf{T}}\mathbf{M} \text{ is positive semi definite} \\
&= (Z^* - \Lambda^*) \cdot (\mathbf{S} \cdot Z^* + Q - {}^{\mathbf{T}}\mathbf{M} \cdot \Lambda^*) \\
&= (-\Lambda^*) \cdot (\mathbf{S} \cdot Z^* + Q - {}^{\mathbf{T}}\mathbf{M} \cdot \Lambda^*) \\
&\leq 0 \text{ because } \Lambda^* \geq 0, \mathbf{S} \cdot Z^* + Q - {}^{\mathbf{T}}\mathbf{M} \cdot \Lambda^* \geq 0
\end{aligned}$$

Thus $Z^* \cdot (\mathbf{M} \cdot Z^* + Q) = 0$. □

Lemma 6.2.1 *Let* $\mathbf{M}$ *be a positive definite matrix, then there exists* $Z = (z_i)$ *such that for all components* i, $z_i > 0$, *i.e.,* $Z > 0$ *and* $(\mathbf{M} \cdot Z)_i > 0$, *i.e.,* $\mathbf{M} \cdot Z > 0$.

Proof. Let $U = (1, 1, \dots, 1)$. It is enough to prove the more general assertion: there exists $Z = (z_i)$ such that for all components i, $z_i \geq 0$ and $(MZ)_i > 0$: if there exists a such Z, we consider $Z' = Z + tU$ with small $t > 0$; we have $Z' > 0$ and by continuity $\mathbf{M} \cdot Z' > 0$.

Let us suppose the contrary: for all $X = (x_i)$ with $x_i \geq 0$, there exists a component j such that $\mathbf{M} \cdot X)_j > 0$.

We consider the linear program: Find (X, Z) that minimizes $U \cdot Z$ subject to $\mathbf{M} \cdot X + Z \geq U$,$X \geq 0$, $Z \geq 0$. As it is bounded below, it has a solution: (X^*, Z^*). According to the Karush-Kuhn-Tucker theorem, there are multipliers Λ^* such that:

$$\begin{aligned}
&X^* \perp \; -{}^{\mathbf{T}}\mathbf{M} \cdot \Lambda^* \\
&Z^* \perp U - \Lambda^* \\
&\Lambda^* \perp \mathbf{M} \cdot X^* + Z^* - U
\end{aligned}$$

Then, we have $\Lambda^* \geq 0$ and $-{}^{\mathbf{T}}\mathbf{M} \cdot \Lambda^* \geq 0$. Thus

$$\Lambda^* \cdot {}^{\mathbf{T}}\mathbf{M} \cdot \Lambda^* \leq 0$$

As ${}^{\mathbf{T}}\mathbf{M}$ is positive definite, this implies that $\Lambda^* = 0$, thus that $Z^* \perp U$, thus that $Z^* = 0$. Thus $\mathbf{M} \cdot X^* - U \geq 0$. This is a contradiction with the above hypothesis about the existence of a non positive component of $\mathbf{M} \cdot X$ for all $X > 0$. □

Theorem 6.2.8 *Let* $\mathbf{M}$ *be a positive definite matrix; then* $\mathcal{LCP}(\mathbf{M}, Q)$ *has a unique solution.*

Proof. Existence. We firstly prove that there exists Z such that $Z > 0$ and $\mathbf{M} \cdot Z > 0$; such a vector exists because of previous lemma. Then we have $t\mathbf{M} \cdot (tZ) = t\mathbf{M} \cdot (Z) \geq -Q$ for enough large t. Thus $\mathcal{LCP}(\mathbf{M}, Q)$ is feasible, and it has a solution.

Proof. Unicity. We use the equivalence between a solution of $\mathcal{LCP}(\mathbf{M}, Q)$ and $\mathcal{MIN}(f, K)$. If $\mathbf{M}$ is positive definite, then f is strictly convex and solution of $\mathcal{MIN}(f, K)$ is unique. Whence the result. □

There are many other relationships between classes of matrix and existence theorems for linear complementarity problems. They are summarized in definition 6.2.4 and figure 6.4. Proofs are given in Cottle et al. (2009), chapter 3 or Facchinei and Pang (2003), chapter 2.

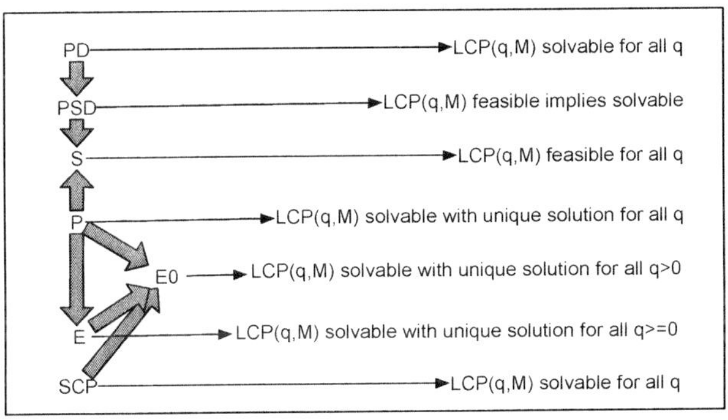

Figure 6.4: Theorem about linear complementarity problems $\mathcal{LCP}$. On the left, different classes of matrix (definition 6.2.4) and their inclusion relationships. On the right, known characteristics of the solution set.

Definition 6.2.4 *A matrix* $\mathbf{M}$ *is said*

- *a* $S-$ *matrix if there exists* $X > 0$ *such that* $\mathbf{M} \cdot X > 0$
- *a* $P-$ *matrix if all its principal minors are positive; recall that the principal sub matrices of* $\mathbf{M}$ *are obtained by removing a same subset in the common set for indices of rows and lines* $\{1, 2, \ldots, n\}$*; recall that the principal minors are the determinants of the principal sub matrices.* $\mathbf{M}$ *is a* P *matrix if and only if the real eigenvalues of* M *and its principal sub-matrices are positive.*
- *a* $E-$*matrix (strictly semi monotone) if for* $X \neq 0, X \geq 0$*, there exists* k *such that* $x_k > 0$ *and* $(M \cdot X)_k > 0$
- *a* E_0-*matrix (semi monotone) if for* $X \neq 0, X \geq 0$*, there exists* k *such that* $x_k > 0$ *and* $(M \cdot X)_k \geq 0$
- *a* SCP *matrix (strictly co positive) if for all* $X > 0$*,* $X \cdot \mathbf{M} \cdot X > 0$

6.2.7 General Complementarity Problems

Definition 6.2.5 *Given a function $F : \mathbb{R}^n \to \mathbb{R}^n$, the general complementarity problem $\mathcal{CP}(F)$ consists in finding X such that $X \perp F(X)$.*

Theorem 6.2.9 *If F is pseudo monotone (cf. definition 6.1.7), then the solution set of $\mathcal{CP}(F)$ is convex. If the $\mathcal{CP}(F)$ problem is feasible, then the solution set is non empty, compact and convex.*

Proof is given in Facchinei and Pang (2003).

6.2.8 Algorithms for Complementarity Problems

Here, we provide a short description of two quite efficient kinds of algorithms. They are extracted from Facchinei and Pand (2003) where one find the presentation of many algorithms for variational inequalities and complementarity problems. They are the ones we use in forthcoming parts.

Projection method. Let us mention firstly that the previous projection algorithm for variational inequality is there very easy to implement because of the simplicity of the projection function on $K = \mathbb{R}^n_+$. If matrix $\mathbf{M}$ is a definite positive matrix, this algorithm converges to the unique solution of $\mathcal{LCP}(\mathbf{M}, Q)$ for all vectors Q.

Newton's method. We present now two algorithms for solving the complementarity problem. They are based on the Newton's method. Let us recall its principles. In dimension 1 (figure 6.5), the algorithm finds a zero of a function $f : \mathbb{R} \to \mathbb{R}$, i.e., a value of x such that $f(x) = 0$ by starting from an arbitrary x_0 and then, until changes are very small, iterating:

$$x_{n+1} = x_n - f(x_n)/f'(x_n)$$

In dimension n, given a function $F : \mathbb{R}^n \to \mathbb{R}^n$, we find a zero of F, i.e., a vector X such that $F(X) = 0$ by starting from an arbitrary X_0 and then, Until changes are very small, iterating:

$$X_{n+1} = X_n - (DF(X_n))^{-1} \cdot F(X_n)$$

The Netwton's method implies the non singularity of matrix $DF(X_n)$ and is effective for some identified classes of functions.

Smoothing algorithm for the complementarity problems. The principle of smoothing algorithm is based on the characterization of solutions complementarity slackness with auxiliary functions. We use the function

$$\phi(a, b) = a + b - \sqrt{a^2 + b^2}$$

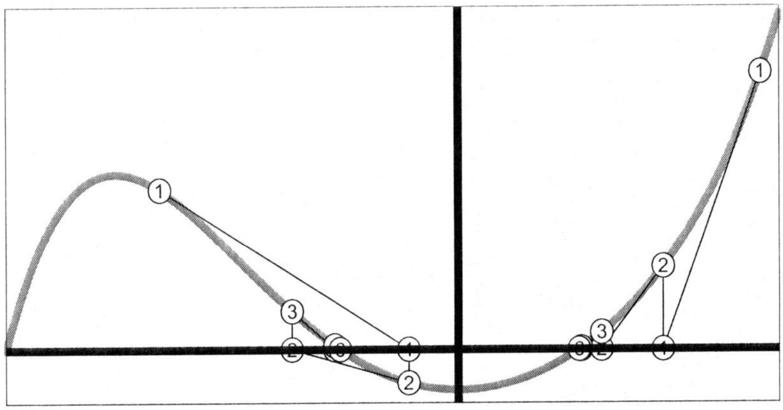

Figure 6.5: Netwon method. Finding a zero of a differentiable function. Start from a point x. The next point is the intersection with the x-axis of the tangent to the graph at point $x, f(x)$. Iterate.

It is such; $\phi(a,b) = 0$ if and only if $a \geq 0$, $b \geq 0$, $ab = 0$, that is if $a \perp b$. We define:

$$\Phi(A,B) = (\phi(a_1,b_1),\ldots,\phi(a_n,b_n))$$
$$\Phi(X) = \Phi(X,F(X))$$

Then X is solution de $LCP(F)$ is equivalent to $\Phi(X) = 0$. Whence the idea of using a Newton algorithm. But, the function ϕ is not continuously derivable in $(0,0)$. To avoid subsequent problems, we consider instead the following function ϕ_μ for small values of μ.

$$\phi_\mu(a,b) = a + b - \sqrt{a^2 + b^2 + \mu^2}$$

We define in the same way:

$$\Phi_\mu(A,B) = (\phi_\mu(a_1,b_1),\ldots,\phi_\mu(a_n,b_n))$$
$$\Phi_\mu(X) = \Phi_\mu(X,F(X))$$

We define, for the computing of the derivatives of Φ_μ, the following functions:

$$\phi_\mu^a(a,b) = 1 - \frac{a}{\sqrt{a^2 + b^2 + \mu^2}}$$

$$\phi_\mu^b(a,b) = 1 - \frac{b}{\sqrt{a^2 + b^2 + \mu^2}}$$

$$\Phi_\mu^a(A,B) = \left(\phi_\mu^a(a_1,b_1),\ldots,\phi_\mu^a(a_n,b_n)\right)$$
$$\Phi_\mu^b(A,B) = \left(\phi_\mu^b(a_1,b_1),\ldots,\phi_\mu^b(a_n,b_n)\right)$$

We have:

$$D\Phi^a_\mu(X) = Diag(\Phi^a_\mu(X, F(X)))$$

$$D\Phi^b_\mu(X) = Diag(\Phi^a_\mu(X, F(X)))$$

We use the algorithms given in algorithm 6.3:

1. Initialization:
   - $\sigma = 1$, $\gamma = 0.5$, $\epsilon = 10^{-5}$, $\eta_k = 3^{-k}$,
   - $U = (1, 0, 0, \ldots, 0)$
2. Step 0:
   - $X^0 =$ Random; $\mu^0 = 1$;
   - $Z^0 = (\mu^0, X^0)$
   - $k = 0$
3. Step 2:
   - $(\mu, X) \leftarrow Z^k$
   - $W = (\mu, \Phi_\mu(X))$
   - If $\|W\| < \epsilon$ Stop
4. Step 3:
   - $\rho = \gamma \|W\|$
   - $DG = \begin{pmatrix} 1 & 0 \\ D_\mu & D_{ab} \end{pmatrix}$
     With $D_\mu = \Phi_\mu(X)$
     and $D_{ab} = D\Phi^a_\mu(X) + D\Phi^b_\mu(X) \cdot DF(X)$
   - Newton step: Solve: $DG \cdot DZ = (\rho U - W)$.
5. Step 4:
   - If $\|Z + DZ\| \geq \|Z\| - \sigma \|DZ\|^2$, then i is the smallest non negative integer such that $\|Z + 2^{-i} DZ\| \leq (1 + \eta_k)\|Z\| - \sigma \|DZ\|^2$
   - Else $i = 0$
6. Step 5:
   - $Z^{k+1} = Z^k + 2^{-i} DZ$
   - $k \leftarrow k + 1$
7. Goto Step 2

Algorithm 6.3: Smoothing algorithm

Interior point algorithm for the complementarity problems. Interior point algorithms are iterative algorithms. They start from an interior point of the constrained set and then maintain the sequence of iterated points away from the boundary of the constrained set, where convergence problems may occur. In the case of the complementarity problem, an effective interior point algorithm is as follows.

Given two vectors $X = (x_1, \ldots, x_n)$, $Y = (y_1, \ldots, y_n)$, we consider the vector obtained with the products of their components: $X \circ Y = (x_1y_1, \ldots, x_ny_n)$. Then it is obvious that $X \in \mathbb{R}^n_+$ is a solution of the complementarity problem $X \perp F(X)$ if and only if there exists $Y \in \mathbb{R}^n_+$ such that $G(X, Y) = 0$, where:

$$G(X,Y) = \begin{pmatrix} Y \circ X \\ Y - F(X) \end{pmatrix}$$

Let $d(X)$ the diagonal matrix associated to X. The Jacobian matrix of G is:

$$DG((X,Y)) = \begin{pmatrix} d(Y) & d(X) \\ -DF(X) & \mathbb{I} \end{pmatrix}$$

A step of the corresponding Newton algorithm is

$$Z^k \to Z^k + DZ$$

where DZ is a solution of: $G(Z^k) + DG(Z^k) \cdot DZ = 0$. That is:

$$\begin{pmatrix} Y \circ X \\ Y - F(X) \end{pmatrix} + \begin{pmatrix} d(Y) & d(X) \\ -JF(X) & \mathbb{I} \end{pmatrix} \begin{pmatrix} DX \\ DY \end{pmatrix} = \begin{pmatrix} 0 \\ 0 \end{pmatrix}$$

We adapt this algorithm in order that, at every step, Z^k stays in the interior of $\mathbb{R}^{2n}_+$. We define the subset:

$$X_I = \{(X,Y) \mid X > 0, Y > 0, Y > F(X)\}$$

We put:

$$U = \begin{pmatrix} \mathbb{I} \\ 0 \end{pmatrix}$$

We solve:

$$\begin{pmatrix} Y \circ X \\ Y - F(X) \end{pmatrix} + \begin{pmatrix} d(Y) & d(X) \\ -DF(X) & \mathbb{I} \end{pmatrix} \begin{pmatrix} DX \\ DY \end{pmatrix} = \sigma U$$

We also use the evaluation function defined by:

$$\psi((X,Y)) = n \log(\|X \circ Y\|^2 + \|Y - F(X)\|^2) - \sum_i \log(X_i Y_i)$$

We use algorithm given in algorithm 6.4:

1. Initialization:
   - Choose $\zeta > n/2$, $\gamma \in (0,1)$, and a sequence $\sigma_k \in (0,1)$
2. Step 0:
   - $Z^0 = (X^0, Y^0) \in X_I$
   - $k = 0$
3. Step 1: If $G(Z^k) = 0$ stop
4. Step 2: Newton step. Solve $G(Z^k) + DG(Z^k) \cdot DZ = (\sigma_k Z \cdot U)U/n$.
5. Step 3: i is the smallest non negative integer such that

$$\psi(Z^k + 2^{-i}DZ) - \psi(Z^k) < \gamma 2^{-i} D\psi(Z^k) \cdot DZ$$

$$Z^k + 2^{-i}DZ > 0, G(Z^k + 2^{-i}DZ) > 0$$

6. Step 4: Put
   - $Z^{k+1} = Z^k + 2^{-i}DZ$
   - $k \leftarrow k+1$
7. Goto Step 1

Algorithm 6.4: Interior point algorithm

6.3 Complementarity Problems and Variational Inequality Problems

We have enunciated two equivalence theorems between variational inequalities and complementarity problems (theorems 6.2.5 and 6.2.6). We have seen also existence and uniqueness theorems for variational inequalities and complementarity problems. Sometimes, emphasis is put on constraints and this leads to use variational inequality. Sometimes, emphasis is put on complementarity relationships and this leads to use complementarity problems.

Personal experience has shown that it is often more natural to express economic systems with $\mathcal{LCP}$ and that in many times the algorithms to solve them work quite well; we formulate an ecological or an economic problem in terms of variational inequality $\mathcal{VI}$, when we need an existence result, and for that we use the theorem 6.1.5: in many cases, the assumption of the compacity of the constrained set is altogether realistic and easy to formulate.

6.4 Nash Equilibrium Problem

In this section, we briefly introduce game theory into the context of variational inequalities or complementarity. The aim of the mathematical theory of games lies in analyzing the interplay between several optimizing agents. In this section, we only consider situations where agents are not cooperative and there is no possibility of coalitions, or cooperation towards a common goal. This may appear to be a significant restriction. However, it is clear that the theory of non-cooperative games provides a suitable framework for mathematical theories of economic equilibrium and is a relevant tool for the analysis of common resource exploitation. It focuses on competition states such that a player gets a lesser reward if she changes her own strategy (assuming other players do not change their strategy). Such a state is called a non-cooperative (or Nash) equilibrium. There are many texts that support this principle, that illustrate its meaning (Farrell, 1988; Myerson, 1999). It has been both extended and generalized (Myerson, 1978) and it offers a framework for the analysis of non-cooperative behavior (Myerson, 1999). One may also keep it with a part of mystery.

6.4.1 An Example

Let us consider a lake with two fishermen, both of them are free to catch as much fish as they like. Their total catch impacts the stock of fish and costs rise when the stock of fish is low.

This kind of problem is mathematically expressed and solved as follows. Let a and b be the two fishermen; y_a and y_b be their catches. The stock of fish depend on the total of catches $s(y_a + y_b)$, with $s' < 0$. The costs of fishing depend on stock for both fishermen: $c_a(s)$, $c_b(s)$ with $c'_a < 0$. Let p be the price of fish. The reward of a is $r_a = y_a(p - c_a)$; that of b is $r_b = y_b(p - c_b)$. At the Nash equilibrium, we have together $\partial r_a / \partial y_a = 0$ and $\partial r_b / \partial y_b = 0$. This gives a pair of equations with two unknown, whose solution, if it exists, is a candidate to be a Nash' equilibrium. Let us consider the elementary case where all functions are linear:

$$c_a(s) = d_a - e_a s$$
$$c_b(s) = d_b - e_b s$$
$$s(y) = u - vy$$

Then

$$\partial r_a / \partial y_a = -2e_a v y_a + (p - d_a - e_a(u - v y_b))$$
$$\partial r_b / \partial y_b = -2e_b v y_b + (p - d_b - e_b(u - v y_a)).$$

We get a linear system:

$$2e_a v y_a + e_a v y_b = p - d_a - e_a u$$
$$e_b v y_a + 2e_b v y_b = p - d_b - e_b u$$

Solutions are obviously:

$$y_a = (p(2e_b - e_a) + d_b e_a + e_a e_b u - 2d_a e_b)/(3e_a e_b v)$$
$$y_b = (p(2e_a - e_b) + d_a e_b + e_a e_b u - 2d_b e_a)/(3e_a e_b v)$$

The conditions on parameters that ensure the positivity of existing solutions are related to the possibility of a competitive co-existence.

6.4.2 Nash Equilibrium Problems

Now we can define the Nash equilibrium of a general non-cooperative game involving an arbitrary number of players.

Definition 6.4.1 *A Nash equilibrium problem $\mathcal{NEP}(C_i, R_i)$ is defined by:*

- *n players: $1, 2, \ldots, n$*
- *the strategy $X_i \in C_i \subset \mathbb{R}^{n_i}$ of player i*
- *the loss R_i of player i; it functionally depends on its own strategy, but also on the strategies of other players: $R_i(X_1, \ldots, X_i, \ldots, X_n)$*

Definition 6.4.2 *A set of strategies $X^* = (X_1^*, \ldots, X_n^*)$ is a solution of the Nash equilibrium problem $\mathcal{NEP}(C_i, R_i)$ if no player can unilaterally change its strategy without getting bigger loss: for all i and $X_i \in C_i$*

$$R_i(X_1^*, \ldots, X_i^*, \ldots, X_n^*) \leq R_i(X_1^*, \ldots, X_i, \ldots, X_n^*)$$

It is common to call a Nash equilibrium the solution of a Nash equilibrium problem. Let $D_{X_i} R_i$ be the gradient of R_i, the loss of agent i, according to his own strategy X_i knowing the strategies of other players $(X_1, \ldots, X_{i-1}, X_{i+1}, \ldots)$:

$$D_{X_i} R_i = \left(\frac{\partial R_i}{\partial x_{i1}}, \ldots, \frac{\partial R_i}{\partial x_{il}}, \ldots \right)$$

Let $C = \prod_i C_i \subset \prod_i \mathbb{R}^{n_i}$ be the set of all combinations of possible strategies. Let $F : C \to \prod_i \mathbb{R}^{n_i}$ be the vector field defined by

$$F(X_1, \ldots, X_i, \ldots) = (D_{X_1} R_1, \ldots, D_{X_i} R_i, \ldots)$$

Then

Theorem 6.4.1 *Suppose that the loss functions of all players $R_i(X_1,\dots,X_i,\dots,X_n)$ are convex functions of the variable X_i. A set of strategies $X^* = (X_1^*,\dots,X_n^*)$ is a solution of the Nash equilibrium problem $\mathcal{NEP}(C_i,R_i)$ if and only if it is a solution of the variational inequality $\mathcal{VI}(C,F)$.*

Proof. Knowing the strategies of other players X_1^*, ..., X_j^*, the losses of player i with strategy X_i will be given by formula: $S_i(X_i) = R_i(X_1^*, X_2^*,\dots,X_i,\dots,X_n^*)$. Function S_i being convex, according to theorem 6.1.2, there will be equivalence between the solution of this optimization problem and the solution of the variational inequality problem $\mathcal{VI}(D_{X_i}S_i,C_i)$. As all players do the same at the equilibrium of the game, we put together the corresponding inequalities and get the proof of the theorem. □

6.4.3 Generalized Nash Equilibrium Problem

In a generalized Nash equilibrium problem (Facchinei et al., 2007) the set of the possible strategies of a player depend on the strategies chosen by other players.

Definition 6.4.3 *A generalized Nash equilibrium problem, denoted $\mathcal{GNEP}(C_i,R_i)$, is defined by:*

- *n players: $1,2,\dots,n$*
- *a set of possible strategies of player i, denoted C_i, that may depend on the strategies of other players $X^{-i} = (x_1,\dots,x_{i-1},x_{i+1},\dots,x_n)$*
- *the loss R_i of player i; it depends on its own strategy $x_i \in C_i$, but also on the strategies of other players: $R_i(x_1,\dots,x_i,\dots,x_n)$*

The most studied Generalized Nash equilibrium problem is the jointly convex one: there exist linear conditions restricting the set of strategies of all players. For example, each player may choose a non negative number x_i and the sum of the strategies of all players is bounded: $\sum_i x_i \leq m$. We may think to a market where the sum of bargains (strategies of buyers) is limited by the sum of sales (strategies of sellers). We have the following theorem.

Theorem 6.4.2 *Let a jointly convex $\mathcal{GNEP}(C_i,R_i)$ be given with continuous and differentiable function R_i. Let X the of possible strategies $X = \{(X_i) \mid X_i \in K_i, L(X_i) = 0\}$. Let, as usual, $F(X) = F(F_i(X)) = \partial R_i(X)/\partial X_i$. Then, every solution of the variational inequality $\mathcal{VI}(X,F)$ is also a solution of the generalized Nash equilibrium problem $\mathcal{GNEP}(C_i,R_i)$.*

Proof is given in Facchinei et al. (2007).

6.5 Inverse Modelling

6.5.1 Introduction

As already mentioned, the equilibrium of a network is expressed in terms of conservation and complementarity equations. Both types of equations may depend on parameters. $X = (x_1, \ldots, x_n)$ denoting a state of the system, $\Omega = (\omega_1, \ldots, \omega_p)$ the set of parameters, the problem to solve appears as a highly non linear mathematical equation $F(X, \Omega) = 0$ (non linearity is an immediate consequence of complementarity).

A direct modelling approach consists of studying the sensitivity of equilibrium states X^* to parameters Ω. There are ways to calibrate (assess values of parameters) and validate models (assess the credibility of predictions obtained with the model according to calibrated parameters); these are, as usual in complex modelling matter, difficult ways.

An inverse modelling approach consists of: (1) getting from observations allowing to define a reference state $\overline{X}$, (2) assuming that this reference state is an equilibrium state, and (3) using the equations $F(\overline{X}, \Omega) = 0$ to assess the reference values of parameters Ω.

The following explains several inverse modeling methods that will be used in the case studies.

6.5.2 Inverse Modelling and Complementarity

Let us assume that the equilibrium of a network appears a the solution of a system of a complementarity problem $\mathcal{CP}(F)$, with a state vector $X = (x_1, \ldots, x_n)$ and a function $F(X, \Omega)$. We have $F(X) = (f_1, \ldots, f_n)$ with:

$$f_n = f_n(x_1, \ldots, x_n, \omega_1, \ldots, \omega_p)$$

We assume that $\overline{X}$ corresponds to a reference equilibrium state. We have $\overline{X} \perp F(\overline{X}, \lambda)$. If $\overline{X}_i > 0$, then $f_i(\overline{X}, \Lambda) = 0$. If $\overline{X}_i = 0$, then $f_i(\overline{X}, \Lambda) \geq 0$. Programming provide simple heuristics to find solutions of this system of equalities and inequalities. For example, we may consider the following program:

(1) find (Ω, Λ)

(2) that minimize

$$A\|\Omega\|^2 + B\|\Lambda\|^2 + C \sum_{i; \overline{X}_i = 0} f_i(\overline{X}, \Lambda)^2$$

(3) subject to

$$\lambda_i \geq 0$$
$$f_i(\overline{X}, \Lambda) = 0 \text{ if } \overline{X}_i > 0$$
$$f_i(\overline{X}, \Lambda) \geq 0 \text{ if } \overline{X}_i = 0$$

6.5.3 Inverse Modelling and Variational Inequality

Let us assume that the equilibrium of the network appears a the solution of a system of a variational inequality problem $\mathcal{VI}(F, K)$, with a state vector $X = (x_1, \ldots, x_n)$, a function $F(X, \Omega)$ depending on unknown parameters $\Omega = (\omega_1, \ldots, \omega_p)$ and a convex set K defined by $x_i \geq 0$ and linear equations $G(X, \Phi) \geq 0$ depending on unknown parameters $\Phi = (\phi_1, \ldots, \phi_q)$. Let us express $F = (f_n)$ and $G = (g_l)$ component wise:

$$f_i = f_i((x_1, \ldots, x_n); (\omega_1, \ldots, \omega_p))$$
$$g_j = g_j((x_1, \ldots, x_n); (\phi_1, \ldots, \phi_q))$$

We know that solving this variational inequality is equivalent to solving the complementarity problem $Z \perp H(Z)$ where

$$Z = (X, \Lambda) = ((x_1, \ldots, x_n), (\lambda_1, \ldots, \lambda_l))$$
$$H(Z) = ((h_1, \ldots, h_n), (\mu_1, \ldots, \mu_l))$$
$$h_i = f_i - \sum_j \frac{\partial g_j}{\partial x_i} \lambda_j$$
$$\mu_j = g_j$$

We assume that $\overline{X}$ corresponds to a reference equilibrium state. This implies that there exists multipliers Λ such that $(\overline{X}, \Lambda) \perp H(\overline{X}, \Lambda)$. This is similar to what has been previously discussed.

6.6 Variational Inequality, Complementarity and Dynamics

In chapter 3.1, we discussed the meaning of the terms statics and dynamics in a network context. Then we have looked more closely at how variational inequality or complementarity tools can be used to define and compute the equilibrium of several kinds of networks. In this section, we give a very short introduction to how these tools can be used to analyze the dynamics of networks. We only illustrate a few options.

6.6.1 Projected Dynamical Systems

The goal of projected dynamical systems consists in keeping the trajectory of a differential equation inside the boundaries of a convex set.

Definition. The specific framework for studying dynamics with network economic tools is that of projected dynamical systems (Nagurney and Zhang, 1996). The idea is related to the projection algorithm for variational inequality. Let $F : K \subset \mathbb{R}^n \to \mathbb{R}^n$ a vector field on the convex set K. The trajectories of the associated projected dynamical system, in the interior of K, follows the vector field and, on the boundary of K follows the projection of the vector field F on this boundary. More precisely:

Definition 6.6.1 *Let K be a convex subset of $\mathbb{R}^n$. Let $F : K \to \mathbb{R}^n$ be a continuous function. The projected dynamical system associated to F and K, denoted $PDS(V, K, X_0)$, is the differential equation*

$$\dot{X} = \Pi_K(X, -F(X))$$
$$X(0) = X_0$$

Where

$$\Pi_K(X, V) = \lim_{\delta \to 0} \frac{P_K(X + \delta V) - X}{\delta}$$

And we have:

Theorem 6.6.1 *Let K be a convex subset of $\mathbb{R}^n$. Let $F : K \to \mathbb{R}^n$ be a continuous function. A solution of the variational inequality $VI(F, K)$ is a stationary point of the projected dynamical system $PDS(V, K, X_0)$*

Proof is obvious. □

Implementation. In economic and ecological applications, we use the projected dynamical systems in the following context. The system is characterized by two kinds of variables. Let X the set of variables related to stocks and flows, resulting from network equilibrium and S the other parameters of the system. Given values of parameters, equilibration mechanism results in the projected dynamical system:

$$\dot{X} = \Pi_K(X, -F(X, S))$$

Given values of flows, X parameters evolve according to a differential equation:

$$\dot{S} = G(X, S)$$

We thus have to consider the full system:

$$\dot{X} = \Pi_K(X, -F(X, S))$$
$$\dot{S} = G(X, S)$$

The key point lies in synchronizing both types of equations. In most of cases, equilibration of network and evolution of parameters are not occurring at the same rhythm. We have to consider constants Θ_X and Θ_S related to this difference and the equations:

$$\dot{X} = \Theta_X \, \Pi_K(X, -F(X, S))$$
$$\dot{S} = \Theta_S \, G(X, S)$$

The key point lies in choosing the values of Θ_X and Θ_S. This is discussed in section 3.3.

6.6.2 Representation of Time In Network Economics

To illustrate the possibilities of network economics for the representation of dynamics systems, we show how it allows combining different conceptions of time. Mathematics has familiarized us with the idea that there are two ways to represent time: (1) discrete and (2) continuous. Economic theory has familiarized us with the idea that there are two kinds of relationship with time: (1) learning (choosing according to past events) and (2) anticipation (choosing according to a projected future). We denote $\mathcal{CP}(F(X, S))$ the complementarity problem associated to a function $F(X, S)$ that depends on parameters S. We have the following alternatives.

A discrete representation of time and the hypothesis of a learning behaviour of entities. X_t^* is the solution of the equation $\mathcal{CP}(F(X, S_t))$, where the set of parameters S_t depends (1) on time t and (2) on the values of previous equilibrium states $X_0^*, \ldots, X_{t-1}^*$. This is easily implemented.

A discrete representation of time and the hypothesis of an anticipative behaviour of entities. X_t^* is the solution of the equation $\mathcal{CP}(F(X, S_t))$, where the set of parameters S_t depends (1) on time t and (2) on the values of previous and ulterior equilibrium states $X_0^*, \ldots, X_T^*$. To solve it, we have to consider a set of complementarity problems:

$$\mathcal{CP}F(X, S(1, X_0, \ldots, X_T)$$
$$\mathcal{CP}F(X, S(2, X_0, \ldots, X_T)$$
$$\ldots$$
$$\mathcal{CP}F(X, S(T, X_0, \ldots, X_T)$$

This can be shown as being equivalent to a unique complementarity problem with state vector $Z = (X_1, X_2, \ldots, X_T)$ and function $G(Z)$ such that:

$$G(Z) = (F(X_1, S(1, X_0, \ldots, X_T)), F(X_2, S(2, X_0, \ldots, X_T)), \\ \ldots, F(X_T, S(T, X_0, \ldots, X_T))$$

A continuous representation of time and the hypothesis of a learning behaviour of entities. Then we use the results from the theory of projected dynamical systems, whose principle consists in keeping the trajectory of a differential equation inside a given polyhedron and that will be presented further.

A continuous representation of time and the hypothesis of a learning behaviour of entities. Previous approach can be generalized to infinite dimensional vector spaces. This is out of the scope of this book. One may find an introduction to this subject and interesting examples in Daniele (2006).

Part III

Network Economics and the Equilibrium of Ecological and Economic Systems

Chapter 7

About the Equilibrium of Trophic and Ecological Networks

7.1 Trophic Networks

Earlier we represented trophic systems as networks where nodes correspond to species and links to a predatory relationship (figure 2.9). These are examples of stock/flow systems. It should be noted that these models are extremely simplified: for example, they take no account of space, age or size that are obviously important factors in the predatory relationship.

Trophic networks are directed weighted networks. Weight relates to the trophic flow between species and the direction is from prey to predator.

The usual way to represent stock/flow systems in ecology consists of defining a function (a so-called functional response) relating predation to the abundance of prey and predators. These functions are extensions of the mass-action principle. Encounters between individuals are related to their abundance, which in turn determines levels of predation. Here we show how network economics may provide an alternative approach based on economic principles. Predation levels operate according to a trade-off between the costs and benefits of predation for the predator.

Presently, there is a need for studying trophic systems that are altogether (1) realistic and (2) accessible to computation and analysis. Especially interesting are networks where the number of nodes is greater than 10 and less than 100.

7.2 Conservation Equations in Marine Ecosystems

Let us look in detail at what happens in a marine ecosystem representing the entire population of a given species (figure 7.1). We can express quantitative relationships between biomass, inflow and outflow. We denote b_i the biomass of species i and x_{ij} the trophic flows from species j to species i. Note the direction; as everywhere in this book, first index correspond to the origin (the prey), second index to the destination (the predator) of the trophic flow. Biomass b_i is expressed in tons of living matter and x_{ij} in tons of living matter consumed per year; trophic inflow is $\sum_j x_{ji}$, trophic outflow is $\sum_j x_{ij}$.

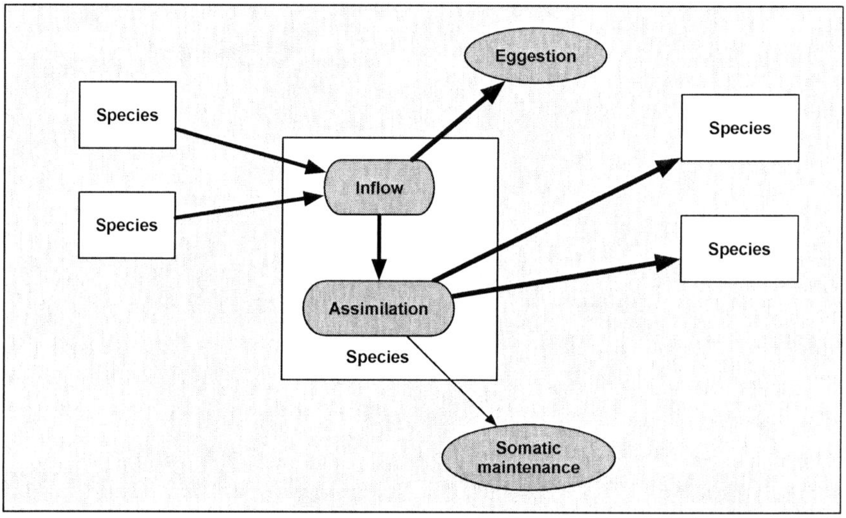

Figure 7.1: Network representation of a trophic system. A compartment.

A part of inflow is assimilated; it is assumed that assimilation rate γ_i of species i is constant. There are losses for species i that are due to natural mortality (other than predation) and somatic maintenance (respiration); it is assumed that these losses are proportional to biomass, with a constant proportionality rate μ_i: $\mu_i b_i$ are the losses of species i. There is a given import e_i (energy) for autotrophic species i.

If production is the assimilated biomass, if consumption is the biomass predated by other species and if losses are due to respiration or somatic maintenance, then the conservation equation expresses the equality between (1) production and (2) consumption and losses. According to previous equations, production is $\gamma_i(\sum_j x_{ji} + e_i)$, consumption is $\sum_j x_{ij}$ and losses are $\mu_i b_i$. This gives the mass balance equations of a marine ecosystem. For autotrophic species:

$$\gamma_i e_i = \mu_i b_i + \sum_j x_{ij}$$

For heterotrophic species:

$$\gamma_i \sum_j x_{ji} = \mu_i b_i + \sum_j x_{ij}$$

This second equation can be rearranged to show biomass as a linear function of inflows and outflows.

$$b_i = \frac{1}{\mu_i}\left[\gamma_i\left(\sum\nolimits_j x_{ji} + e_i\right) - \left(\sum\nolimits_{ji} x_{ij}\right)\right] \tag{7.2-1}$$

We will draw heavily on this equation. Biomass is the result of trophic exchanges, assimilation and somatic maintenance rates. This equation can also be interpreted using an analogy from accounting where it expresses the relationship between capital accounts (biomass) and financial accounts (trophic flows).

These principles are close to those of the Ecopath models (Pauly et al., 2000). It is possible to get above equation from the Ecopath main equation:

$$p_i b_i = \sum_j d_{ji} q_j b_j + (1 - e_i) p_i b_i$$

Where p_i is the production per biomass, q_i the consumption per biomass, d_{ij} are the diet coefficient, e_i the ecotrophic efficiency coefficient. According to definitions, we have $x_{ij} = d_{ji} q_j b_j$. We put $\gamma_i = p_i / q_i$ and $\mu_i = (1 - e_i) p_i$. We get the equality

$$\gamma_i q_i b_i = \sum_j x_{ij} + \mu_i b_i$$

As $\sum_j d_{ij} = 1$, we may write $\gamma_i q_i b_i = \gamma_i \sum_j d_{ij} q_i b_i = \gamma_i \sum_j x_{ji}$. Whence

$$\gamma_i \sum_j x_{ji} = \sum_j x_{ij} + \mu_i b_i$$

7.3 Complementarity Equations

The economic approach to predation leads to the hypothesis that, for a predatory species, there is a cost/benefit trade-off between the hunt or pursuit of prey, and its consumption[1]. If prey are abundant, predation is more advantageous. However, if predators are abundant, then predation is costly because of intra-species competition. The nature of the relationship between prey and predator takes account of such characteristics as preferences for a particular prey or the mechanical impossibility of capturing a prey of a very different size, etc.

We define excess function f_{ij} as the cost of predator j feeding upon prey i. It is equal to the cost of predation minus its benefits. When prey is abundant and easy to catch, excess function decreases $\partial f_{ij}/\partial b_i \leq 0$. It is assumed to increase with the abundance of predator, due to competition between predators: $\partial f_{ij}/\partial b_j \geq 0$. This example does not take account of collective predatory behavior in hunting for, and catching prey.

The system tends to equilibrium (figure 7.2). If $f_{ij} > 0$, predation is costly for the predator, flows x_{ij} decrease, consequently b_i increases, and f_{ij} decreases. However, if $f_{ij} < 0$, predation is beneficial for the predator and flows x_{ij} increase; consequently b_i decreases, and f_{ij} increases. Therefore, the system tends towards equilibrium characterized by complementarity equations: $x_{ij} > 0$ and excess $f_{ij} = 0$, or $x_{ij} = 0$ and excess $f_{ij} \geq 0$. Let us remark that in this last situation, biomass $b_i = 0$.

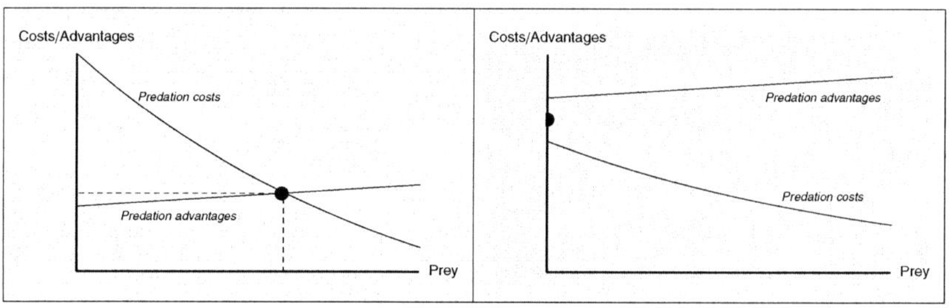

Figure 7.2: Ecological equilibrium. Excess and flows.

7.4 Some Simple Trophic Systems

Here we express both conservation and complementarity equations for some very simple ecosystems, whose network structures are shown in figure 7.3, and examine how their conservation and complementarity equations can be solved.

[1]Most important references in that direction are related to optimal foraging theory (MacArthur and Pianka, 1966), to the marginal value theorem (Charnov, 1976), to the ideal free distribution (Fretwell and Lucas, 1970a).

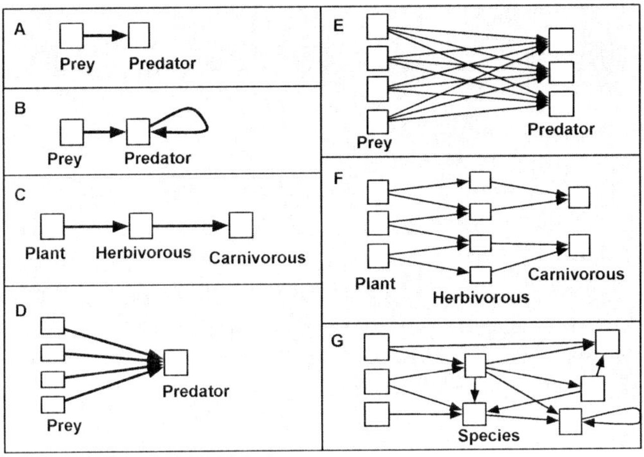

Figure 7.3: The network structure of simple trophic systems

7.4.1 Notations

In what follows, we use the following notations: i, j, k for compartments, x_{ij} for trophic flows from i (prey) to j (predator), b_i for biomass of i. We will use the reverse equation 7.2-1. We will express predation costs for j feeding on i as ϕ_{ij}.

7.4.2 Prey Predator

Figure 7.4 represents a system with only one prey i and a predator j. Conservation equations are:

$$\begin{aligned} \gamma_i e_i &= x_{ij} + \mu_i b_i \\ \gamma_j x_{ij} &= \mu_j b_j \end{aligned} \tag{7.4-2}$$

To express the excess function, we make the following linearity assumptions: for predator the ease of predation is $\kappa_i b_i$ where κ_i denotes the accessibility of prey, and there is no intra specific competition. The excess functions is $f_{ij} = \phi_{ij} - \kappa_i b_i$.

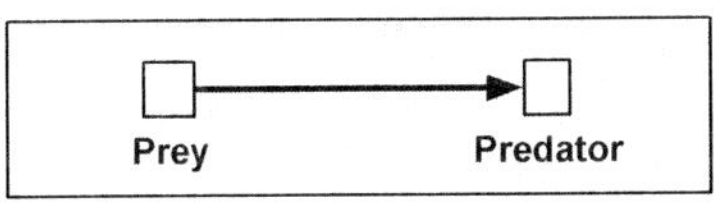

Figure 7.4: The network structure of simple trophic systems. Prey predator

Complementarity equations mean that if flows are > 0, then:

$$\kappa_i b_i = \phi_{ij} \tag{7.4-3}$$

The three equations of (7.4-2), (7.4-3) contain three unknowns: x_{ij}, b_i, b_j depending on parameters $\mu_i, \mu_j, \gamma_i, \gamma_j, \phi_{ij}$. They can be solved by the following. If flows are > 0, then biomass and flows are given by:

$$b_i = \frac{\phi_{ij}}{\kappa_i}$$

$$b_j = \frac{\gamma_i \gamma_j \kappa_i e_i - \mu_i \gamma_j \phi_{ij}}{\mu_j \kappa_i}$$

$$x_{ij} = \gamma_i e_i - \frac{\mu_i \phi_{ij}}{\kappa_i}$$

This result leads to the observation that if flows are > 0:

$$\frac{\partial b_i}{\partial \phi_{ij}} > 0, \ \frac{\partial b_j}{\partial \phi_{ij}} < 0, \ \frac{\partial x_{ij}}{\partial \phi_{ij}} < 0$$

This simply means that predation costs impact positively prey biomass and negatively predator biomass.

7.4.3 Cannibalism

Figure 7.5 shows a simple trophic system in which the predator feeds upon itself. In this case, conservation equations are:

$$\gamma_i e_i = x_{ij} + \mu_i b_i \qquad (7.4\text{-}4)$$

$$\gamma_j (x_{ij} + x_{jj}) = x_{jj} + \mu_j b_j$$

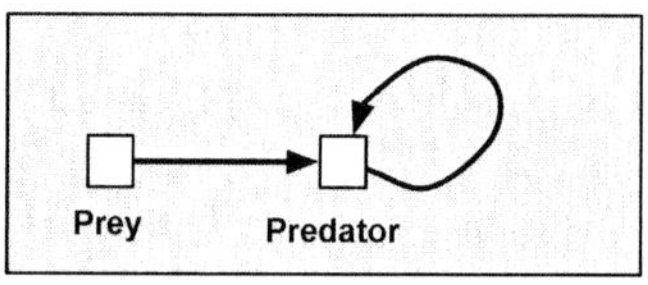

Figure 7.5: The network structure of simple trophic systems. Prey predator and cannibalism of the predator.

In this case we assume that excess function takes the same form as before: $f_{ij} = \phi_{ij} - \kappa_i b_i$ and $f_{jj} = \phi_{jj} - \kappa_j b_j$. If flows are > 0, complementarity equations are:

$$x_{ij} = (\gamma_i \gamma_j e_i - \gamma_i \mu_i \beta_{ij} - \mu_j \beta_{jj})/(1 - \gamma_j)$$

The system of equations (7.4-4) and (7.4-5) can be solved. There is, between parameters, a relationship that allows cannibalism:

$$\gamma_i \gamma_j e_i - \gamma_i \mu_i \beta_{ij} - \mu_j \beta_{jj} \geq 0$$

that is

$$\beta_{jj} \leq \frac{\gamma_i \gamma_j e_i}{\mu_j} - \frac{\gamma_i \mu_i \beta_{ij}}{\kappa_i}$$

We get a numerical condition relating the existence of cannibalism to the costs of alternative predations, including predation on it self.

7.4.4 A Short Food Chain

Figure 7.6 represents a short food chain linking a plankton population i, a planktivorous population j and a carnivorous population k. Lets denote b_i the biomass of i, b_j the biomass of j, b_k the biomass of k, x_{ij} the trophic flow from i to j, x_{jk} the trophic flow from j to k.

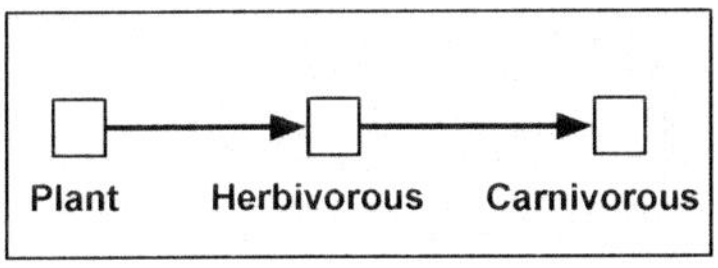

Figure 7.6: The network structure of simple trophic systems. A short food chain.

Conservation equations are:

$$\begin{aligned} \gamma_i e_i &= x_{ij} + \mu_i b_i \\ \gamma_j x_{ij} &= x_{jk} + \mu_j b_j \\ \gamma_k x_{jk} &= \mu_k b_k \end{aligned} \tag{7.4-5}$$

The excess function remains linear and we can write: $f_{ij} = \phi_{ij} - \kappa_i b_i$ and $f_{jk} = \phi_{jk} - \kappa_j b_j$. Complementarity equations are:

$$x_{ij} \perp f_{ij} \tag{7.4-6}$$

$$x_{jk} \perp f_{jk} \tag{7.4-7}$$

When there are positive flows: $x_{ij} > 0$, $x_{jk} > 0$, we have

$$\kappa_i b_i = \phi_{ij} \tag{7.4-8}$$

$$\kappa_j b_j = \phi_{jk} \tag{7.4-9}$$

Taking equations (7.4-5) and (7.4-8), together we have a linear system with five equations and five unknowns (b_i, b_j, b_k, x_{ij}, x_{jk}). Solutions are:

$$b_i = \frac{\phi_{ij}}{\kappa_i}$$

$$b_j = \frac{\phi_{jk}}{\kappa_j}$$

$$b_k = \frac{\gamma_k \gamma_j \gamma_i e_i}{\mu_k} - \frac{\gamma_j \gamma_k \mu_i \phi_{ij}}{\kappa_i \mu_k} - \frac{\gamma_k \mu_j \phi_{jk}}{\kappa_j \mu_k}$$

$$x_{ij} = \gamma_i e_i - \frac{\mu_i \phi_{ij}}{\kappa_i}$$

$$x_{jk} = \gamma_j \gamma_i e_i - \frac{\gamma_j \mu_i \phi_{ij}}{\kappa_i} - \frac{\mu_j \phi_{jk}}{\kappa_j}$$

The effects of the parameters on the solution: $\partial b_i / \partial \phi_{ij}$, $\partial b_k / \partial \phi_{jk}$, $\partial x_{jk} / \partial \phi_{ij}$, are easy to determine, and are all significant. For example, changing energy input e_i has no effect on b_i or b_j. Accessibility parameters have the most impact on solutions.

7.4.5 Several Prey and a Predator

Figure 7.7 represents a system with several prey i and a predator j. Let us denote b_i, the biomass of prey i, b_j the biomass of predator j and x_{ij} the trophic flow from i to j. Conservation equations are:

$$\gamma_i e_i = x_{ij} + \mu_i b_i$$

$$\gamma_j \sum_i x_{ij} = \mu_j b_j$$

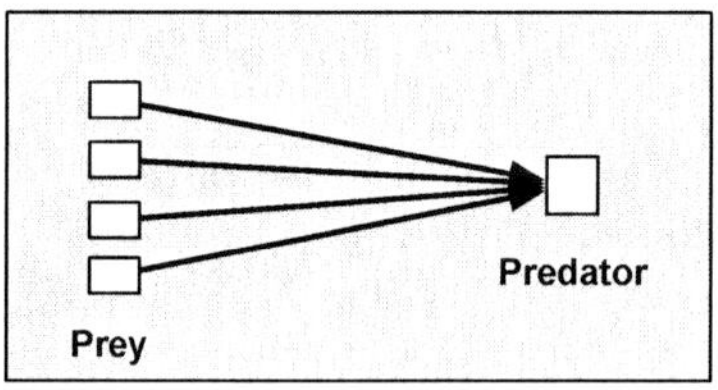

Figure 7.7: The network structure of simple trophic systems. Several prey and a predator.

From which we obtain:

$$b_i = \frac{\gamma_i e_i - x_{ij}}{\mu_i}$$

$$b_j = \frac{\gamma_j \sum_i x_{ij}}{\mu_j}$$

Excess functions are linear: $f_{ij} = \kappa_i b_i - \phi_{ij}$. Complementarity equations are thus:

$$x_{ij} \perp f_{ij}$$

Constraints are obvious. Flows x_{ij} must result in positive biomass: $b_i \geq 0$, $b_j \geq 0$. All together, we get when all flows are positive:

$$\gamma_i e_i = x_{ij} + \mu_i b_i$$
$$\gamma_j \sum_i x_{ij} = \mu_j b_j$$
$$\kappa_i b_i - \phi_{ij} = 0$$

This is easy to solve:

$$b_i = \frac{\phi_{ij}}{\kappa_i}$$
$$x_{ij} = \gamma_i e_i - \mu_i \frac{\phi_{ij}}{\kappa_i}$$
$$b_j = \sum_i \frac{\gamma_j \gamma_i e_i}{\mu_j} - \sum_i \frac{\gamma_j \gamma_i e_i \phi_{ij}}{\mu_j \kappa_i}$$

This provides a simple mathematical approach to the ideal free distribution theory (Fretwell and Lucas, 1970b).

7.4.6 Several Prey and Several Predators

Figure 7.8 represents a system with several prey i and several predators j. Let us denote b_i, the biomass of prey i, b_j the biomass of predator j and x_{ij} the trophic flow from i to j. Conservation equations are:

$$\gamma_i e_i = \sum_j x_{ij} + \mu_i b_i$$
$$\gamma_j \sum_i x_{ij} = \mu_j b_j$$

From which we obtain:

$$b_i = \frac{\gamma_i e_i - \sum_j x_{ij}}{\mu_i}$$
$$b_j = \frac{\gamma_j \sum_i x_{ij}}{\mu_j}$$

The excess functions remains linear: $f_{ij} = \kappa_i b_i - \phi_{ij} + \lambda_j b_j$ and complementarity equations are therefore:

$$x_{ij} \perp f_{ij}$$

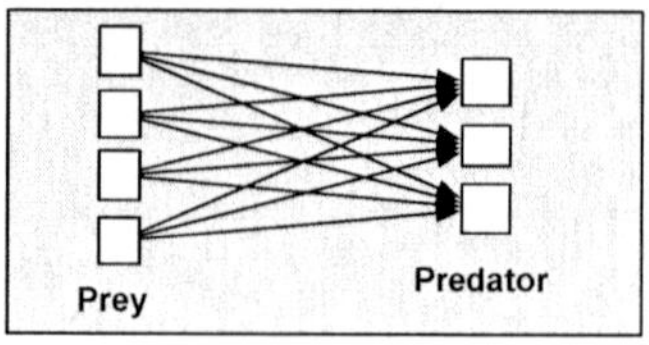

Figure 7.8: The network structure of simple trophic systems. Several prey and several predators.

Constraints are obvious. Flows x_{ij} must result in positive biomass: $b_i \geq 0$, $b_j \geq 0$.

Let us introduce the function $X = (x_{ij}) \to F(X) = (f_{ij})$, defined by:

$$\begin{aligned} f_{ij} &= \phi_{ij} - \kappa_i b_i + \lambda_j b_j \\ &= \phi_{ij} - \kappa_i (\frac{\gamma_i e_i - \sum_k x_{ik}}{\mu_i}) + \lambda_j \frac{\gamma_j \sum_i x_{ij}}{\mu_j} \end{aligned}$$

Then, we have:

Theorem 7.4.1 *$X^* \in K$ is an equilibrium state if and only if it is a solution of the variational inequality $\mathcal{VI}(F, K)$.*

Proof. Let $X^* \in K$ an equilibrium state. Let $S = \{(ij) \mid x^*_{ij} = 0\}$. Then we have $f^*_{ij} = 0$ for $(ij) \notin S$. We deduce:

$$\begin{aligned} F(X^*) \cdot (X - X^*) &= \sum_{(ij) \in S} f^*_{ij}(x_{ij} - x^*_{ij}) + \sum_{(ij) \notin S} f^*_{ij}(x_{ij} - x^*_{ij}) \\ &= \sum_{(ij) \in S} f^*_{ij} x_{ij} \geq 0 \end{aligned}$$

Now, suppose that X^0 is not an equilibrium. Then, either (1) there exists $(i'j')$ such $f^0_{i'j'} < 0$, or (2) there exists $(i'j')$ such $f^0_{i'j'} > 0$ and $x^0_{i'j'} > 0$.

In the first case, we consider a state X such that $x_{i'j'} = x^0_{i'j'} + \epsilon$ and $x_{ij} = x^0_{ij}$ for $(ij) \neq (i'j')$. Such a state is in K: a slight reduction in predation by j' diminishes its biomass, but leaves it positive. This gives:

$$\begin{aligned} \sum_{ij} f^0_{ij}(x_{ij} - x^0_{ij}) &= f^*_{i'j'}(x_{i'j'} - x^0_{i'j}) \\ &= -f^0_{i'j'}\epsilon \\ &< 0 \end{aligned}$$

Thus variational inequality is not verified.

In the second case: there exists $(i'j')$ such $f^0_{i'j} > 0$ and $x^0_{i'j}$. We consider a state X such that $x_{i'j'} = x^0_{i'j'} - \epsilon$ and $x_{ij} = x^0_{ij}$ for $(ij) \neq (i'j')$. Such a state is in K, because $f^0_{i'j'} = \phi_{i'j'} - \kappa_{i'} b^0_{i'} + \lambda_{j'} b^0_{j'} < 0$ and thus $b^0_{i'} > 0$ and can be reduced by a slightly higher level of predation by j'. This gives:

$$\begin{aligned}\sum_{ij} f^0_{ij}(x_{ij} - x^0_{ij}) &= f^0_{i'j'}(x_{i'j'} - x^0_{i'j'}) \\ &= f_{i'j'}\epsilon < 0\end{aligned}$$

We remark that:

Theorem 7.4.2 *F is an affine function $F(X) = M \cdot X + N$, with symmetric associated matrix M: $M_{ij,i'j'} = 0$ if $i \neq i'$, $M_{ij,ij'} = \kappa_i/\mu_i$ and $N_{ij} = \phi_{ij} - \kappa_i \gamma_i e_i/\mu_i$*

Proof. We develop:

$$\begin{aligned}f_{ij} &= \phi_{ij} - \kappa_i(\frac{\gamma_i e_i - \sum_k x_{ik}}{\mu_i}) + \lambda_j \frac{\gamma_j \sum_k x_{kj}}{\mu_j} \\ &= \phi_{ij} - \frac{\kappa_i \gamma_i e_i}{\mu_i} + \sum_k \frac{\kappa_i}{\mu_i} x_{ik} + \sum_k \frac{\lambda_j \gamma_j}{\mu_j} x_{kj} \\ &= N_{ij} + \sum_{i'j'} M_{ij,i'j'} x_{i'j'}\end{aligned}$$

□

We remark that:

Theorem 7.4.3 *F is strongly monotone.*

Proof:

$$\begin{aligned}(F(X) - F(X')) \cdot (X - X') &= \sum_{ij} (f_{ij} - f'_{ij})(x_{ij} - x'_{ij}) \\ &= \sum_{ijj'} \frac{\kappa_i}{\mu_i}(x_{ij'} - x'_{ij'})(x_{ij} - x'_{ij}) \\ &= \sum_i \frac{\kappa_i}{\mu_i}(O_i - O'_i)^2\end{aligned}$$

□

We conclude that:

Theorem 7.4.4 *Equilibrium exists and can be computed.*

Proof. A simple consequence of theorem 6.1.9. □

7.4.7 Pelagic Food Chains

Figure 7.9 represents a similar system with several plankton species i, several planktivorous species j, several carnivorous species k. Conservation equations are:

$$\gamma_i e_i = \sum_j x_{ij} + \mu_i b_i$$

$$\gamma_j \sum_i x_{ij} = \sum_k x_{jk} + \mu_j b_j$$

$$\gamma_k \sum_j x_{jk} = \mu_k b_k$$

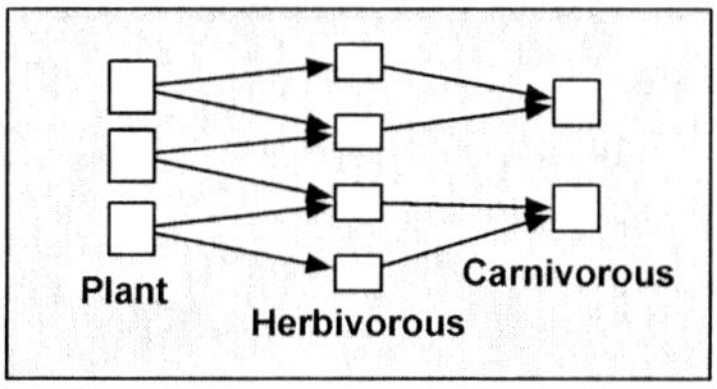

Figure 7.9: The network structure of simple trophic systems. Pelagic food chain.

From which we obtain:

$$b_i = (\gamma_i e_i - \sum_j x_{ij})/\mu_i$$

$$b_j = (\gamma_j \sum_i x_{ij} - \sum_k x_{jk})/\mu_j \qquad (7.4\text{-}10)$$

$$b_k = \gamma_k \sum_j x_{jk})/\mu_k$$

Excess functions are as before:

$$\begin{aligned} f_{ij} &= \kappa_i b_i - \phi_{ij} \\ f_{jk} &= \kappa_j b_j - \phi_{jk} \end{aligned} \qquad (7.4\text{-}11)$$

Complementarity equations are:

$$x_{ij} \perp \phi_{ij} - \kappa_i b_i$$

$$x_{jk} \perp \phi_{jk} - \kappa_j b_j$$

There are obvious constraints on flows. They must be positive and such: $b_i \geq 0$, $b_j \geq 0$, $b_k \geq 0$. We observe that flows determine biomass according to equation 7.4-10 and that biomass determine excess functions according to equation 7.4-11.

Let $X = (x_{ij}, x_{jk})$ a state of the system. We can define the following function: $X \rightarrow F(X) = (f_{ij}, f_{jk})$ and

$$f_{ij} = \phi_{ij} - \kappa_i b_i$$

$$= \phi_{ij} - \kappa_i(\frac{\gamma_i e_i - \sum_l x_{il}}{\mu_i})$$

$$f_{jk} = \phi_{jk} - \kappa_j b_j$$

$$= \phi_{jk} - \kappa_j(\frac{\gamma_j \sum_i x_{ij} - \sum_l x_{jl}}{\mu_j})$$

Then we have the following theorem:

Theorem 7.4.5 *$X^* \in K$ is an equilibrium state if and only if it is a solution of the variational inequality $\mathcal{VI}(F, K)$.*

Proof. Let $X^* \in K$ an equilibrium state. Let $S_1 = \{(ij) \mid x^*_{ij} = 0\}$ and $S_2 = \{(jk) \mid x^*_{jk} = 0\}$. Then $f^*_{ij} = 0$ for $(ij) \notin S_1$ and $f^*_{jk} = 0$ for $(jk) \notin S_2$.

$$F(X^*) \cdot (X - X^*) = \sum_{(ij) \in S_1} f^*_{ij} x_{ij} + \sum_{(jk) \in S_2} f^*_{jk} x_{jk}$$

$$\geq 0$$

Now, suppose that X^0 is not an equilibrium. Equilibrium conditions are not verified on a path (ij) or on a path (jk). In the last case, proof is exactly the same as the one regarding a bilayered network.

Therefore we can assume that equilibrium conditions are satisfied on paths jk and that either (1) there exists (i', j') such $f^0_{i'j'} < 0$, or (2) there exists $(i'j')$ such $f^0_{i'j'} > 0$ and $x^0_{i'j'} > 0$.

In the first case: there exists (i', j') such $f^0_{i'j'} < 0$, we consider a state X such that $x_{i'j'} = x^0_{ij}(1 + \epsilon)$ with $0 < \epsilon$, $x_{ij} = x^0_{ij}$ for $(i, j) \neq (i'j')$, $x_{j'k} = (1 + \epsilon)x^0_{j'k}$ and $x_{jk} = x^0_{jk}$ for $j \neq j'$. There exists a value of ϵ such that all resulting biomass are positive, thus such that $X \in K$. And we have

$$S = \sum_{ij} f^0_{ij}(x_{ij} - x^0_{ij}) + \sum_{jk} f^0_{jk}(x_{jk} - x^0_{jk})$$

$$= f^0_{i'j'}(x_{i'j'} - x^0_{i'j'}) + \sum_k f^0_{j'k}(x_{j'k} - x^0_{j'k})$$

$$= f^0_{i'j'} \epsilon x^0_{i'j'} + \sum_k f^0_{j'k} \epsilon x^0_{j'k}$$

But complementarity conditions are satisfied on paths $j'k$ and thus $f^0_{j'k} \perp x^0_{j'k} = 0$. Fianlly

$$S = f^0_{i'j'}(1-\epsilon)x^0_{i'j'} < 0$$

Thus variational inequality is not verified for X^0.

In the second case: there exists $(i'j')$ such $f^0_{i'j'} > 0$ and $x^0_{i'j'} > 0$, we consider a state X such that $x_{i'j'} = (1-\epsilon)x^0_{ij}$ with a small ϵ such that $b_{j'} > 0$, $x_{ij} = x^0_{ij}$ for $(i,j) \neq (i'j')$, and $x_{jk} = x^0_{jk}$. Then, this state is in K: $b_{i'} \geq b^0_{i'} > 0$, $b_{j'} > 0$. And:

$$\begin{aligned}
S &= \sum_{ij} f^0_{ij}(x_{ij} - x^0_{ij}) + \sum_{jk} f^0_{jk}(x_{jk} - x^0_{jk}) \\
&= \sum_{ij} f^0_{ij}(x_{ij} - x^0_{ij}) \\
&= f^0_{i'j'}(x_{i'j'} - x^0_{i'j'}) \\
&= -\epsilon f^0_{i'j'} x^0_{i'j'} \\
&< 0
\end{aligned}$$

Thus variational inequality is not verified for X^0. □

Theorem 7.4.6 *F is an affine function: $F(X) = M \cdot X + N$, with associated matrix M such that:*

$$M = \begin{pmatrix} A & 0 \\ C & D \end{pmatrix}, \ N = \begin{pmatrix} E \\ F \end{pmatrix}$$

- $A_{ij,i'j'} = 0$ *if* $i \neq i'$, $A_{ij,ij'} = \kappa_i/\mu_i$
- $C_{jk,i'j'} = 0$ *if* $j \neq j'$, $C_{jk,i'j} = -\kappa_j\gamma_j/\mu_j$
- $D_{jk,j'k'} = 0$ *if* $j \neq j'$, $D_{jk,jk'} = \kappa_j/\mu_j$
- $E_{ij} = \phi_{ij} - \kappa_i\gamma_i E_i/\mu_i$, $f_{jk} = \phi_{jk}$

Proof. We develop

$$f_{ij}(X) = \phi_{ij} - \frac{\kappa_i\gamma_i E_i}{\mu_i} + \sum_l \frac{\kappa_i}{\mu_i} x_{il}$$

$$f_{jk}(X) = \phi_{jk} - \sum_i \frac{\kappa_j\gamma_j}{\mu_j} x_{ij} + \sum_l \frac{\kappa_j}{\mu_j} x_{jl}$$

□

7.5 General Trophic Networks

Finally we consider how the network economic approach may be applied to the overall marine ecosystem. In this system there are autotrophic, herbivorous, specialized carnivorous, and generalized carnivorous species (figure 7.10).

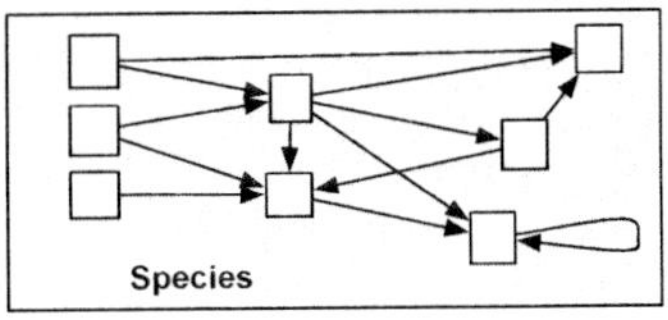

Figure 7.10: The network structure of simple trophic systems. General trophic networks.

7.5.1 Formulation

We denote species i, j, k, b_i the biomass of i and x_{ij} the trophic flow from i to j. Species i is characterized by its assimilation rate γ_i, its somatic maintenance rate μ_i, its energy import e_i; we have $E_i > 0$ only for autotrophic species. As before, we get conservation equations

$$\gamma_i \left(\sum_j x_{ji} + E_i\right) = \left(\sum_{ji} x_{ij}\right) + \mu_i b_i$$

and inverse equations

$$b_i = \frac{1}{\mu_i}\left[\gamma_i \left(\sum_j x_{ji} + e_i\right) - \left(\sum_{ji} x_{ij}\right)\right]$$

Let L the number of effective links. The state of the system is fully determined by $x_{ij} \in \mathbb{R}^L$. We consider constraints $b_i \geq 0$ and $x_{ij} \geq 0$. They define a compact polyhedron $K \subset \mathbb{R}^L$.

We consider the excess function: $f_{ij} = \phi_{ij} - \kappa_i b_i + \lambda_j b_j$. Complementarity equations are:

$$x^*_{ij} \perp f^*_{ij}$$

We consider the function: $X = (x_{ij}) \to F(X) = (f_{ij})$.

Theorem 7.5.1 *$X^* \in K$ is an equilibrium state if and only if it verifies the variational inequality $\mathcal{VI}(F, K)$.*

Proof. Let us prove first that an equilibrium is a solution of the variational inequality. Let $C = \{ij \mid x^*_{ij} = 0\}$. Then

$$\begin{aligned}
\sum_{ij} f^*_{ij}(x_{ij} - x^*_{ij}) &= \sum_{ij\in C} f^*_{ij}(x_{ij} - x^*_{ij}) + \sum_{ij\notin C} f^*_{ij}(x_{ij} - x^*_{ij}) \\
&= \sum_{ij\in C} f^*_{ij}(x_{ij} - 0) + \sum_{ij\notin C} 0(x_{ij} - x^*_{ij}) \\
&= \sum_{ij\in C} f^*_{ij} x_{ij} \\
&\geq 0
\end{aligned}$$

Now let us prove that a solution of the variational inequality is an equilibrium. Let us assume that it is not. In this case, there are three possibilities:

(1) there exists no pair i_0, j_0 such that $x^*_{i_0,j_0} > 0$ and $f^*_{i_0,j_0} < 0$,

(2) there exists no pair i_0, j_0 such that $x^*_{i_0,j_0} > 0$, $b_{j_0} > 0$ and $f^*_{i_0,j_0} > 0$,

(3) there exists no pair i_0, j_0 such that $x^*_{i_0,j_0} > 0$, $b_{j_0} = 0$ and $f^*_{i_0,j_0} > 0$.

If there exists no pair i_0, j_0 such that $x^*_{i_0,j_0} > 0$ and $f^*_{i_0,j_0} < 0$, we consider X such that $x_{i_0 j_0} = x^*_{i_0 j_0} + \epsilon$, $x_{ij} = x^*_{ij}$ otherwise. Then $X \in K$ is obvious. It is equally obvious that X is a counter example to variational inequality.

If there exists no pair i_0, j_0 such that $x^*_{i_0,j_0} > 0$, $b_{j_0} > 0$ and $f^*_{i_0,j_0} > 0$ we consider X such that $x_{i_0 j_0} = x^*_{i_0 j_0} - \epsilon$ and $x_{ij} = x^*_{ij}$ otherwise.Then $X \in K$ is obvious. It is equally obvious that X is a counter example to variational inequality.

Now, suppose that there exists no pair i_0, j_0 such that $x^*_{i_0,j_0} > 0$, $b_{j_0} = 0$ and $f^*_{i_0,j_0} > 0$. We define S the set of nodes i such that $i = i_0$ or there exists a trophic path starting in i_0 and ending in i. We consider X such that $x_{ij} = (1-\epsilon)x^*_{ij}$ if $i \in S$ and otherwise $x_{ij} = x^*_{ij}$.

Let us prove that for all i, corresponding biomass b_i are positive. If $i \in S$,

$$\begin{aligned}
b_i &= \frac{1}{\mu_i}(\gamma_i(\sum_{j \in S} x_{ji} + \sum_{j \notin S} x_{ji}) - \sum_j x_{ij}) \\
&= \frac{1}{\mu_i}(\gamma_i(\sum_{j \in S}(1-\epsilon)x^0_{ji} + \sum_{j \notin S} x^0_{ji}) - \sum_j (1-\epsilon)x^0_{ij}) \\
&= (1-\epsilon)b^0_i + \frac{1}{\mu_i}\epsilon \sum_{j \notin S} x^0_{ji} \\
&> 0
\end{aligned}$$

If $i \notin S$,

$$\begin{aligned}
b_i &= \frac{1}{\mu_i}\gamma_i(\sum_j x_{ji} - \sum_{j \in S} x_{ij} - \sum_{j \notin S} x_{ij}) \\
&= \frac{1}{\mu_i}\gamma_i(\sum_j x_{ji} - \sum_{j \in S}(1-\epsilon)x^0_{ij} - \sum_{j \notin S} x^0_{ij}) \\
&= b^0_i + \frac{1}{\mu_i}\epsilon \sum_{j \in S} x^0_{ij} \\
&> 0
\end{aligned}$$

Then $X \in K$. It is obvious that X is a counter example to variational inequality. □

Chapter 8

About the Equilibrium of Economic Networks

8.1 About Economic Equilibrium

8.1.1 Introduction

This chapter is an introduction to the theories of spatial economic equilibrium and supply chains with, of course, an emphasis on the underlying network structure of the system. It includes a digression that is more specifically addressed to ecologists, in which we expand on the conventional economic approach to equilibrium, its relation to utility functions and introduce the theory of the general equilibrium. We also show how these ideas can be expressed using the framework of network economics.

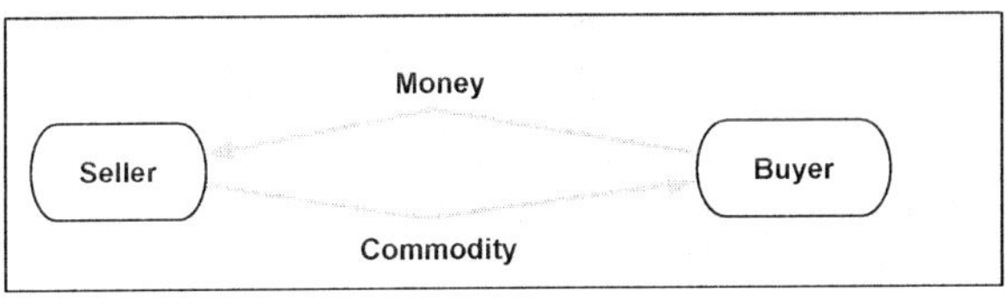

Figure 8.1: Economic exchange. With commodity and money flows.

The most important factor when studying economic systems with a network approach is clearly the fact that flows have a dual component: a material or service flows in one direction, while money flows in the opposite direction (figures 8.1 and 8.2). Of course, the two flows are related. The ratio between money flow m and commodity flow c is obviously the price of the commodity $p = m/c$. The situation where prices functionnaly depend on the relationship between the behaviours of the sellers and the buyers is described according to the law of supply and demand that is detailed in the next paragraph.

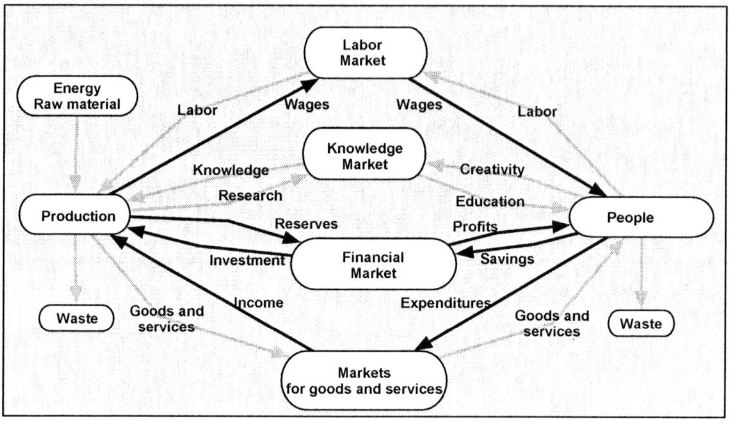

Figure 8.2: A general economic network. Including material (grey), services (grey) and money flows (black).

8.1.2 The Law of Supply and Demand

In a simple model such as the one of figure 8.1, the law of supply and demand is illustrated in figure 8.3 and can be expressed as follows:

- We assume that the producers have production constraints that make them charging a higher price for a higher quantity: there is an increasing supply function.
- We assume that the consumers have money constraints that make them to seek a lower price for a higher quantity: there is a decreasing demand function.
- Then we conclude: (1) there is an economic exchange between both agents when supply price equals demand price; (2) if there is a surplus of supply price on demand price, there is no exchange, no market price; and (3) there are no other alternatives.

Now let us consider an economic system where several agents exchange various commodities. We make the (risky) assumption that the excess price (the demand price minus the supply price) of any commodity depends on all exchanged commodities, i.e., there is a function such that:

$$Q = (q_1, \ldots, q_n) \to E(Q) = (e_1, e_2, \ldots e_n)$$

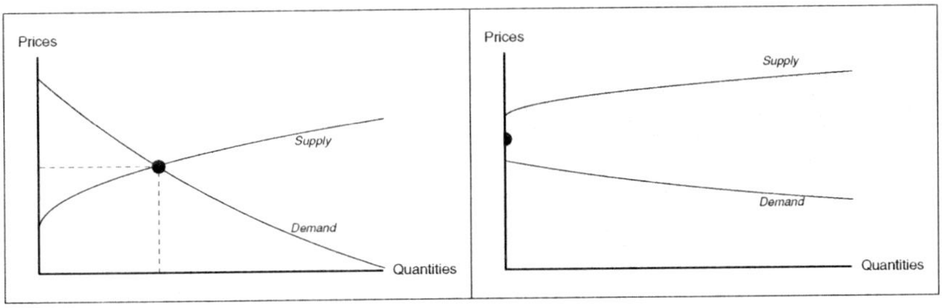

Figure 8.3: The law of supply and demand. Left: equality of supply and demand: there is an exchange and a price. Right: excess of supply on demand; there is no exchange and no price.

We express the law of supply and demand in this case as $q_i \geq 0, e_i \geq 0, q_i e_i = 0$, that is $q_i \perp e_i$; this is a new example of a complementarity relationship: equilibrium is a state in which two related characteristics are positive, one of them being zero.

8.2 Spatial Economics

In spatial economics (figure 8.4), there are (1) a single commodity, (2) several producers i, (3) several commodity markets j and (4) commodity flows x_{ij} between producers and markets[1]. Producers are characterized by their production q_i, which is greater than their sales: $q_i \geq \sum_j x_{ij}$ and their production costs function: $q_i \rightarrow c_i(q_i)$; costs functions are supposed increasing. Markets are characterized by their deals s_j which are such that $s_j \leq \sum_i x_{ij}$ and by a function $s_j \rightarrow p_j(s_j)$, relating the prices to the volume of their supply; such a function is called an inverse demand function[2]; it is supposed to be decreasing. Links are characterized by transportation costs function $X = (x_{ij}) \rightarrow t_{ij}(X)$. It should be noted that transportation costs on a link may depend on the load on all links, and not simply on its own load.

Definition 8.2.1 *An equilibrium state $X^* = (x^*_{ij})$ of the above spatial economics system is defined by the complementarity relationship:*

$$x^*_{ij} \perp (c^*_i + t^*_{ij} - p^*ij)$$

The above definition is an expression of the law of supply and demand. The mechanism is as follows. If prices and costs are such that $c_i + t_{ij} - p_j > 0$ (prices

[1] Spatial economics originated in the work of Samuelson (1952); see Nagurney (1999) for details and an historical sketch.

[2] Demand and inverse demand function, and their relationship with the idea of utility will be discussed in section 8.5. The suitability of using inverse demand function for fish markets is discussed in Anderson (1980); Asche et al. (1997); Nielsen (1999).

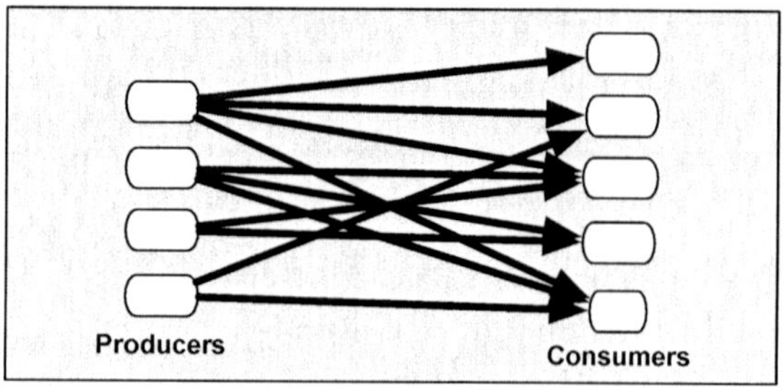

Figure 8.4: Spatial equilibrium. Several producers (left), several consumers (right). The structure is that of a bi layered network.

are lower than production and transportation costs), the flow x_{ij} decreases, there is a consequent decrease of supply to the market j and an increase in price p_j. Then there is a decrease in production by producer i and a decrease in costs c_i. This occurs until either $c_i + t_{ij} - p_j = 0$ or $x_{ij} = 0$.

The function of spatial equilibrium is: $F : X = (x_{ij}) \to F(X) = (f_{ij})$ where $f_{ij} = c_i + t_{ij} - p_j = 0$. The set of possible flows is: $K = \mathbb{R}_+^{ni \times nj}$.

Main characterization theorems are:

Theorem 8.2.1 *$X^* = (x^*_{ij})$ is an equilibrium if and only if it is a solution of the complementarity problem $\mathcal{CP}(F)$.*

Proof: it is enough to rephrase the definition of the equilibrium of a spatial economical system. □

Theorem 8.2.2 *$X^* = (x^*_{ij})$ is an equilibrium if and only if it is a solution of the variational inequality $\mathcal{VI}(F, K)$.*

Proof. Let us prove first that an equilibrium state is a solution of the variational inequality $\mathcal{VI}(K, F)$. We develop

$$\begin{aligned}
(X - X^*) \cdot F(X^*) &= \sum_{ij} (x_{ij} - x^*_{ij}) f^*_{ij} \\
&= \sum_{ij|x^*_{ij}=0} (x_{ij} - x^*_{ij}) f^*_{ij} \\
&\quad + \sum_{ij|x^*_{ij}>0} (x_{ij} - x^*_{ij}) f^*_{ij} \\
&= \sum_{ij|x^*_{ij}=0} x_{ij} f^*_{ij} \\
&\geq 0
\end{aligned}$$

Let us now prove first that a solution of the variational inequality $\mathcal{VI}(K, F)$ is an equilibrium state. Suppose that X^* is not an equilibrium state of the system. Then either there exists a path a path i_0j_0 such that $x^*_{i_0j_0} > 0$ and $f^*_{i_0j_0} \neq 0$, or there exists a path a path ij such that $x^*_{x^*_{i_0j_0}} = 0$ and $f^*_{i_0j_0} < 0$. We consider $x_{ij} = x^*ij$ for $(ij) \neq (i_0j_0)$. We have: $(X - X^*) \cdot F(X^*) = (x_{i_0j_0} - x^*_{i_0j_0})f^*_{i_0j_0}$. If $x^*_{i_0j_0} > 0$ and $f^*_{i_0j_0} > 0$, we take $0 < x_{i_0j_0} < x^*_{i_0j_0}$. If $x^*_{i_0j_0} > 0$ and $f^*_{i_0j_0} < 0$, we take $x_{i_0j_0} > x^*_{i_0j_0}$. If $x^*_{x^*_{i_0j_0}} = 0$ and $f^*_{i_0j_0} < 0$, we take $x_{i_0j_0} > 0$. In all cases, we get a contradiction. □

8.2.1 Extensions

We draw heavily upon inverse demand functions in this presentation: in all markets j, the function $q_j \rightarrow p_j = F_j(q_j)$, relating prices to quantities, plays an important role. We do not take into account the role of consumers in arbitrating between several markets and the fact that the inverse demand functions on these markets are related: prices on a given market depend on the quantities traded on an other market. This can be achieved by considering general inverse demand functions $(q_1, \ldots, q_n) \rightarrow p_j = F_j(q_1, \ldots, q_n)$. In this case, all the results of spatial economics remain valid.

8.2.2 An Elementary Example

Let us illustrate the principles of spatial economics theory with an example consisting of a system consisting of two producers a and b and two consumers u and v depicted in figure 8.5. These actors produce, exchange and consume a single commodity. Producers are characterized by their production costs that relate the quantity they produce q_a and q_b to the production costs of one unit of the commodity. We assume that production costs are linear and increasing: $c_a(q_a) = 100 + q_a$ and $c_b(q_b) = 200 + 5q_b$. Consumers are characterized by an inverse demand function that relates the quantity they purchase (s_u and s_v) to the price they agree to pay. We assume that inverse demand functions are linear and decreasing: $p_u(s_u) = 900 - 4s_u$, $p_v(s_v) = 1200 - 8s_u$.

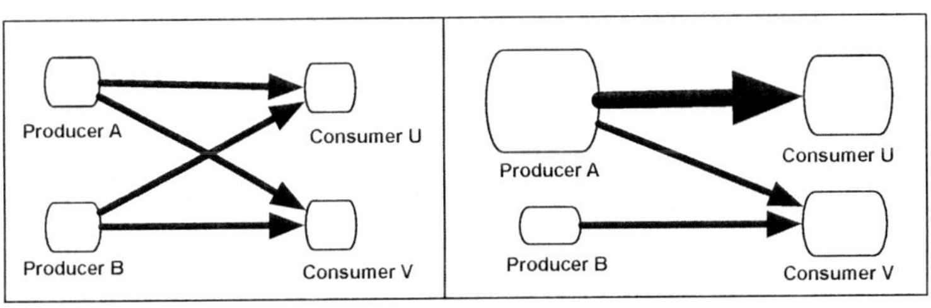

Figure 8.5: Spatial equilibrium. Two producers, two consumers. Left: the structure of the network. Right: a representation of the network with sizes of nodes and thickness of links are proportional to consumption, production or sales.

Let $x_{au}, x_{av}, x_{bu}, x_{bv}$ be the quantity exchange between producers and consumers. The model can be called a spatial definition as it takes into account transportation costs functions. Here we assume that transportation costs functions are constant $t_{au} = 25$, $t_{av} = 135$, $t_{bu} = 144$, $t_{bv} = 12$. Obviously, productions are given by: $q_a = x_{au} + x_{av}$, $q_b = x_{bu} + x_{bv}$ and consumptions are given by: $s_u = x_{au} + x_{bu}$, $s_v = x_{av} + x_{bv}$.

Then an equilibrium of the system is a set of exchanges between producers and consumers: $x_{au}, x_{av}, x_{bu}, x_{bv}$ such that any producer cannot unilaterally change its behavior without lowering their per-unit benefits. This is expressed as:

$$x_{au} \perp (c_a + t_{au} - p_u)$$

$$x_{av} \perp (c_a + t_{av} - p_v)$$

$$x_{bu} \perp (c_b + t_{au} - p_u)$$

$$x_{bv} \perp (c_b + t_{av} - p_v)$$

It is not possible to get an equilibrium with all flows being positive; with above values, we would have altogether:

$$5x_{au} + x_{av} + 4x_{bu} + 0x_{bv} = 775$$

$$x_{au} + 9x_{av} + 0x_{bu} + 8x_{bv} = 965$$

$$4x_{au} + 0x_{av} + 9x_{bu} + 5x_{bv} = 556$$

$$0x_{au} + 8x_{av} + 5x_{bu} + 13x_{bv} = 988$$

This equation has no solution (add first and fourth equalities, second and third equalities, compare).

We must now examine situations in which one flow is zero (there are four), two flows are zero (six), three flows are zero (four), and all flows are zero (one). We begin by checking what happens when one flow is zero, and then proceed flow by flow. Although it is always possible to solve the restricted equations, in all cases (bar one) the result is a negative value for one flow, or a positive value of F for the zero flow. The only correct solution is:

$$x_{au} = 144.579$$

$$x_{av} = 52.103$$

$$x_{bu} = 0$$

$$x_{bv} = 43.936$$

Productions, productions costs, trade and prices depend immediately of these values of flows. Solution is depicted in figure 8.5 (right).

8.2.3 Concluding Remark

It has been very simple to compute the equilibrium of the system described here because there are only four possible flows. In the worst case, there are only $2^4 = 16$ situations to examine. In the case of a more complex system, with n possible flows, we would have to examine 2^n situations, which rapidly becomes infeasible to compute. This is one of the reasons why we use variational inequality or complementarity problems in the examples that follow.

8.3 Supply Chains: Limitation of Paths Capacity

Supply chains are generalizations of the short chain relating producers and markets that we looked at in the previous section. In this case, there are intermediate nodes (transformers) between producers and consumers (figure 8.6). In this section we analyze supply chains that have capacity constraints, i.e., flows are limited on some links. Of course, there are many other constraints that can characterize a supply chain, such as limits on production, consumption, or costs and in the next section we will examine supply chains with production constraints. All these supply chains can be analyzed with the network economics approach.

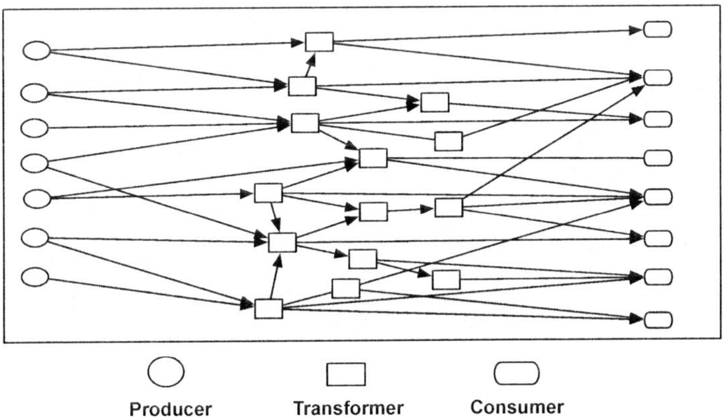

Figure 8.6: The structure of a supply chain. Left part: producers. Middle part: transformers. Right part: consumers.

8.3.1 Network Structure

Nodes n are characterized by an observed price p_n; for producers, p_n equals the production costs c_n; for consumers, p_n is the market price given by an inverse demand function $p_n(q_n) = a_n - b_n q_n$; for transformers, p_n results from the network complementarity equations.

Links l are characterized by the observed flows x_l, transportation costs c_l, and a capacity m_l, that is a limit on effective flows.

We denote $o(l)$ the node origin and $d(l)$ the node destination of a link l. We define usual incidence functions: $\delta^o_{nl} = 1$ if $o(l) = n$, $\delta^d_{nl} = 1$ if $d(l) = n$.

Inflow of node n is $i_n = \sum_l \delta^o_{nl} x_l$. Outflow of node n is $o_n = \sum_l \delta^d_{nl} x_l$. The excess is $f_n = i_n - o_n = \sum_l (\delta^o_{nl} - \delta^d_{nl}) x_l$.

On link l, the excess between (1) inflow price and transportation costs and (2) outflow prices is: $f_l = p_{o(l)} + c_l - p_{d(l)}$.

Network equilibrium. A network equilibrium of the supply chain is a state made of prices and trade flows such that:

- for a transformer, either inflow is greater than outflow, and there is no price (according to the law of supply and demand, there is not enough demand, supply is too high); or inflow equals outflow, and there is a price;
- for a link, either price at origin plus transportation costs are greater than price at destination and there is no flow; or price at origin plus transportation costs equals price at destination and there is flow smaller than capacity; or price at origin plus transportation costs are greater than price at destination and flow equals capacity.

Thus we have:

Definition 8.3.1 *A network equilibrium of the supply chain is a state* $X = ((p_n), (x_l))$ *such that:*

- *for a transformer* n*, either* $p_n \perp f_n 0$
- *for a link* l*, either* $x_l = 0$ *and* $f_l \geq 0$*; or* $0 < x_l < m_l$ *and* $f_l = 0$*; or* $x_l = m_l$ *and* $f_l \geq 0$.

Variational inequality. Now we formulate the equilibrium of a supply chain in the framework of variational inequalities. A state is a combination of flows quantities for all links and prices for all transformers:

$$Z = ((x_l), (p_n))$$

Flows are limited by the capacity of links. Prices must be positive. The constrained set is: $K = \prod_l [0, m_l] \times \mathbb{R}_+^{N_G}$.

Functional relationship is defined as: $F : K \to \mathbb{R}_+^{N_L} \times \mathbb{R}_+^{N_G}$,

$$Z = ((x_l), (p_n)) \to F(Z) = ((f_l), (f_n))$$

Then, we have:

Theorem 8.3.1 $Z^* \in K$ *is an equilibrium state if and only if it is a solution of the variational inequality* $VI(K, F)$.

Proof. Let us prove first that an equilibrium state is a solution of the variational inequality. Let $Z^* = ((x_l^*), (p_n^*)) \in K$ be an equilibrium state. Then we write:

$$(Z - Z^*) \cdot F(Z) = \sum_l (x_l - x_l^*) f_l^* + \sum_n (p_n - p_n^*) f^* n$$

- If $p_n^* = 0$ and $f_n^* \geq 0$, then $(p_n - p_n^*) f_n^* \geq 0$.
- If $p_n^* > 0$ and $f_n^* = 0$, then $(p_n - p_n^*) f_n^* = 0$.
- If $x_l^* = 0$ and $f_l^* \geq 0$ then $(x_l - x_l^*) f_l^* \geq 0$.
- If $0 < x_l^* < m_l$ and $f_l^* = 0$ then $(x_l - x_l^*) f_l^* = 0$.
- If $x_l^* = m_l$ and $f_l^* \geq 0$ then $(x_l - x_l^*) f_l^* \geq 0$ because $x_l \leq x_l^* = m_l$.

Thus the sum involved in the variational inequality is positive.

Let us prove now that a solution of the variational inequality is an equilibrium state. Suppose that Z^* is not an equilibrium state. Then, it is immediate that $(Z - Z^*) \cdot F(Z^*)$ can be made positive in all possible alternatives:

- If m is such $p_m^* \geq 0$ and $o_n^* < i_n^*$, we take Z identical to Z^* except $p_m > p_m^*$.
- If m is such that $p_m^* > 0$ and $o_n^* > i_n^*$, we take Z identical to Z^* except $p_m < p_m^*$.
- If k is such that $x_k^* = 0$ and $f_k^* < 0$, we take Z identical to Z^* except $0 = x_k^* < x_k < m_k$.
- If k is such that $0 < x_k^* < m_k$ and $f_k^* > 0$, we take Z identical to Z^* except $0 < x_k^* < x_k < m_k$.
- If k is such that $0 < x_k^* < m_k$ and $f_k^* < 0$, we take Z identical to Z^* except $0 < x_k < x_k^* < m_k$.
- If k is such that $x_k^* = m_k$ and $f_k^* > 0$, we take Z identical to Z^* except $0 < x_k < x_k^* = m_k$.

Let P the set of producers; let D the set of consumers. We have:

Theorem 8.3.2 *$F(Z) = \mathbf{M} \cdot Z + N$ with vector N and matrix $\mathbf{M}$ defined as follows.*

$$\mathbf{M} = \begin{pmatrix} \mathbf{M}_{LL} & \mathbf{M}_{LN} \\ -^{\mathbf{T}}\mathbf{M}_{LN} & 0 \end{pmatrix}, N = \begin{pmatrix} N_L \\ 0 \end{pmatrix}$$

- *$N_L = (n_l)$ is such that: $n_l = -c_l - \sum_{n \in P} \delta_{nl}^o c_n$.*
- *$\mathbf{M}_{LL} = (m_{ll'})$ is such that $m_{ll'} = \sum_{n \in D} b_n (\delta_{nl}^d - \delta_{nl'}^d)$.*
- *$\mathbf{M}_{LN} = (m_{ln})$ is such that $m_{ln} = \delta^d nl - \delta_{nl}^o$.*

Theorem 8.3.3 *$\mathbf{M}$ is a skew symmetric matrix.*

Proof: it is enough to remark that $m_{ll'} = -m_{l'l}$

8.4 Supply Chains: Limitation of Production Capacity

We now consider a slightly different supply chain structure. As in the previous section, nodes $n \in N$ can be producers ($p \in P$), transformers ($g \in G$), consumers ($d \in D$). Links $l \in L$ support the flows of a commodity between nodes. For a subset Q of transformers G, there is a constraint on inflow that they can process. Nodes $q \in Q$ are subject to:

$$\sum_l \delta_{ql}^d f_{ql} x_l \leq g_q$$

To such a constraint, is associated a shadow price λ_q.

Links are characterized by the observed flows x_l, transportation costs c_l. Nodes are characterized by an observed price p_n; for producers p_n represent production costs; for consumers p_n is the market price given by an inverse demand function $p_n(q_n) = a_n - b_n q_n$; for transformers p_n results from the network complementarity equations.

8.4.1 Network Equilibrium

A network equilibrium is a combination of flows, prices, shadow prices such that:

- For all transformers, either ingoing flow is greater than outgoing flow and there is no price; or ingoing flow equals outgoing flow and there is a price;
- For all constrained transformers, constraint on inflow is satisfied; and there is a shadow price; this shadow price is non zero only if the corresponding constraint is saturated;
- For all links that do not end in a constrained transformer, either price at origin plus transportation cost is greater than price at destination and there is no flow; or price at origin plus transportation cost equals price at destination and flow is smaller than capacity;
- For links that end to a constrained transformer, either the excess cost equals the shadow price and there is a flow; or the excess cost is more than the shadow price and there is no flow.

These alternatives are shown in figure 8.7. With mathematical notations, it gives:

- To a transformer node $n \in G$, we associate the excess $e_n = i_n - o_n$; then, either $p_n^* = 0$ and $e_n^* \geq 0$; or $p_n^* > 0$ and $e_n^* = 0$: we have $p_n^* \perp e_n^*$.

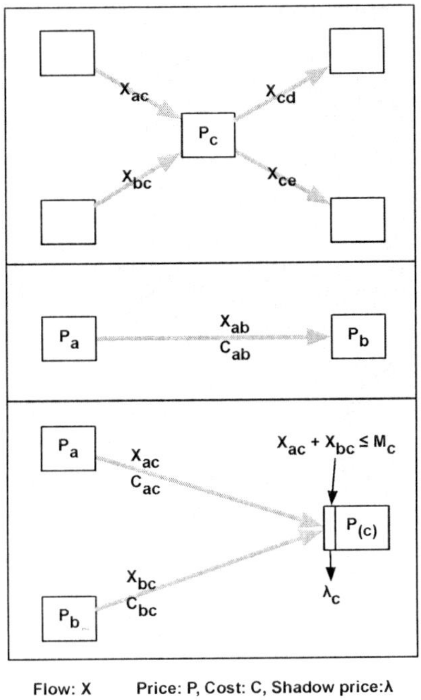

Figure 8.7: Supply chain and shadow costs.

- To a constrained node $q \in Q$, we associate the satisfaction of constraint $e_q = g_q - \sum_l \delta^d_{ql} f_{ql} x_l$; then, either $e^*_q \geq 0$ and $\lambda^*_q = 0$, or $e^*_q = 0$ and $\lambda^*_q \geq 0$: we have $\lambda^*_q \perp e^*_q$.
- To a link l that does not end in a constrained node, we associate the excess $a_l = p_{o(l)} + c_l - p_{d(l)}$; then either $x^*_l = 0$ and $a^*_l \geq 0$; or $0 < x^*_l$ and $a^*_l = 0$: we have $x^*_l \perp a^*_l$.
- To a link l that ends in a constrained node q, we associate the excess $a_l = p_{o(l)} + c_l - p_{d(l)} + \lambda_q$; then either $x^*_l = 0$ and $a^*_l \geq 0$, or $0 < x^*_l$: we have $x^*_l \perp a^*_l$.

A state of the system is a vector $Z = ((x_l), (p_n), (\lambda_q))$. Let $F : \mathbb{R}^{N_L}_+ \times \mathbb{R}^{N_G}_+ \times \mathbb{R}^{N_Q}_+ \to \mathbb{R}^{N_L} \times \mathbb{R}^{N_G} \times \mathbb{R}^{N_Q}$ be defined by:

$$Z = ((x_l), (p_n), (\lambda_q)) \to F(Z) = ((b_l), (e_n), (e_q))$$

Then, we may rephrase the above definition in mathematical terms:

Definition 8.4.1 *$Z^* \in H$ is an equilibrium state of the system if and only if it is a solution of the complementarity problem $\mathcal{CP}(F)$.*

Theorem 8.4.1 *F is an affine function: $F(Z) = \mathbf{M} \cdot Z + N$ with*

$$\mathbf{M} = \begin{pmatrix} \mathbf{M}_{LL} & \mathbf{M}_{LN} & \mathbf{M}_{LQ} \\ \mathbf{M}_{NL} & 0 & 0 \\ \mathbf{M}_{QL} & 0 & 0 \end{pmatrix}, \; N = \begin{pmatrix} N_L \\ 0 \\ N_Q \end{pmatrix}$$

Where:

- $N_l = c_l + \sum_{n \in P} \delta^o_{nl} p_n - \sum_{n \in D} \delta^d_{nl} a_n, \; N_q = g_q,$
- $M_{ll'} = \sum_{n \in D} (\delta^d_{nl'} \delta^d_{nl}) b_n;$
- $M_{ln} = \delta^o_{nl} - \delta^d_{nl}, \; M_{nl} = -M_{ln};$
- $M_{ql} = \delta^d_{ql} f_{ql}, \; M_{lq} = \delta^d_{ql}.$

Proof. We write

$$\begin{aligned} a_l &= \sum_n \delta^o_{nl} p_n + c_l - \sum_n \delta^d_{nl} p_n + \sum_q \delta^d_{nq} \lambda_q \\ &= \sum_{n \in P} \delta^o_{nl} p_n + \sum_{n \in G} \delta^o_{nl} p_n + c_l \\ &\quad - \sum_{n \in D} \delta^d_{nl} p_n - \sum_{n \in G} \delta^d_{nl} p_n + \sum_q \delta^d_{nq} \lambda_q \end{aligned}$$

Using, $p_n = a_n - b_n i_n = a_n - b_n(\sum_l x^d_{nl} x_l)$, for $n \in D$, and developing, we get:

$$\begin{aligned} a_l = &\sum_{l'} (\sum_{n \in D} \delta^d_{nl} \delta^d_{nl'} b_n) x_{l'} + \sum_{n \in G} (\delta^o_{nl} - \delta^d_{nl}) p_n + \sum_q \delta^d_{nq} \lambda_q \\ &+ \sum_{n \in P} \delta^o_{nl} p_n + c_l - \sum_{n \in D} a_n \delta^d_{nl} \end{aligned}$$

We have

$$e_n = i_n - o_n = \sum_l \delta^d_{nl} x_l - \sum_l \delta^o_{nl} x_l = \sum_l (\delta^d_{nl} - \delta^o_{nl}) x_l$$

And

$$e_q = g_q - \sum_l \delta^d_{ql} f_{ql} x_l$$

□

Theorem 8.4.2 *If constraints are on total inflow: $f_{ql} = 1$, then* **M** *is a non negative matrix.*

Proof. We have then $\mathbf{M}_{QL} = -{}^{\mathbf{T}}\mathbf{M}_{LQ}$. Then:

$$X \cdot \mathbf{M} \cdot X = \begin{pmatrix} X_L \\ X_N \\ X_Q \end{pmatrix} \cdot \begin{pmatrix} \mathbf{M}_{LL} & \mathbf{M}_{LN} & \mathbf{M}_{LQ} \\ -{}^{\mathbf{T}}\mathbf{M}_{LN} & 0 & 0 \\ -{}^{\mathbf{T}}\mathbf{M}_{LQ} & 0 & 0 \end{pmatrix} \cdot \begin{pmatrix} X_L \\ X_N \\ X_Q \end{pmatrix}$$
$$= X_L \cdot \mathbf{M}_{LL} \cdot X_L$$

and

$$\begin{aligned} X_L \cdot \mathbf{M}_{LL} \cdot X_L &= \sum_{ll'} x_l M_{ll'} x_{l'} \\ &= \sum_{ll'} x_l x_{l'} \sum_{n \in D} (\delta^d_{nl'} \delta^d_{nl}) b_n \\ &= \sum_{n \in D} b_n \sum_{ll'} x_l x_{l'} (\delta^d_{nl'} \delta^d_{nl}) \\ &= \sum_{n \in D} b_n (\sum_{l} x_l \delta^d_{nl})^2 \geq 0 \end{aligned}$$

□

8.5 Digression: Utility, Prices and Equilibrium

The idea of utility plays a major role in economic theory. On one hand, utility represents a simple, general but hypothetical idea. On the other hand, it can be represented mathematically with specific functions. These functions are very important as they make it possible to reason in terms of an economic system consisting of rational agents ('rational' here meaning optimizing some utility). Utility is a difficult idea to model, due to the gap between the basic mathematical definitions and the many different behaviors it is supposed to represent. Here, we simply present the basics of the theory of marginal utility[3]. We then give a short introduction to the theory of economic equilibrium and formulated in the context of network economics.

8.5.1 Utility Functions

In theories of marginal utility, an agent is associated with a utility function u. Let $i = 1, \ldots, n$ denote the goods that are produced and consumed in an economics system. The utility function of an agent associates to its own consumption vector: $Q = (q_1, q_2, \ldots, q_n) \in \mathbb{R}^n_+$, a measure of its satisfaction $u(Q) \in \mathbb{R}$. In most of cases, u is assumed to be a concave function (i.e., $-f$ is a convex function), which may be justified and provides better results from a mathematical point of view.

[3]The points developed here can be found in any introduction to micro economics, e.g., Mankiw (2011).

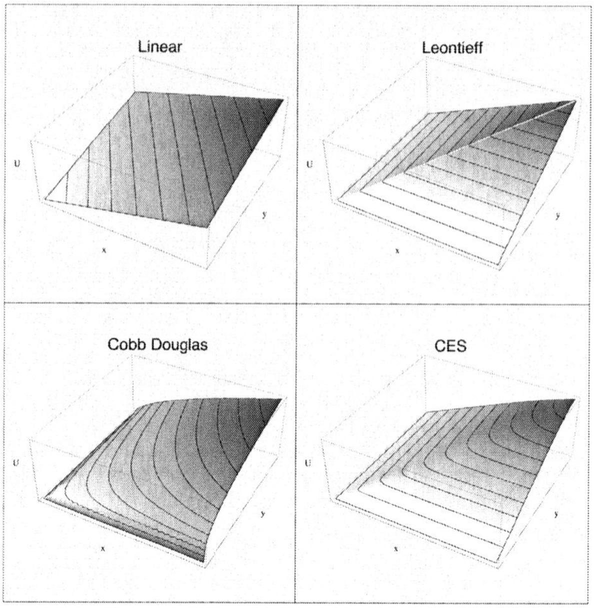

Figure 8.8: Utility Functions. Linear, Cobb Douglas, Leontief and Constant elasticity substitution functions. Utility U is expressed in terms of the consumption of two goods X and Y.

Most used utility functions are depicted in the case of two goods in figure 8.8. For all of them, we will derive the mathematical expressions that are of interest for their characterization (for example their concavity).

(1) Linear utility functions are defined as $u_L(Q) = \sum_j \beta_j q_j$.

(2) Cobb Douglas utility functions are defined as: $u_{CB}(Q) = \prod_j (q_j)^{\beta_j}$. Coefficients β are elasticity coefficients. Normally, we suppose that $\beta_j < 1$ (in this case, u_{CB} is a concave function).

(3) Leontief utility functions are defined as $u_{LF}(Q) = \min_j q_j$. Of course they are not linear, not differentiable. They represent a very interesting case of utility functions, even if they are more related to constraints on the set of possible consumptions.

(4) Constant elasticity substitution utility functions are defined as: $u_{CES}(Q) = \left[\sum_i \alpha_i q_i^{\theta}\right]^{\frac{1}{\theta}}$.

Theorem 8.5.1 *Constant elasticity substitution utility functions are generalizations of the other common utility functions.*

We give a sketch of the proof. At this point, we recall consequence of the L'Hopital's rule about the ratio of two functions: let f and g two functions such that $\lim_{\theta \to \theta_0} f(\theta) = \lim_{\theta \to \theta_0} g(\theta) = 0$. Then

$$\lim_{\theta \to \theta_0} \frac{f(\theta)}{g(\theta)} = \frac{f'(\theta_0)}{g'(\theta_0)}$$

Let $u_{CES}(Q) = \left[\sum_i \alpha_i q_i^\theta\right]^{\frac{1}{\theta}}$ be constant elasticity substitution utility functions. We assume that $\sum_i \alpha_i = 1$.

Proof for a linear utility function. When $\theta = 1$ a constant elasticity substitution function becomes a linear utility function: $\lim_{\theta \to 1} u_{CES}(Q) = \sum_i \alpha_i q_i$.

Proof for a Cobb Douglas utility function. We apply the L'Hopital's rule with $g(\theta) = \theta$, $f(\theta) = \ln(\sum_i \alpha_i q_i^\theta)$. We get:

$$\lim_{\theta \to 0} \ln u_{CES}(Q) = \frac{\sum_i \alpha_i \ln q_i}{\sum_i \alpha_i} = \sum_i \alpha_i \ln q_i$$

That is:

$$\lim_{\theta \to 0} u_{CES}(Q) = \prod_i q_i^{\alpha_i}$$

Thus, when $\theta \to 0$ a constant elasticity substitution function becomes a Cobb Douglas utility function.

Proof for a Leontief utility function. Let $q_m = min(q_1, \ldots, q_n)$. We write:

$$u_{CES}(Q) = \left(q_m^\theta \sum_i \alpha_i (q_i/q_m)^\theta\right)^{\frac{1}{\theta}}$$

$$\log u_{CES}(Q) = \ln q_m + \frac{1}{\theta} \log \left(\sum_i \alpha_i (q_i/q_m)^\theta\right)$$

$$= \ln q_m + \frac{1}{\theta} \log \left(\alpha_m \sum_{i \neq m} \alpha_i (q_i/q_m)^\theta\right)$$

But $\lim_{\theta \to -\infty} (q_i/q_m)^\theta = 0$. Thus, $\lim_{\theta \to -\infty} \log u_{CES}(Q) = \ln q_m$. When $\theta \to -\infty$ a constant elasticity substitution function becomes a Leontief utility function. □

8.5.2 Prices and Utility

Utility functions and prices are closely related. Let us consider an economic system with n goods. An agent has an utility function $(q_1, \ldots, q_n) \to u(q_1, \ldots, q_n)$. He has an initial endowment d. Commodities have known prices: $(p_1, \ldots, p_n) \in \mathbb{R}_+^n$. They are given. Then the agents problem consists in choosing a consumption vector $(q_1, \ldots, q_n) \in \mathbb{R}_+^n$ that maximizes

$u(q_1, q_2, \ldots, q_n)$ given the constraints $q_i \geq 0$, $p_1 q_1 + p_2 q_2 + \cdots + p_n q_n \leq d$. From the Karush-Kuhn-Tucker theorem (5.6.3), we deduce:

Theorem 8.5.2 *Consumption vector $(q_1, q_2, \ldots, q_n)$ is a solution if there exists λ such that:*

$$\lambda \perp (d - p_1 q_1 + p_2 q_2 + \cdots + p_n q_n)$$

$$q_i \perp (-\partial u(q_1, q_2, \ldots, q_n)/\partial q_i + \lambda p_i)$$

This allows to develop relationship between prices and marginal utility. For two goods that are consumed, i.e., such that $q_i > 0, q_j > 0$, then:

$$\frac{1}{p_i}\frac{\partial u}{\partial q_i}(q_1, q_2, \ldots, q_n) = \frac{1}{p_j}\frac{\partial u}{\partial q_j}(q_1, q_2, \ldots, q_n) = -\lambda$$

Suppose that $\lambda = 0$; some part of endowment may be not used.

- $\partial u(q_1, q_2, \ldots, q_n)/\partial q_i = 0$ for consumed goods ($q_i > 0$),
- $\partial u(q_1, q_2, \ldots, q_n)/\partial q_i \leq 0$ for unconsumed goods ($q_i = 0$).

In the case $\lambda > 0$, if we apply theorem 8.5.1:

- to a Cobb Douglas utility function, we get the equality: $\frac{\beta_i}{q_i p_i} = \frac{\beta_j}{q_j p_j}$. It suggests inverse demand functions of kind

$$p_j \simeq \frac{1}{q_j}$$

- to a linear utility function, we get: $\frac{\beta_i}{p_i} = \frac{\beta_j}{p_j}$. It suggests that there are no inverse demand functions.
- to a constant elasticity substitution utility function, we get: $\frac{\alpha_i q_i^{\theta-1}}{p_i} = \frac{\alpha_j q_j^{\theta-1}}{p_j}$. It suggests an inverse demand function of kind:

$$p_i \simeq q_i^{\frac{1}{\theta-1}}$$

8.5.3 Consumers and Markets: Hedonist Consumers

Now let us use the network economics framework to express the relationship between utilities, prices and equilibrium for a simple economic system with consumers and markets operating under the assumptions: (a) there are limits on traded goods; and (b) given these constraints, a pricing mechanism ensures the utility maximization of consumers. This is called a Fisher's market as it was defined by Irving Fisher in 1891[4]. We represent the network structure of this system (see figure 8.9) as follows:

[4]It is interesting here that in Fisher's thesis, the equilibrium on a market was defined and computed by analogies with a mechanistic system

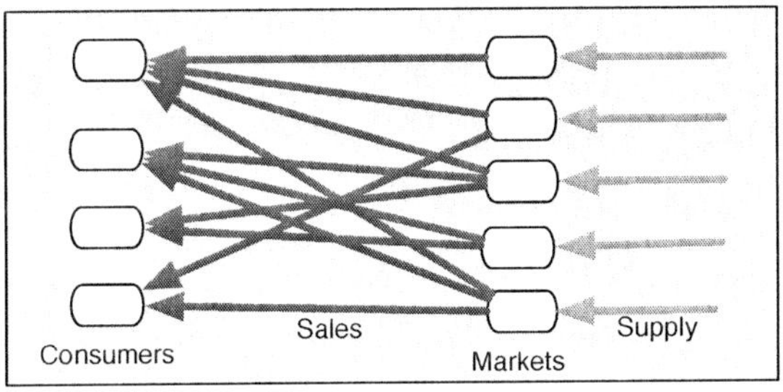

Figure 8.9: A network representation of an economic system with consumers and commodity markets. Black arrows indicate limitations on the quantities delivered by markets.

Consumers are denoted a. Goods are denoted i. Prices are denoted p_i. Maximum quantity put on market i is m_i. Consumption vector of consumer a is $X^a = (x_i^a)$. Consumer a has an utility function $X^a = (x_i^a) \to u^a(X^a)$. Initial endowment of a is denoted d^a.

An equilibrium is a state of prices (p_i) and traded quantities (x_i^a). It is characterized by the optimization of utilities of all consumers and the clearing of markets. With multipliers $\lambda^a \geq 0$ associated to the constraint $\sum_i p_i x_i^a \leq d^a$, we get using the Karush-Kuhn-Tucker theorem:

$$x_i^a \perp \left(-\frac{\partial}{\partial x_i^a} u^a(X^a) + \lambda^a p_i\right) \text{ for all } a, i$$

$$\lambda^a \perp \left(d^a - \sum_i p_i x_i^a\right) \text{ for all } a$$

$$p_i \perp \left(m_i - \sum_a x_i^a\right) \text{ for all } i$$

We define the following function. To $Z = (x_i^a, \lambda^a, p_i)$ we associate $F(Z) = (f_i^a, f^a, f_i)$

$$f_i^a = -\frac{\partial}{\partial x_i^a} u^a(X^a) + \lambda^a p_i$$

$$f_a = d^a - \sum_i p_i x_i^a$$

$$f_i = m_i - \sum_a x_i^a$$

Let $K = \{Z = (x_i^a, \lambda^a, p_i) \mid x_i^a \geq 0, \lambda^a \geq 0, p_i \geq 0\}$. It is clear that Z is an equilibrium state if an only if it is a solution of the complementarity problem $\mathcal{CP}(F)$.

Theorem 8.5.3 *The derivative of F is given by:*

$$JF(Z) = \begin{pmatrix} -\delta_{aa'}\frac{\partial^2 u^a}{\partial x_i^a \partial x_{i'}^{a'}} & \delta_{aa'}p_{i'} & \delta_{ii'}\lambda_a \\ -\delta_{aa'}p_{i'} & 0 & -\delta_{ii'}x_{i'}^a \\ -\delta_{ii'} & 0 & 0 \end{pmatrix}$$

Proof. Elementary computation. □

Even if, for all a, u^a is a concave function, i.e., if

$$-Ju^a = -\frac{\partial^2 u^a}{\partial x_i^a \partial x_{i'}^a}$$

is a non negative matrix, there is no reason to infer that F is a monotone function. Let $V = (v_i^a, v^a, v_i)$, then:

$$\begin{aligned} V \cdot JF(Z) \cdot V = &- \sum_{ai,a'i'} \delta_{aa'}\frac{\partial^2 u^a}{\partial x_i^a \partial x_{i'}^{a'}} v_i^a v_{i'}^{a'} + \sum_{ai,a'} \delta_{aa'}p_{i'}v_i^a v^{a'} \\ &+ \sum_{ai,i'} \delta_{ii'}\lambda_a v_i^a v_{i'} - \sum_{a,a'i'} \delta_{aa'}p_{i'}v^a v_{i'}^{a'} \\ &- \sum_{a,i'} \delta_{ii'}x_{i'}^a v^a v_{i'} - \sum_{i,a'i'} \delta_{ii'}v_i v_{i'}^{a'} \end{aligned}$$

And there are vectors V that are such that $V \cdot JF(Z) \cdot V < 0$. For example, with $v_i^a = 0$ for all a and all i, we get:

$$V \cdot JF(Z) \cdot V = -\sum_{ai} x_i^a v^a v_i$$

This obviously can be easily made negative as soon as there are non null values of x_i^a. This demonstrates the difficulty of the economic equilibrium problem, and a similar issue arises in the general equilibrium problem that we will examine later.

8.5.4 Consumers and Markets: Avaricious Constrained Consumers

Still using the network economics framework let us now formulate a slightly different economic system made up of producers and markets. In this case we assume: (a) there are limitations on traded goods, (b) there are consumption constraints on consumers and (c) a pricing mechanism ensures the minimization of consumers expenses, given these constraints.

Consumers are denoted a. Goods are denoted i. Prices are denoted p_i. Maximum quantity put on market i is m_i. Consumption vector of a is $X^a = (x_i^a)$. It must satisfy constraints $\sum_i f_{ci}^a x_i^a \geq g_c^a$. Consumer a minimizes its expenses $e^a = \sum_i p_i x_i^a$.

Clearly, there are situations where supply and consumption constraints cannot both be satisfied. We are analyzing an economic system that may have no solutions.

An equilibrium is a state of prices (p_i) and traded quantities (x_i^a). It is characterized by the minimization of the expenses of all producers and the clearing of all markets. When multipliers $\lambda_c^a \geq 0$ are associated with consumers constraints, we obtain:

$$x_i^a \perp (p_i - \sum_c \lambda_c^a f_{ci}^a) \text{ for all } a, i$$

$$\lambda_c^a \perp (\sum_i f_{ci}^a - g_c^a) \text{ for all } a, c$$

$$p_i \perp (m_i - \sum_a x_i^a) \text{ for all } i$$

A state vector of the economic system is: $Z = (x_i^a, \lambda_c^a, p_i)$.

Theorem 8.5.4 *Z is an equilibrium state if an only if it is a solution of the complementarity problem $\mathcal{LCP}(M, N)$ with*

$$\mathbf{M} = \begin{pmatrix} 0 & -\delta_{aa'} f_{ci'}^a & \delta_{ii'} \\ \delta_{a'a} f_{c'i}^a & 0 & 0 \\ -\delta_{i'i} & 0 & 0 \end{pmatrix}, \qquad N = \begin{pmatrix} 0 \\ -g_c^a \\ m_i \end{pmatrix}.$$

Proof. Elementary computation. □

If $Z = (x_i^a, \lambda_c^a, p_i)$ is a solution, then $Z_t = (x_i^a, t\lambda_c^a, tp_i)$ is also a solution for all $t > 0$. We may impose the usual constraint on prices: $\sum_i p_i = 1$

It is obvious that $\mathbf{M}$ is a skew symmetric matrix : $\mathbf{M} + {}^{\mathbf{T}}\mathbf{M} = 0$. This has implications for the solutions of $\mathcal{LCP}(M, N)$. For all Z, we have $Z \cdot \mathbf{M} \cdot Z = 0$, thus $Z \cdot (\mathbf{M} \cdot Z + N) = Z \cdot N$. That means that all vectors Z such that $Z \cdot N = 0$, $\mathbf{M} \cdot Z + N \geq 0$ and $Z \geq 0$ are solutions of $\mathcal{LCP}(M, N)$. As for all Z, $Z \cdot \mathbf{M} \cdot Z = 0$, $\mathbf{M}$ is a positive semi-definite matrix. We have seen (theorem 6.2.7) that a linear complementarity problem with a positive semi-definite matrix has a solution as soon as it is feasible, i.e., as soon as there exists $Z \geq 0$ such that $\mathbf{M} \cdot Z + N \geq 0$. Suppose that there exist a consumptions vector$X^a = (x_i^a)$ such as supply and consumption constraints are satisfied. Then it is obvious that, for all $p_i \geq 0$, $Z = (x_i^a, 0, p_i)$ is such that $Z \geq 0$ and $\mathbf{M} \cdot Z + N \geq 0$. The linear complementarity problem is feasible. The solution set is non empty, and it is obvious that it is convex.

Suppose now that i denote a producer who sells y_i under constraint $y_i \leq m_i$. It maximizes $p_i y_i$. This constraint adds a multiplier μ_i. Complementarity equations are now:

$$
\begin{aligned}
x_i^a &\perp (p_i - \sum_c \lambda_c^a f_{ci}^a) \\
\lambda_c^a &\perp (\sum_i f_{ci}^a - g_c^a) \\
p_i &\perp (y_i - \sum_a x_i^a) \\
y_i &\perp -p_i + \mu_i \\
\mu_i &\perp (m_i - y_i)
\end{aligned}
$$

We still get a linear complementarity problem with as skew symmetric matrix. Sate vector is $Z = (x_i^a, \lambda_c^a, p_i, y_i, \mu_i)$.

$$
\mathbf{M} = \begin{pmatrix} 0 & -\delta_{aa'} f_{ci'}^a & \delta_{ii'} & 0 & 0 \\ \delta_{a'a} f_{c'i}^a & 0 & 0 & 0 & 0 \\ -\delta_{i'i} & 0 & 0 & 1 & 0 \\ 0 & 0 & -1 & & 1 \\ 0 & 0 & 0 & -1 & 0 \end{pmatrix}, \qquad N = \begin{pmatrix} 0 \\ -g_c^a \\ 0 \\ 0 \\ m_i \end{pmatrix}.
$$

The linear complementarity problem has a non empty convex solution set as soon as it is feasible. And just as before, it is feasible on an economic point of view as soon as it is feasible on a production point of view, i.e., if there exists y_i and x_i^a such that $\sum_i f_{ci}^a x_i^a \geq g_c^a$ and $\sum_a x_i^a \leq y_i \leq m_i$. It is elementary to show that, at equilibrium, when a good produced by i is exchanged at a positive price, there is equality of maximum production m_i and consumption $\sum_a x_i^a$.

We mention this result to highlight the possibility that complementarity problems can have many or no solutions. This kind of result is not usually welcomed in economic theory. Nevertheless, the assumptions are realistic, and the mathematical formulation and developments are very simple. This raises the question of why it is considered preferable for economic systems to have a unique equilibrium.

8.5.5 General Equilibrium

General equilibrium theory aims at explaining prices formation[5]. The framework is as follows. We consider a system in which producers and consumers exchange commodities on markets: producers optimize their

[5] The problem has been formulated in Walras (1874); it has been solved in Arrow and Debreu (1954); see Ginsburgh and Keyzer (2002) for a general presentation.

income while consumers optimize the utility of the goods they consume. Consumers share the benefits of producers income and cannot spend more than they earn. Market prices depend on demand and supply, which themselves depend on anticipated income, which depend on prices.

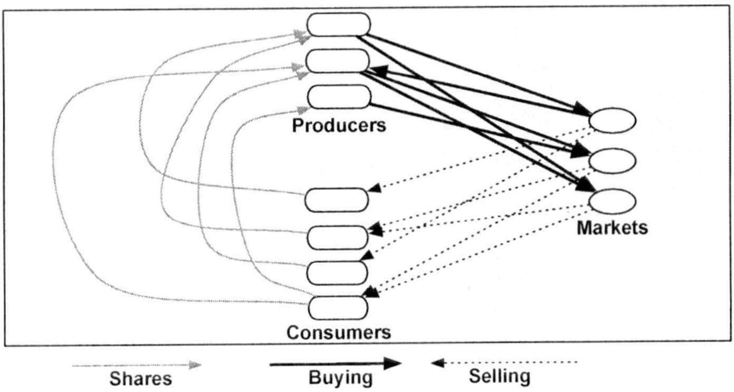

Figure 8.10: A network representation of a general economic system. Top: producers; bottom: consumers; right: markets.

Mathematical definition. Commodities are denoted k. A commodity k is exchanged at a price p_k between consumers denoted i and producers denoted j. Let $P = (p_k)$ be the price vector.

Let $Y^j = (y^j_k)$ the production vector of producer j; constraints of production imply that this vector belongs to a subset K_j of $\mathbb{R}^{nk}_+$. Profit of j is $r^j = \sum_k p_k y^j_k$.

Let θ_{ji} be the share of producer j owned by consumer i; we have $\sum_i \theta_{ji} = 1$. Income of i is $r^i = \sum_j \theta_{ji} r^j$.

Consumption vector of i is a vector $X^i = (x^i_k)$; consumption are limited by income; we have: $\sum_k x^i_k p_k \leq r^i$. Consumer i chooses the consumption vector X^i that maximizes an utility function: $X^i = (x^i_k) \to u^i(X^i) \in \mathbb{R}$.

Definition 8.5.1 *A general equilibrium is defined as a vector of prices P, productions vectors $X^i = (x^i_k)$, consumption vectors $Y^j = (y^j_k)$ such that*

- *For all producers j, production vector $Y^j \in K_j$ maximizes r^j.*
- *For all consumer i, consumption vector $X^i = (x^i_k)$ maximizes it utility function $u^i(X^i)$ under constraint $\sum_k x^i_k p_k \leq r^i$,*
- *For all markets k, there is clearing, i.e., with complementarity notations: $p_k \perp (\sum_i x^i_k - \sum_j y^j_k)$..*

The Arrow Debreu theorem. Arrow and Debreu (1954) gave after fifty years of research an elegant solution and proved the existence of an equilibrium. Let us recall two basic definitions.

Definition 8.5.2 *An utility function u is said strictly concave if, for all X and X' such that $X \neq X'$,then*

$$u(\frac{X+X'}{2}) > \frac{u(X)+u(X')}{2}$$

Definition 8.5.3 *An utility function u is said strictly decreasing if, for all $X = (x_k)$ and $X = (x_k)'$ such that $x_k < x'_k$ for all k, then*

$$u(X) < u(X')$$

Theorem 8.5.5 *If production sets K_j are compact and convex, if $0 \in K_j$, if utility functions $X^i \to u^i(X^i)$ are continuous, strictly concave, increasing, then there exists an equilibrium price P^*.*

Almost all proofs are based on Brouwer's or Kakutani's theorems (cf. section 5.8). Although they conclude that a solution exists they are non-constructive: they don't give the solution. A simple proof is given in Ginsburgh and Keyzer (2002).

8.5.6 General Equilibrium and Network Economics

The general equilibrium theory can be expressed in the network economics framework as follows.

For producer j, we suppose that the production set K^j is defined be equations $\sum_c a^j_{ck} y^j_k \leq b^j_c$. According to Karush-Kuhn-Tucker theorem (5.6.3), there exists λ^j_c such that

$$y^j_k \perp (-p_k + \sum_c \lambda^j_c a^j_{ck})$$

$$\lambda^j_c \perp (b^c_j - \sum_k a^j_{ck} y^j_k)$$

Consumer i has consumption vector $X^i = (x^i_k)$ that maximizes $u^i(X^i)$ with budget constraint $\sum_k x^i_k p_k \leq r^i$. There exists μ^i such that

$$x^i_k \perp (-\frac{\partial u^i}{\partial x^i_k} + \mu_i p_k)$$

$$\mu^i \perp (\sum_{jk} \theta_{ij} p_k y^j_k - \sum_k x^i_k p_k)$$

A state of the system is an element $Z = (x^i_k, y^j_k, p_k, \lambda^j_c, \mu_i)$. Components of Z are supposed non-negative:

$$Z \in K = \mathbb{R}^{ni \times nk}_+ \times \mathbb{R}^{nj \times nk}_+ \times \mathbb{R}^{nk}_+ \times \mathbb{R}^{nc}_+ \times \mathbb{R}^{ni}_+$$

For $Z \in K$, we define the function:

$$F(Z) = (f_k^i, f_k^j, f_k, f_c^j, f^i)$$

Where:

$$f_k^i = -\frac{\partial u^i}{\partial x_k^i} + \mu_i p_k$$

$$f_k^j = -p_k + \sum_c \lambda_c^j a_{ck}^j$$

$$f_k = \sum_j y_k^j - \sum_i x_k^i$$

$$f_c^j = b_j^c - \sum_k a_{ck}^j y_k^j$$

$$f^i = \sum_{jk} \theta_{ij} p_k y_k^j - \sum_k x_k^i p_k$$

We have $F(Z) \in R^{ni \times nk} \times R^{nj \times nk} \times R^{nk} \times R^{nc} \times R^{ni}$.

Theorem 8.5.6 (X^*, Y^*, P^*) *is an equilibrium if and only if there exists* λ^*, μ^* *such that* $Z^* = (X^*, Y^*, P^*, \lambda^*, \mu^*) \in K$ *is a solution of the complementarity problem* $\mathcal{CP}(F)$, *that is of the variational inequality* $\mathcal{VI}(F, K)$.

Proof. Obvious. □

Due to quadratic terms $p_k y_k^j$, $X \to F(X)$ is not linear. More over, it is not a monotone function. It is fastidious but elementary to show that, if $S = (Z - Z') \cdot (F - F')$, then

$$\begin{aligned} S = &(x_k^i \mu_i p_k - x_k^i \mu_i' p_k' - x_k'^i \mu_i p_k + x_k'^i \mu_i' p_k') \\ &- (x_k^i p_k - x_k^i p_k' - x_k'^i p_k + x_k^i p_k') \\ &+ \sum_j \theta_{ij} (\mu_i p_k y_k^j - \mu_i p_k' y_k'^j - \mu_i' p_k y_k^j + \mu_i' p_k' y_k'^j) \end{aligned}$$

It has no reason to be positive. This the reason why the network economics approach to general equilibrium is not straightforward; it has a positive solution (Jofré et al., 2007), but it involves tools that are beyond the ones presented in this book.

8.5.7 General Equilibrium: The Network Perspective

We have discussed the application of network economics to general equilibrium theory. This makes it possible to address an important question related to the analysis of an economic system: What are the relationships between the possible equilibriums and structural characteristics of the network shown in figure 8.10.

Chapter 9

Migrations

In this chapter, we represent a migration system in which individuals make decisions according to what they know about the place they are living in, and what they know about the places they can go, following network links. In chapters 17 and 18 we illustrate how repeating this process creates a dynamical system.

9.1 Migration of Similar Individuals

Let us consider the system with several locations i that is shown in figure 9.1. In each location, is living a population with p_i individuals. People in a location i evaluate cost of living there according to density population: there is a cost of life function $p_i \to u_i(p_i)$. A given number of individual x_{ij} migrate from i to j. Migration costs from location i to location j are denoted c_{ij}. They may be constant or depend upon the number of individuals migrating $x_{ij} \to c_{ij}(x_{ij})$. If before migration, population in location i is $\overline{p_i}$, then after migration, it is $p_i = \overline{p_i} - \sum_i x_{ij} + \sum_i x_{ji}$. There are constraints on migration flows: $x_{ij} \geq 0$ and $\sum_j x_{ij} \leq \overline{p_i}$ must always be satisfied. Thus the constrained set is defined as the set K of migration vectors $\{(x_{ij})\}$ such that $x_{ij} \geq 0$ and $\sum_j x_{ij} \leq \overline{p_i}$.

To a migration vector $X = \{(x_{ij})\}$, we associate $F(X) = \{(f_{ij})\}$ where the balance between costs and advantages of migrations $f_{ij} = u_i + c_{ij} - u_j$.

Definition 9.1.1 *An equilibrium of the migration system is a migration vector* $\{(x^*_{ij})\} \in K$ *such that there exists* $\lambda^*_i \geq 0$*:*

$$x^*_{ij} \perp (f^*_{ij} + \lambda^*_i)$$

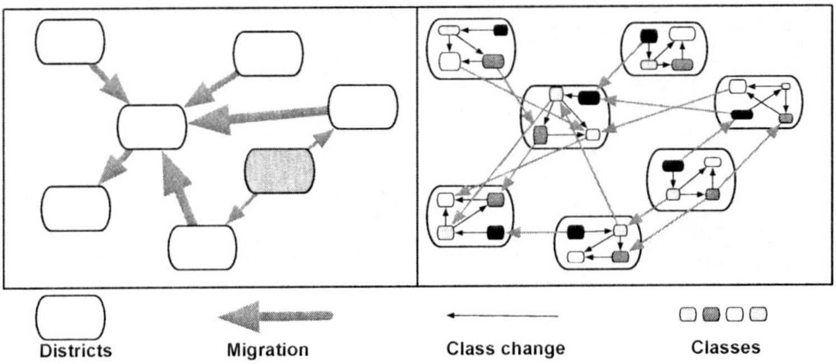

Figure 9.1: A migration network. Left: migration without classes. At each time step, a part of the population of a given area moves to another area depending on migration costs and standard of living. Right: migration with classes. In a given area the population is split into several classes; improved living conditions depend on class and population distribution.

and

$$\lambda_i^* \perp (\overline{p_i} - \sum_j x_{ij}^*)$$

This may be interpreted as follows. If there are individuals starting from i and going to j and j', the migration balance is the same for both destinations. Otherwise there would be more individuals going to the most advantageous place. There is a equality of the advantages and the disadvantages of migration, unless all individuals from a given location have migrated because they find better elsewhere.

Theorem 9.1.1 *A migration vector $X^* \in K$) is a migration equilibrium if and only if it satisfies the variational inequality $\mathcal{VI}(F, K)$: $(X - X^*) \cdot F(X^*) \geq 0$ for all $X \in K$.*

Proof. Let $(x_{ij}^*) \in K$ be a migration equilibrium. Let H the set of pairs of indices (ij) such that $x_{ij}^* > 0$, i.e., such that $f_{ij}^* = -\lambda_i^*$. Let L the set of indices (i) such that $\sum_j x_{ij}^* < \overline{p_i}$, i.e., such that $\lambda^* i = 0$. Then we have, by definition of H, we successively may write:

$$\begin{aligned}
S &= \sum_{ij} f_{ij}^*(x_{ij} - x_{ij}^*) \\
&= \sum_{ij \in H} f_{ij}^*(x_{ij} - x_{ij}^*) + \sum_{ij \notin H} f_{ij}^*(x_{ij} - x_{ij}^*) \\
&= \sum_{ij \in H} (-\lambda_i^*)(x_{ij} - x_{ij}^*) + \sum_{ij \notin H} f_{ij}^* x_{ij} \\
&\geq \sum_{ij \in H} (-\lambda_i^*)(x_{ij} - x_{ij}^*) + \sum_{ij \notin H} (-\lambda_i^*) x_{ij}
\end{aligned}$$

$$\geq \sum_{ij\in H}(-\lambda_i^*)(x_{ij}-x_{ij}^*) + \sum_{ij\notin H}(-\lambda_i^*)(x_{ij}-x_{ij}^*)$$
$$= \sum_i(-\lambda_i^*)\sum_j(x_{ij}-x_{ij}^*)$$

By definition of L, using $\sum_j x_{ij} \leq \overline{p_i}$, we obtain:

$$\begin{aligned}
S &= \sum_{ij} f_{ij}^*(x_{ij}-x_{ij}^*)\\
&\geq \sum_{i\in L}(-\lambda_i^*)\sum_j(x_{ij}-x_{ij}^*) + \sum_{i\notin L}(-\lambda_i^*)\sum_j(x_{ij}-x_{ij}^*)\\
&\geq \sum_{i\in L} 0 \times \sum_j(x_{ij}-x_{ij}^*) + \sum_{i\notin L}(-\lambda_i^*)(\sum_j x_{ij} - \overline{p_i})\\
&\geq 0
\end{aligned}$$

Let $(x_{ij}^*) \in K$ a solution of the variational inequality. We define λ_i^* as follows. If $x_{ij}^* = 0$ for all j then $\lambda_i^* = 0$, else $-\lambda_i^*$ is the maximum value of (f_{ij}^*) for all j such that $x_{ij}^* > 0$.

Let us prove that $\lambda_i \geq 0$. Suppose that it is not true. Let i_0 such that $\lambda_{i_0} < 0$. There exists j_0 such that $f_{ij}^* > 0$. With X such that $x_{ij} = x_{ij}^*$, except $x_{i_0 j_0} = 0$, we get a contradiction:

$$\sum_{ij} f_{ij}^*(x_{ij}-x_{ij}^*) = f_{i_0j_0}^*(-x_{i_0j_0}^*) < 0$$

Let us prove that $\lambda_i \perp (\overline{p_i} - \sum_j x_j)$. Suppose that it is not true. For some i_0, we have $\lambda_{i_0} > 0$ and $\sum_j x_{i_0 j} < \overline{p_{i_0}}$. Let j_0 such that $f_{i_0j_0}^* < 0$. With X such that $x_{ij} = x_{ij}^*$, except $x_{i_0j_0} = x_{ij}^* + \overline{p_{i_0}} - \sum_j x_{i_0 j}^*$, we get a contradiction:

$$\sum_{ij} f_{ij}^*(x_{ij}-x_{ij}^*) = f_{i_0j_0}^*(\overline{p_{i_0}} - \sum_j x_{i_0j}^*) < 0$$

Let us prove that $x_{ij}^* \perp (f_{ij}^* + \lambda_i^*)$. Suppose that it is not true. There exists a pair (i_0, j_0) such that $x_{i_0j_0}^* > 0$ and $f_{i_0j_0}^* + \lambda_{i_0}^* > 0$. By definition of $\lambda_{i_0}^*$, there exists j_1 such that $x_{i_0j_1}^* > 0$ and $f_{i_0j_1}^* + \lambda_{i_0}^* = 0$. We obviously have $f_{i_0j_0}^* < f_{i_1j_1}^*$. With X such that $x_{ij} = x_{ij}^*$, except $x_{i_0j_0} = x_{i_0j_0}^* - \epsilon$, $x_{i_0j_1} = x_{i_0j_1}^* + \epsilon$, we get a contradiction:

$$\begin{aligned}
\sum_{ij} f_{ij}^*(x_{ij}-x_{ij}^*) &= f_{i_0j_0}^*(x_{i_0j_0}-x_{i_0j_0}^*) + f_{i_0j_1}^*(x_{i_0j_1}-x_{i_0j_1}^*)\\
&= f_{i_0j_0}^*(-\epsilon) + f_{i_0j_1}^*(\epsilon)\\
&= \epsilon(f_{i_0j_1}^* - f_{i_0j_0}^*)\\
&< 0
\end{aligned}$$

Thus X^* is an equilibrium state. □

We also have a formulation of the migration equilibrium using complementarity problem. We define the function $Z = (x_{ij}, \lambda_i) \rightarrow G(Z) = (f_{ij}, \mu)$ where:

$$f_{ij} = u_i + c_{ij} - u_j + \lambda_i$$

$$\mu_i = \overline{p_i} - \sum_j x_{ij}$$

Theorem 9.1.2 *A migration vector $X^* = (x^*_{ij})$ is a migration equilibrium if and only there exists $\lambda^* i$ such that $Z^* = (x^*_{ij}, \lambda^*_i)$ is a solution of $\mathcal{CP}(G)$.*

Proof is immediate according to previous theorem. □

9.2 Migration of Individuals Belonging to Several Classes

Now, we consider that there are several classes, for example age, size, or social level; they are denoted k. We denote p^k_i the population of class k in location i: u^k_i the cost of living in location i for individuals of class k: x^k_{ij} the flow of population of class k from i to j, c^k_{ij} the cost of migrating from i to j for class k.

Just as before, the constrained set is: $K = \{x^k_{ij}\}$ such that $x^k_{ij} \geq 0$ and $\sum_j x^k_{ij} \leq \overline{p^k_i}$. To a migration vector $X = \{(x^k_{ij})\}$, we associate $F(X) = \{(f^k_{ij})\}$ where $f^k_{ij} = u^k_i + c^k_{ij} - u^k_j$ with above formulations for c^k_{ij} u^k_i, u^k_j. And we have:

Definition 9.2.1 *An equilibrium of the migration system is a migration vector $X^* = \{x^{k*}_{ij}\} \in K$ such that there exists $\lambda^{k*}_i \geq 0$ verifying:*

$$x^{k*}_{ij} \perp (f^{k*}_{ij} + \lambda^{k*}_i)$$

and

$$\lambda^{k*}_i \perp (\overline{p^k_i} - \sum_j x^{k*}_{ij})$$

Theorem 9.2.1 *A migration vector $X^* \in K$ is a migration equilibrium if and only if it satisfies the variational inequality $\mathcal{VI}(F, K)$: $(X - X^*) \cdot F(X^*) \geq 0$ for all $X \in K$.*

Proof is similar to previous one. □

Part IV

Illustrations

Chapter 10

Approach and Method

The following chapters illustrate the method. Almost all of them relate to published research where the emphasis was on a scientific problem, and the network economics approach was used to tackle it. Here, we take the opposite approach: we anchor the methods of network economics in actual ecological or economic problems. Our conclusions relate to the appropriateness of networks economics in solving problems related to double exposure, and the different perspective it provides.

Several methodological issues arise in the modeling and simulation of complex systems. They appear as a result of differences between initial expectations and actual results. Excess of confidence in modeling can be hazardous. A now famous diagram of Koomey demonstrates the issue (Koomey and Holdren, 2008). It articulates the steps in the publication of a scientific paper that involves modeling, namely: (1) sparse and infrequent observations, observational errors, and incorrect interpretation of observations; (2) theoretical misunderstanding, and oversimplified models; (3) computer models and coding errors; (4) unrealistic assumptions, refinement of unimportant details and coincidental agreement between theory and observation, and (5) publication.

We suggest a more modest approach to economic or ecological modeling (see figure 10.1). Building a model, especially a simulation, is like building a house of cards: it consists of exploring the possibility of a new hypothesis, choosing between alternative hypotheses, testing their feasibility, and demonstrating their effectiveness. This is exactly the nature of the models presented in the following chapters. In the same way that falsifiability is a requirement to ensure the scientific quality of a theory, we suggest that collapsibility ensures the quality of a model in economics or ecology.

Figure 10.1: A modeler. According to the French painter Chardin: The Son of M. Le Noir amusing himself in making a house of cards. National Gallery. London.

10.1 Pattern-Oriented Approach

In the following chapters we are careful to use methodologies that have been developed in order to obtain, with simulation experiments, results that are applicable to ecological theory. The illustrations given here are based on the Pattern Oriented approach to modeling (Railsback, 2001; Grimm et al., 2005a; Amberg, 2007). This makes it possible to combine specific procedures (based on the idea of the pattern) to design the model, compare and contrast alternative theories, and handle uncertainty parameters: *In Pattern Oriented Modeling, we explicitly follow the basic research program of science: the explanation of observed patterns. Patterns are defining characteristics of a system and often, therefore, indicators of essential underlying processes and structures. Patterns contain information on the internal organization of a system, but in a coded form. The purpose of Pattern Oriented Modeling is to decode this information* (Railsback, 2001). Its principles are as follows.

(1) The goal of the model is not to produce a detailed and exact representation of the system. Economic and ecological systems

incorporate many processes and are described by many datasets, from very diverse points of view. Attempts to achieve exact representations are irrelevant.

(2) A first step consists in extracting patterns from knowledge and data about a given system (stylized facts).
(3) Hypotheses are expressed in terms of processes and patterns. A necessary balance between the details of processes and patterns is respected.
(4) A model is created and simulations are carried out in order to represent and test the hypothesis.
(5) Results are interpreted in terms of patterns.
(6) Only then can scientific conclusions be drawn.

10.2 Overview-Design-Details Protocol

This protocol was developed to model complex systems, individual-based models, and multi-agent models (Grimm et al., 2005b; Kramer-Schadt et al., 2007; Grimm et al., 2010). We have adapted it for the presentation of a network model: the common outline of the forthcoming cases studies will be according to the sequence:

(1) The Overview phase consists in making explicit, without any mathematical symbols, the initial question and the purpose of modeling.
(2) The Design phase consists in making explicit the strategy used to get results. How will the analysis of network processes and structure be used to represent the patterns identified from observations. What are the initial assumptions? What are the hypotheses to test? How is the model able to analyze relationships between phenomena both at the level of the network and at the level of nodes and links? How will network simulations be analyzed?
(3) The Details phase consists in presenting the model in such a way that a reader can implement the model and reproduce the simulations. How is the initial state of the system defined at the beginning of a simulation? What parameters will be used for sensitivity analysis? Present the mathematical skeleton of the model. Give a full model description.

These recommendations can be seen as simple common sense. However, the authors own experience has shown that the recent ODD protocol has already had a positive impact on modeling practices in ecology.

Chapter 11

About the Gause's Exclusion Principle

11.1 Overview

11.1.1 Questioning

Gause's principle plays a major role in ecological theory (Gause, 1936; Levin, 1970). It states that when several similar predator species compete for one prey species, only one of them survives. The principle is supported by empirical evidence (initially found by Gause in experimentation on yeast species, and subsequently confirmed in many other studies) and a mathematical modeling approach. In the modeling approach (developed by Gause himself) the competition process is expressed as a Lotka-Volterra-like model with two, or a small number of species. In this model, relationship coefficients characterize inter- and intra-species competition and it follows directly that (except in exceptional cases) only one species can survive. The approach has been confirmed by similar models that include many species of prey and predators.

Gause's principle has been frequently observed in laboratory studies and occasionally in the field. However, the question of its generality has been widely debated. It has led to the plankton paradox (Hutchinson, 1961): Why are there so many plankton species feeding on the same resource pool? Or the tropical forest paradox: Why are there so many tree species in a homogenous tropical environment (Prance, 1982)? The best-known answer to these paradoxes is given by the theory of non-equilibrium disturbance (Rykiel, 2006). This theory does not contradicts Gause's competitive exclusion principle, but suggests that a mathematical taking into account of dynamics is missing. In particular, it argues that there are disturbances over a short time scale which prevent Gause's equilibrium being reached (Connell, 1978).

Other questions have been raised about whether the competitive exclusion principle is a tautology, and whether the reason for its success and persistence is due to its logical nature (Levin, 1970). Are there logical flaws in the reasoning, related to the vague definition of a niche? If we observe many species there should be many niches, which may lead to a tendency to define a niche according to the need to have many of them.

11.1.2 Modeling

There are many reasons to revisit, explore, and criticize Gause's principle, notably from the perspective of theoretical research into biological diversity. Network economics offers an alternative to the conventional Lotka-Volterra approach. Taking the example of the trophic network with several prey and several predators presented in section 7.4, we analyze Gauses competitive exclusion principle and discuss its generality.

11.2 Design

11.2.1 Goals

They consists in analyzing how the network economics represent the Gause's competetitive exclusion principle.

11.2.2 Network Structure

We consider a bi-layered trophic network with 10 prey and 50 predators (section 7.4).

11.2.3 Rationale

Recall that model described in section 7.4.6 examines the equilibrium state of an ecosystem in which several predators arbitrate between the costs and benefits of predation on several prey. The benefits of predation depend on the abundance of prey, while costs depend partially on predator abundance. There is intra-species competition (the importance of which is underlined in Vandermeer et al. (2002)). Here, we simply examine how the resulting equilibrium depends on the values of the predator intra-species competition parameter.

11.2.4 Summary of Assumptions

Most of the assumptions of this experience, given in box 11.1, are those of section 7.4.6.

Entities	Several prey species. Several predation species. The related trophic flows.
Assumptions	Assimilation of prey is proportional to a given energetic inflow Assimilation of predators is proportional to predation inflow There is a mass balance principle, whence a relationship between biomass and trophic flows for prey and predator Costs of predation are positively related to predator biomass (through an intra specific coefficient) and negatively to prey biomass There is a complementarity relationship between predation costs and predation flows There is a network equilibrium
Statics/Dynamics	Equilibrium analysis
Calibration	Assimilation coefficients, somatic coefficients are given standard values
Observations	The relationship between the intra specify coefficient and the diversity of the predator population, once the network equilibrium has been reached

Box 11.1: Design of a model for analyzing competitive exclusion principle

11.3 Details

11.3.1 A Full Model Description

Recall that the system described in detail in section 7.4.6 represent prey species i and predator species j. Let b_i, b_j, x_{ij} the biomass and the trophic flows. The model is defined by balance equations:

$$\gamma_i e_i = \sum_j x_{ij} + \mu_i b_i$$

$$\gamma_j \sum_i x_{ij} = \mu_j b_j$$

and complementarity equations based on function $F_{ij} = \kappa_i b_i - \phi_{ij} + \lambda_j b_j$, where κ_i is the advantage of feeding on species i, ϕ_{ij} is the cost for species j of feeding on species i, λ_j is the competition coefficient inside species j.

We study how the resulting network equilibrium depends on the intra-species competition coefficient λ_j: we build the simulation experiment that is detailed in the algorithm 11.1.

For values of Λ fixed to $0, 0.02, 0.04, 0.06, 0.08, 0.1$
Repeat 1000 times the following sequence:

1. Randomly choose the intra-specific competition parameter $\lambda_j \in [\Lambda, 2\Lambda]$,
2. Randomly choose the other parameters in a relatively wide range of values: $e_i \in [1000, 2000]$, $\gamma_i \in [1, 2]$, $\mu_i \in [0.5, 1]$, $\gamma_j \in [0.1, 0.2]$, $\mu_j \in [0.2, 0.4]$, $\phi_{ij} \in [0.1, 0.3]$, $\kappa_i \in [0.1, 0.2]$, $\phi_{ij} \in [0.1, 0.3]$;
3. Compute the resulting equilibrium, get the number of each species N_i and its proportion $p_i = N_i/N$, with $N = \sum_i N_i$.

Observe how resulting diversity indices such as S the number of predator species with positive abundance $N_i > 0$, or the Shannon index $H = -\sum_i p_i Log(p_i)$, their mean, their standard deviation depend on the parameter of intra-specific competition.

Algorithm 11.1: Biological diversity and intra specific competition

11.4 Results

Results are given in figure 11.1. We observe that without any intra-specific competition, ($\lambda = 0$) the number of predator species is equal to the number of prey species. This is a crude form of Gauses principle. For each prey i, survive the most efficient of its predators: it put the biomass of i at a level that is not interesting for other predators.

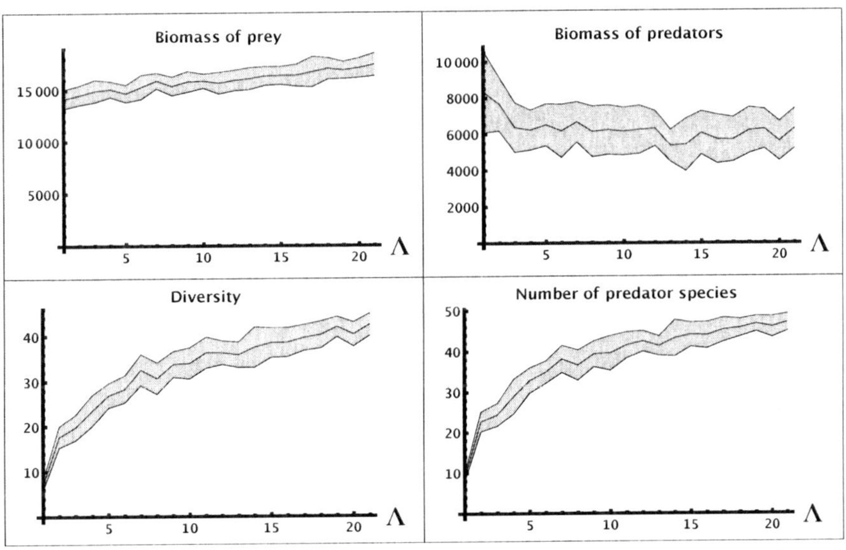

Figure 11.1: Intra-specific competition and diversity. X-axis correspond to the intra-specific values of Λ. Y-axis corresponds to different characteristics of the resulting equilibrium. What is represented is the mean plus or minus standard deviation of a given characteristic.

This is, in a very robust manner, a demonstration of how biodiversity is related to the level of intra-species competition: in this modeling approach, intra-species competition is a limiting factor for the biomass of a given predator, and thus when prey is abundant, most efficient predator leaves food for other predators. This is a well-known feature of the Lotka-Volterra equations of a competitive system. However, the result based on network economics is less drastic. While conventional theory predicts that the special relationship between inter- and intra-species competition allows the coexistence of several competitors, our model suggests that very low levels of intra-species competition are sufficient to allow the coexistence of several competitors.

11.5 Concluding Remark

11.5.1 Features

We have shown how the number of predator species in a simple ecological system is related to the intensity of intra-species competition.

11.5.2 Teachings

Network economics provides a simple demonstration of an ecological principle without using the theory of dynamical systems.

11.5.3 Going Further

The application of this result to the plankton paradox suggests that intra-species competition in plankton in the predation of resources is as important as the niche effect.

This approach may contribute to explain what is called the wasp waist pattern. In the trophic food web of upwelling pelagic systems there are relatively few intermediate small pelagic species with a high biomass (Bakun, 1996). We suggest to relate these features to the propensity of pelagic fish to schooling (Fréon and Misund, 1999; Pitcher, 2001). Let us consider a hierarchy of prey and predator systems. At the first level, there is a common prey (a pool of resources) and many predators (plankton species). We deduce that intra-species competition is significant in plankton. At the second level there are many prey species (plankton), and few predators (small pelagic species). We deduce that inter-species competition between small pelagic fish predominates; schooling is interpreted as a limitation of intra-species competition. At the third level, there are few prey (small pelagic species) and many predators (carnivorous fish). We deduce that intra-species competition is more prevalent. In this way, predation is a factor of biological diversity: there are several predators because of intra-species competition between predators and several prey because of differences in predation costs for predators. This is the most usual explanation of the importance of top-down controls in the maintenance of biological diversity.

Chapter 12

About the Peruvian Puzzle

12.1 Overview

12.1.1 Questioning

Upwelling is a phenomenon found in the eastern regions of oceans where there is an upwelling of cold, nutrient-enriched water (figure 12.1). The phenomenon is related to physical mechanisms: Eckman transport, and Coriolis effects due to the Earths rotation.

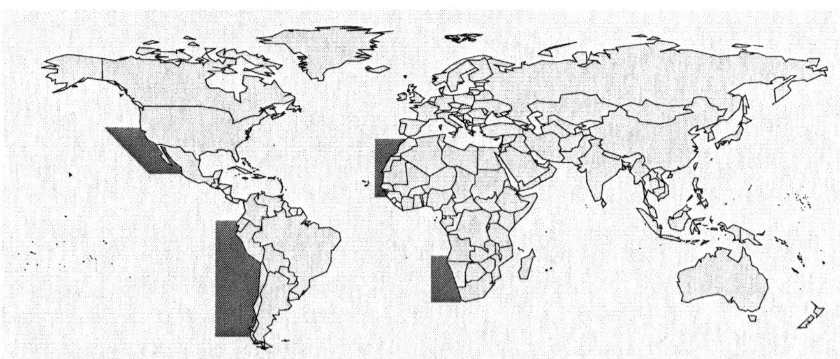

Figure 12.1: Upwelling areas. In dark grey: California, Humboldt, Canary and Benguela.

The main characteristic of these highly productive and varied ecosystems is the dominance of small pelagic fish (anchovies and sardines). While these systems are similar in terms of primary production and species structure, there are several differences between them (figure 12.2). In the Humboldt ecosystem, primary production is relatively low and small pelagic fish are abundant. In the Benguela ecosystem, primary production is very high and pelagic fish are less abundant. In the other upwelling ecosystems, the trophic pyramid has the expected shape. These differences are puzzling (Shannon et al., 2008).

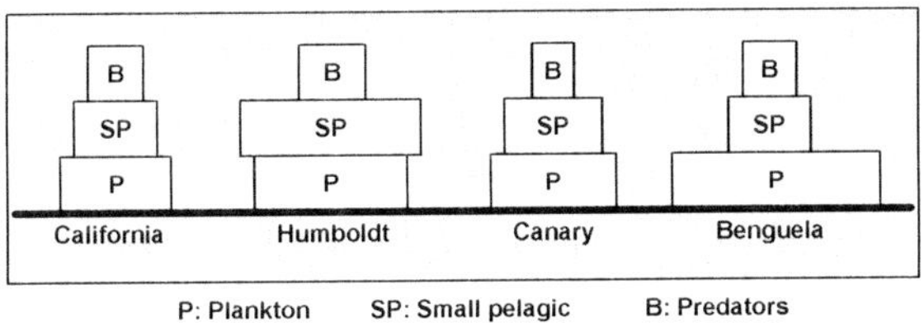

Figure 12.2: Schematic representations of trophic pyramids in the main upwelling systems.

12.1.2 Modeling

We apply the short food chain model presented in section 7.4.4 to compare upwelling ecosystems in the Humboldt, Canary, California, and Benguela currents (Shannon et al., 2008; Rossi et al., 2008). The aim is to provide potential simple, heuristic explanations for the differences between them. In this example there is no explicit modeling. Instead, it focuses on an analysis of how the principles of network economics may help to identify the key ecosystem processes.

12.2 Design

12.2.1 Goals

They consists in analyzing how the network economics can explain the observed differences of the trophic pyramids between several upwelling system.

12.2.2 Rationale

It consists in relating observed patterns of the trophic pyramids to known features of intrinsic accessibility of upwelling ecosystems.

12.2.3 Network Structure

We use the model of short food chain shown in section 7.4.4

Entities	One phytoplankton population. One zooplankton population. One small pelagic population. The trophic flows between these populations.
Assumptions	Assimilation of phytoplankton is proportional to a given energetic inflow Assimilation of zooplankton and of small pelagic fish populations is proportional to their predation inflow Losses in all populations are proportional to biomasses There is a mass balance principle, whence a relationship between biomass and trophic flows for all populations Costs of grazing are negatively to phytoplankton biomass (through an accessibility coefficient) Costs of predation are negatively to zooplankton biomass (through an accessibility coefficient) There is a complementarity relationship between predation costs and predation flows There is a network equilibrium
Statics/Dynamics	Equilibrium analysis
Calibration	There is no need for calibration
Observations	The mathematical relationship between resulting biomass of the three populations and the accessibility coefficients

Box 12.1: Design of a model for analyzing productivity differences between upwelling ecosystems

12.2.4 Summary of Assumptions

We assume that the observed characteristics of systems should be sufficient to explain observed differences in productivity (see box 12.1). We suggest to explain differences in trophic pyramids in terms of accessibility to food at different levels of the trophic chain.

12.3 Details

The equilibrium equations of a short food chain were formulated in section 7.4. Recall that solutions are given by:

$$b_i = \frac{\phi_{ij}}{\kappa_i}$$

$$b_j = \frac{\phi_{jk}}{\kappa_j}$$

$$b_k = \frac{\gamma_k \gamma_j \gamma_i e_i}{\mu_k} - \frac{\gamma_j \gamma_k \mu_i \phi_{ij}}{\kappa_i \mu_k} - \frac{\gamma_k \mu_j \phi_{jk}}{\kappa_j \mu_k}$$

where b_i, b_j, b_k denote the respective biomass of phytoplankton, zooplankton, pelagic fish, $\gamma_i, \gamma_j, \gamma_k$ their assimilation efficiency, γ_i, γ_j, γ_k their maintenance coefficient, κ_i, κ_j the density dependence of predation costs, ϕ_{ij}, ϕ_{jk} the accessibility part of predation costs and e_i the energetic input for primary production. This is the simplest mathematically explicit example of a bottom up control: changes in primary production have an effect of the top levels of the trophic chain, not on the intermediate ones.

In order to make comparisons, we underline the effects of predation costs on biomass at the different level of the trophic chain. According to previous equations, we can qualitatively determine the effects of accessibility parameters on biomass. They are given in table 12.1.

Table 12.1: Short trophic chain. Effects of accessibility parameters on biomass

$\partial b_i/\partial \kappa_i < 0$	$\partial b_j/\partial \kappa_i = 0$	$\partial b_k/\partial \kappa_i > 0$
$\partial b_i/\partial \kappa_j = 0$	$\partial b_j/\partial \kappa_j < 0$	$\partial b_k/\partial \kappa_j > 0$
$\partial b_i/\partial \phi_{ij} > 0$	$\partial b_j/\partial \phi_{ij} = 0$	$\partial b_k/\partial \phi_{ij} < 0$
$\partial b_i/\partial \phi_{jk} = 0$	$\partial b_j/\partial \phi_{jk} > 0$	$\partial b_k/\partial \phi_{jk} < 0$

12.4 Results

To analyze the productivity of the main upwelling systems, we can build the table 12.2 associating their main productivity features (left part of the table) with plausible values for accessibility parameters (right part of the table).

Table 12.2: Main upwelling systems. The characteristics of California, Benguela, Canary and Humboldt Currents. Left: key observed features. Right: potential explanations in terms of accessibility coefficients

System	b_i	b_j	b_k	κ_i	κ_j	ϕ_{ij}	ϕ_{jk}
California Benguela Canary	+	++		−	−	+	
Humboldt	+	+	++	+	+	−	−

12.5 Concluding Remark

12.5.1 Features

We have carried out a qualitative analysis of the solutions of the short food chain model: we have related the patterns observed in the differences of productivity between the main upwelling systems to the parameters of this model. Let us keep in mind (1) that coefficients κ represent the prey density dependence effect of predation costs: when κ is low, there is not much of an advantage for the predator to feed on abundant prey, and (2) coefficient ϕ represent the costs of predation related to environment only.

12.5.2 Teachings

We may conclude that differences in the accessibility of prey at different levels of the trophic chain provides a shared principle for the study of the productivity of ecosystems and a simple explanation for the Peruvian puzzle.

12.5.3 Going Further

We now extend and add further weight to the hypothesis that accessibility is a key parameter in the productivity of upwelling systems (cf. table 12.2). The differences of the accessibility parameters in this table may be related to:

(1) the high hydrodynamic activity of the Benguela that makes the spatial distribution of living organism highly heterogeneous,

(2) the shifts of water masses in the Humboldt Current due to the El Nino-Southern Oscillation (ENSO) event. This has an impact on the level of the oxycline, which is in itself a component of accessibility (Bertrand et al., 2004).

Accessibility may be related to the shape of the continental shelf. A narrow shelf may favor encounters as it reduces the spatial distribution of predators and prey. Accessibility may also be related to hydrodynamic activity and the existence of mesoscale structures that could increase the probability of encounters.

This approach contributes to the debate about the effects of climate change on the structure of marine ecosystems (Bopp et al., 2001). The causal relationship: Global change $\rightarrow$ Change in precipitation $\rightarrow$ Increase of stratification $\rightarrow$ Decrease in winter mixing $\rightarrow$ Decrease in nutrient supply $\rightarrow$ Decrease in phytoplankton $\rightarrow$ Decrease in zooplankton, can be interpreted using our approach. It suggests that attention be paid to changes in food accessibility at different levels (e.g. the effect of a increase of stratification on plankton accessibility, or the development of a mismatch between pelagic fish and their predators due to changes in the vertical column).

Chapter 13

Ecological Controls in the Chilean Coastal Ecosystem

13.1 Overview

13.1.1 Questioning

The Chilean coastal ecosystem (figure 13.1) is one of the worlds richest marine systems. This is related to the coastal upwelling effect (see chapter 12).

The system has been intensely studied and we have a good idea of fish abundance and distribution at different points in time. Some of them are from the pre-fishing era (1920), while others concern different levels of fishing exploitation (Neira and Arancibia, 2004; Neira et al., 2004). From these studies, it appears that the system has switched between several reference states. This switching may be related to the influence of climate (e.g. the effect of the El Nino event) and fishing pressure (some species being over-exploited).

13.1.2 Modeling

In this chapter we describe how to apply network economics to a switching ecosystem. Our goal is to illustrate how a large coastal ecosystem, represented as a network, can be sensitive to both climate and pressure. In this way we tackle the question of ecological controls: i.e., who controls whom in the ecosystem (Cury et al., 2000, 2003).

Figure 13.1: The Chilean coast. The upwelling area.

Entities, initial state, and input parameters are provided by earlier attempts to model the Chilean coastal ecosystem (Neira and Arancibia, 2004). This makes it possible to define the nodes (20 compartments corresponding to the main species) and links (60 trophic links) of a moderately complex network.

13.2 Design

13.2.1 Goals

The goal is to analyze how network economics can explain (in terms of controls) observed switches in the trophic structures of a large ecosystem such as the Chilean coast. The model performs a sensitivity analysis of top-down and bottom-up controls, in order to identify the important system nodes (keystone species).

13.2.2 Network Structure

We use the model of a general trophic system analyzed in section 7.5.

13.2.3 Rationale

We express the conservation and complementarity equations that are characteristic of a supply chain. Given a productivity level (depending on climate) and an exploitation level, the system reaches equilibrium. We study the sensitivity of these equilibrium states to changes in climate and fishing pressure controls.

13.2.4 Summary of Assumptions

These are summarized in box 13.1.

Entities	Several phytoplankton populations, several zooplankton populations, several fish population, one fleet. The trophic flows between these populations. The catches
Assumptions	Assimilation of phytoplankton population is proportional to a given energetic inflow Assimilation of zooplankton and of fish populations is proportional to their predation inflow Losses in all populations are proportional to biomasses There is a mass balance principle, whence a relationship between biomass and trophic flows for all populations Costs of grazing are negatively related to phytoplankton biomass; Costs of predation and costs of fishing are negatively related to prey biomass There is complementarity between grazing costs and grazing flows, predation costs and predation flows, fishing costs and fishing flows There is a network equilibrium
Statics/Dynamics	Equilibrium analysis
Calibration	Assimilation coefficients, somatic coefficients are estimated with an Ecopath model Costs coefficients are estimated with an elementary application of inverse modeling
Observations	The effects of changing the way to compute some accessibility costs on the structure of the resulting network. Top down controls: changing top predators predation costs. Bottom-up: changing grazing costs. Keystone species: changing the accessibility costs from and to a given species

Box 13.1: Design of a model for analyzing the network structure of the Chilean coastal ecosystem

13.3 Details

13.3.1 Entities

Following Neira and Arancibia (2004), we consider a trophic network with the following functional groups: phytoplankton, micro-zooplankton, meso-zooplankton (copepods), macro-zooplankton (euphausiids), jellies (salps and jellyfishes), macrobenthos, anchovy, common sardine, mesopelagics, horse mackerel, large hake, small hake, pelagics I (medium-sized pelagic fish like hoki), pelagics II (large-sized pelagic fish like swordfish), demersal fish I (benthic feeders species), demersal fish II (pelagic feeders species), chondricthyans (mainly skates), cephalopods (squid), sea lion, sea birds (penguins, pelicans, cormorants). We also consider fisheries as a specific node.

13.3.2 Initialization and Input Parameters

Initialization and input parameters are given in table 13.1. They also are extracted from Neira and Arancibia (2004).

Table 13.1: Chilean coast ecosystem. Main characteristics. Values of parameters: γ: trophic assimilation efficiency, μ: somatic maintenance coefficient, κ: accessibility, λ: intra-specific competition, E: energy input, B: reference biomass. Most parameters values are from Neira et al. (2004).

Compartment	γ	μ	κ	λ	E	B
PHYTOPLANKTON	2.0	16.89	0.21		8000	302.50
ZOOPLANKTON I	0.5	96.87	0.23	0.11		11.62
ZOOPLANKTON II	0.29	0.04	0.20	0.05		14.09
ZOOPLANKTON III	0.41	0.01	0.26	0.10		26.35
JELLIES	0.1	0.05	0.24	0.1		44.70
MACROBENTHOS	0.25	0.50	0.23		25	2.0
ANCHOVY	0.1	0.94	0.26	0.02		8.35
COMMON SARDINE	0.1	1.10	0.27	0.02		11.59
MESOPELAGICS	0.2	1.20	0.29	0.05		13.26
HORSE MACKEREL	0.05	0.68	0.29	0.01		13.79
HAKE (LARGE)	0.11	0.13	0.25	0.02		4.28
HAKE (SMALL)	0.30	0.56	0.26	0.08		4.48
PELAGICS I	0.1	0.35	0.21	0.02		13.38
DEMERSAL FISH I	0.2	0.26	0.23	0.04		1.68
DEMERSAL FISH II	0.2	0.24	0.26	0.05		0.78
CHONDRICTHTHYAN	1.3	0.03	0.23	0.30		0.43
PELAGICS II	0.4	0.95	0.25	0.10		0.31
CEPHALOPODS	0.5	1.53	0.24	0.12		1.63
SEA LION	0.02	0.15		0.01		0.09
SEA BIRDS	0.025	0.5		0.01		0.06
FISHERY	0.3	0.08		0.01		12.00

13.3.3 Notations and Equations

We use definitions, notations and results from section 7.5. We denote species i, j, k, b_i the biomass of i and x_{ij} the trophic flow from i to j. Conservation equations are:

$$\gamma_i \left(\sum_j x_{ji} + e_i \right) = \left(\sum_{ji} x_{ij} \right) + \mu_i b_i$$

K is the set of trophic flows (x_{ij}) that are positive and result in positive biomass. We put: $f_{ij} = \phi_{ij} - \kappa_i b_i + \lambda_j b_j$. Complementarity equations are:

$$x_{ij} \perp f_{ij}$$

Defining the functional F by $X = (x_{ij}) \rightarrow F(X) = (f_{ij})$ we know that a state of trophic flows x^* is an equilibrium if and only if it is a solution the variational inequality $\mathcal{VI}(F, K)$.

13.4 Results

13.4.1 Analysis of Controls

A sensitivity analysis of the impact of input parameters makes it possible to analyze the influence of top-down and bottom-up controls on the trophic network. To address the question, we relate the characteristics of equilibrium states to specific changes in predation costs.

13.4.2 Bottom-Up Control

To show the dynamics brought about by a bottom-up control, predation parameters ϕ_{ij} for grazers j of phytoplankton or macrobenthos i are multiplied successively by 0.5, 0.6, 0.7, ..., 2.0. Results are presented in terms of food transfer along the food chain (see figure 13.2 for factors 0.5 and 2.0).

When grazing costs are low (top of the figure), biomass of autotroph species is at medium levels. The trophic network is connected and competition for macro-zooplankton is detrimental for demersal fish and mesopelagics.

When grazing costs are high (bottom of the figure), biomass of autotroph species is at a higher level. Only macro-zooplankton survives the competition with micro- and mezo-zooplankton and jellies for phytoplankton. Small pelagic fish (anchovy and sardines) lose the competition with other fish species for macro-zooplankton. There are also smaller effects on top-level predators.

There is a shift between a regime where micro-zooplankton, jellies, sardines and anchovy are dominant, and phytoplankton biomass is low, and a regime

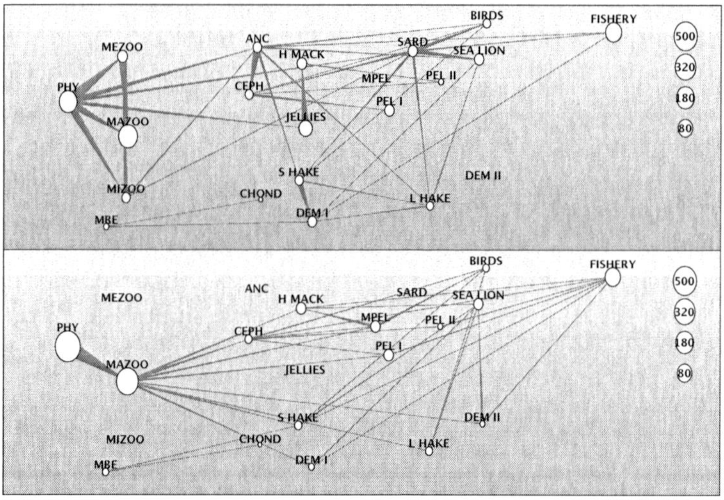

Figure 13.2: Bottom-up control. Effects of changing grazing costs on paths from autotrophic species. Top, with a factor ×0.5. Bottom with a factor ×2 (right).

where macro-zooplankton, mesopelagics and demersal fish are dominant, and phytoplankton biomass is high.

13.4.3 Top-Down Control

To represent the dynamics induced by a top-down control hypothesis predation costs parameters ϕ_{ij} for fisheries and top predators i have been successively multiplied by 0.8, 0.85, ..., 1.25.

We observe from figure 13.3 that the main changes in the dynamics are the consequence of competition between fisheries and all top-level predator species (chondrichtyans, demersal fish I and II, pelagics II, large hake, sea lions and sea birds).

High predation costs for fisheries and top-level predators have a big impact on the connectivity of the network (top of the figure). Fisheries and top-level predators mainly target small pelagic species; this reduces the abundance of other predators of these species and cuts many trophic links.

Low predation costs for fisheries and top-level predators contributes to the health of the ecosystem, with effective trophic flows on all possible predation links (bottom of the figure).

13.4.4 Wasp Waist Control

To represent the dynamics brought about by a wasp waist control (Cury et al., 2000; Bakun, 2006; Jordan et al., 2005), predation cost parameters for small pelagic fish (sardines and anchovy) were systematically increased, while parameters related to the cost of predation on these fish were

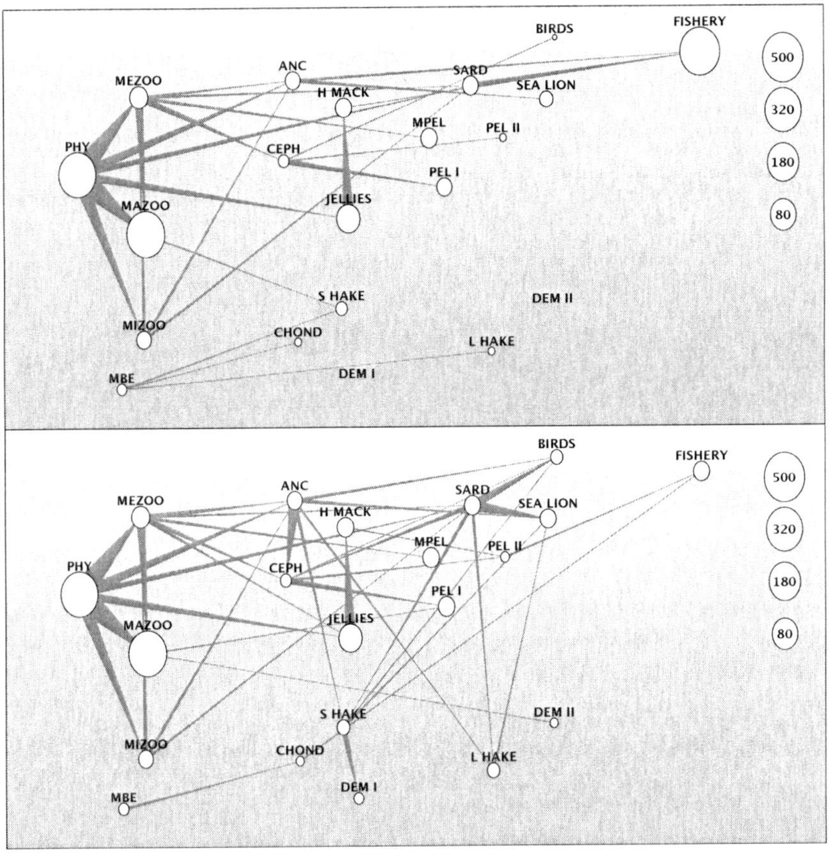

Figure 13.3: Top-down control. The effects of changing predation costs on paths to top predator species and fisheries. Top: with a factor ×0.8. Bottom: with a factor ×1.25

decreased (figure 13.4). When life is easy for small pelagic (top of figure 13.4), their efficiency leads to the disappearance of most of their competitors, especially zooplankton species.

It should be noted that we also carried out a sensitivity analysis of the respective efficiency of sardines and anchovies. This demonstrated the effect of competition between them, and showed the abrupt collapse of anchovy when the environment favors sardines, and vice-versa. This provides a simple representation of the synchronies commonly observed in upwelling ecosystems (Schwartzlose et al., 1999).

13.4.5 Keystone Species

A sensitivity analysis can be used to identify keystones species in the ecosystem. For each species i, we repeat the procedure applied to small

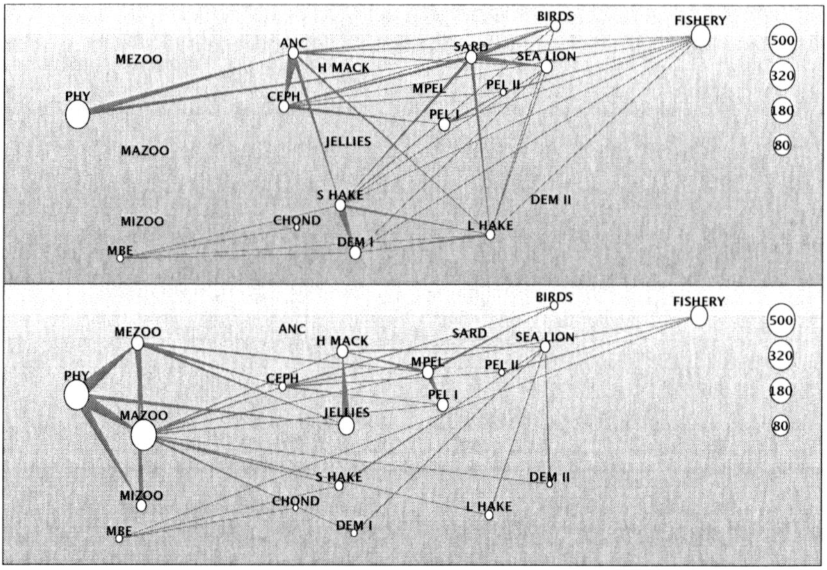

Figure 13.4: Wasp waist controls. The effects of coefficients related to small pelagic. Top: accessibility multiplied by a factor ×0.8, predation costs by a factor 1.25. Bottom: accessibility multiplied by a factor ×1.25, predation costs by a factor 0.8.

pelagic fish described in the previous section. We systematically change the predation costs coefficients (increase of ϕ_{ij} for other species j that are prey of i and decrease of ϕ_{ki} for other species k that are predators of i). We then observe the variations of biomass b_k according to the variation of b_i. When b_i is small and $\Delta b_k/\Delta b_i$ is significant for several species k, i may be considered as a keystone species (Power et al., 1996; Libralato et al., 2006).

13.4.6 Is Macro-zooplankton a Keystone Species?

Figure 13.5 shows how the network equilibrium states change according to changes in the environment of macro-zooplankton.

When the environment is very favorable, macro-zooplankton is the only consumer of phytoplankton and this is origin of most of the food web (figure 13.5, top).

This changes when the environment becomes less favorable for macro-zooplankton. Then there is competition with micro-zooplankton, which causes a slight reduction in the macro-zooplankton biomass. At a certain threshold its predators (such as hake) no longer prey on it, and disappear.

When the environment for macro-zooplankton is even more difficult (figure 13.5, bottom), the biomass of sardine and anchovies increases to a

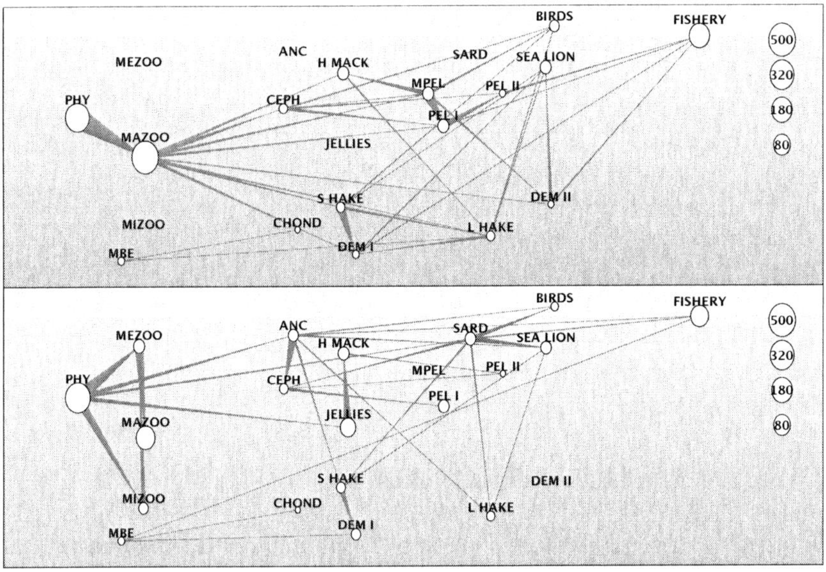

Figure 13.5: Chilean coast: large zooplankton as a keystone species. The effects of changing predation costs on paths from and to large zooplankton. Top: with a factor ×0.4. Bottom: with a factor ×2.

level that can sustain the whole food chain. They become the prey of hake and other predators that, because of high intra-species competition, prefer them to macro-zooplankton. Jellies have an important negative effect on zooplankton species and are an indicator of the health of the ecosystem. Competition with demersal II fish (pelagic feeders) is strong, but indirect. Observation of the structure of the trophic network suggests that the key mechanism is the control of macro-zooplankton by jellies. At a given level jellies consume macro-zooplankton, but this level is below the threshold that would lead to them being preyed upon by mesopelagics, which would annihilate the subsequent food chain (mainly demersal II fish).

It seems that meso-zooplankton correspond to the definition of a key species. It has a relatively low biomass, and its impact on species much further up the food chain is high. This is related to its indirect control of macro-zooplankton.

13.5 Concluding Remark

13.5.1 Features

We have shown how to represent the trophic structure of a large marine ecosystem, such as the Chilean coast, with network economics. The model takes into account the balance of costs and benefits of predation depending on prey abundance. We used a sensitivity analysis to investigate several general hypotheses about the nature of trophic controls in this system and to explain the patterns observed in its switching dynamics.

13.5.2 Teachings

The approach has proven to be feasible in terms of formulation and computation. The definition of entities and the calibration of model parameters was possible using ecosystem models previously defined using the Ecopath approach.

Switches are a natural feature of marine ecosystems when they are represented by network economics. Network economics reproduces switches, not only between alternative supply chains, but also between highly connected and less connected trophic networks.

13.5.3 Going Further

If, in comparison to other modeling approaches, our approach is found to be useful it is easy to imagine the analysis of other, similar ecosystems. It may be possible to compare the results obtained during the calibration phase (there may be general values for coefficients related to predation costs and advantages), to identify typical behaviors and relate them to differences in trophic structures.

We have represented switches in ecosystems in a static way: different external conditions are related to different structures of the trophic network. An extension of this work would involve representing the dynamics of the system. We leave this as an open question.

Chapter 14

The Worldwide Supply Chain for Tuna

14.1 Overview

14.1.1 Questioning

Tuna is one of the world's most important exploited species. It is caught in all the major oceans. Figure 14.1 represents the worldwide distribution of tuna catches.

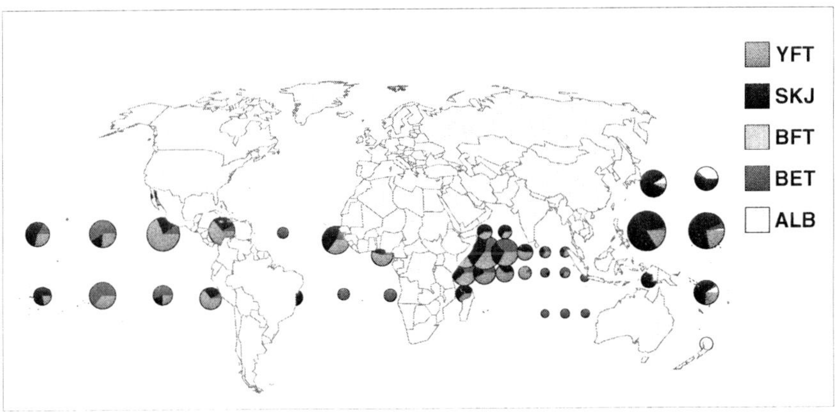

Figure 14.1: Tuna fishing areas. For each selected area, the circle represents the catches of all species. Abbreviations are as follows. YFT: yellow fin tuna, SKJ: skipjack, BFT: Bluefin tuna, BET: blue-eye tuna, ALB: albacore. Data is from the Sardara database. Spatial resolution is higher in the Indian Ocean.

Tuna products are consumed worldwide, either fresh, frozen or canned. One may consider that there is global market for tuna products (Jeon et al., 2008; Jiménez-Toribio et al., 2010; Catarci, 2005). The main actors in this commodity chain are specialized: Japan (fleet, importer of fresh/frozen fish), Spain (fleet, transformer, consumer of prepared fish), Thailand (importer of fresh/frozen fish, transformer, and exporter of prepared fish), etc. Figure 14.2 shows the distribution, production, and consumption of tuna commodities according to country.

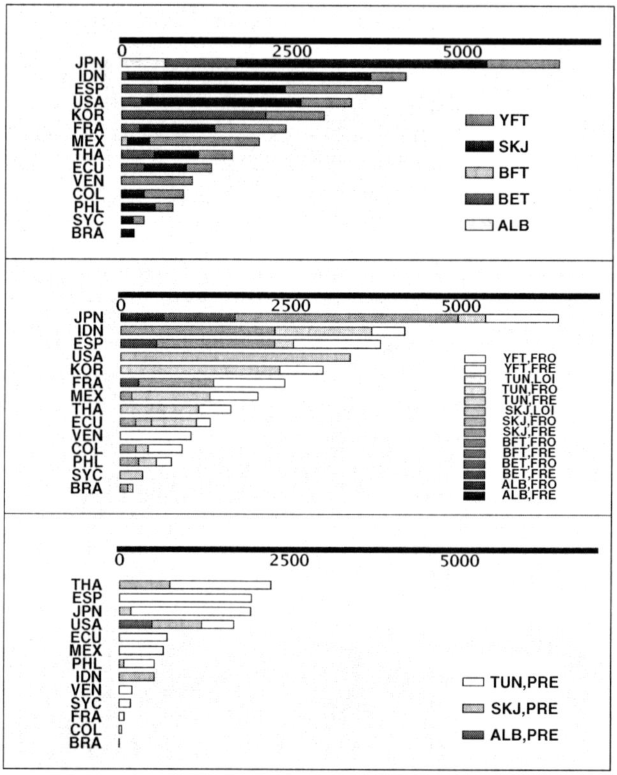

Figure 14.2: Production of tuna. Top: catches. Middle: production of fresh and frozen commodities. Bottom: production of can commodities. Plots represent the mean values for the period 1993–2006.

Some selected characteristics of the global tuna commodity chain are given in table 14.1. Values are taken from Miyake et al. (2010) who provide a general view of this bio-economic system according to biology, fisheries, markets, etc.

The observation of a reduction in the size of large predators (Myers and Worm, 2003) suggests that many tuna stocks are endangered (Majkowski, 2007). At the same time, fishing fleets and tuna markets are affected by economic globalization (Catarci, 2005; Pauly et al., 2005). The increasing

Table 14.1: Some important features of the global tuna supply chain. Magnitudes orders and trends

	1980	2010.
World tuna catches	3 M tons	6 M tons
Sales of can	800 K tons	1500 K tons
Prices of fresh	1400 $ per ton	2000 $ per ton
Prices of can	3000 $ per ton	3000 $ per ton
Average size of catches in Western Pacific Ocean	60 kg	20 kg

demand for tuna products and structural changes in the demand for tuna (Jeon et al., 2008; Jiménez-Toribio et al., 2010) explain the high pressure on stocks.

The global control of tuna exploitation may be a possible solution. Most tuna is caught offshore, beyond Exclusive Economics Zones (EEZ) by long distance fleets (Reid et al., 2003). The management of tuna fishing capacity (Bayliff, 2004) implies international cooperation initiated by specialized international commissions focused on the main oceans: the Atlantic (the International Commission for the Conservation of Atlantic Tunas), the Pacific (the Western and Central Pacific Fisheries Commission), and the Indian (the Indian Ocean Tuna Commission).

14.1.2 Modeling

This chapter presents a model of the global tuna commodity chain. It is based on an integrated representation of the biological and economic processes operating in the global exploitation of a natural resource. The goal is to build future scenarios for a complex bio-economic system in the context of climate change and economic globalization. The results of these scenarios should help to address the question of global governance of a worldwide commodity chain.

Building future scenarios is an important feature of any modeling attempt. It makes it possible to determine time scales, processes, and interactions, which are all key issues in the modeling of complex systems. The following are some potential scenarios for the model.

Collapse of fresh market. What could happen if, following a health problem, there is a huge decrease of the demand for fresh tuna (Bestor, 2004)?

Increase of oil price. What could happen if there is a huge increase in oil prices, affecting both fishing and shipment costs (Tyedmers, 2004; Tyedmers et al., 2005)?

Trade regulation. What could happen if, as it has happened before, there is a decrease in import taxes for all tuna commodities (Grafton et al., 2010)?

Climate change and productivity. What could happen if, in the context of current climate changes, the productivity of marine areas either dramatically increases or decreases (Brander, 2007; Cheung et al., 2010)?

Offshore marine protected areas. What could happen if global governance of tuna fisheries is put in place and results in the application of a system of protected marine areas (Balmford et al., 2004; Scovazzi, 2004; Hyrenbach et al., 2000; Young, 2003; Game et al., 2009; Mills and Carlton, 1998)?

Moratorium on fishing aggregative devices. What could happen if, in the framework of global governance, it is decided to prohibit fish aggregation devices for some years, reducing the quantity of fish that can be caught (Ménard et al., 2000)?

14.2 Design

14.2.1 Goals

The goal is to analyze how network economics can contribute to exploring future scenarios concerning the global tuna supply chain.

14.2.2 Network Structure

The model is based on a network representation of the tuna commodity chain that links stocks and fish markets (figure 14.3). It has been designed using existing statistics concerning stocks (catches, biomass), fleets (effort, fishing capacity, investment behavior), and markets (volumes, prices).

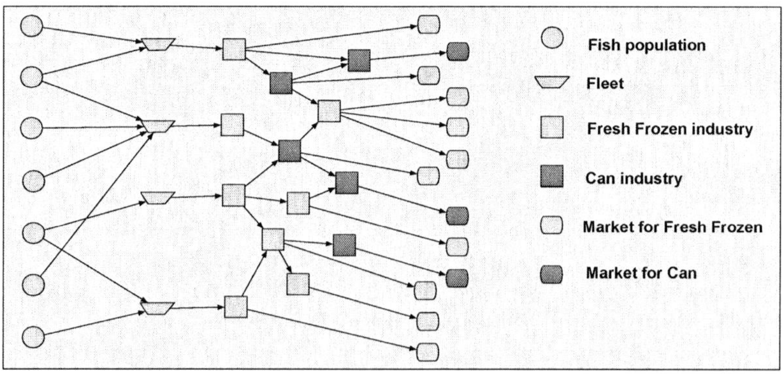

Figure 14.3: Network representation of the tuna commodity chain: type of nodes and links.

14.2.3 Rationale

Scenarios are based on sensitivity analyzes. They specify, a priori the input to the simulation (the parameter to be analyzed as a factor) and the output of simulations. The analysis focuses on the following changes to the network: (1) variation in intensive variables costs on links and prices on nodes; and (2) variation in extensive variables flows on links and quantities on nodes.

14.2.4 Modeling Dynamics

The dynamics of the system is represented as the iteration, during several time steps, of the succession of two kinds of processes: (1) the computation of an economic equilibrium of the system; and (2) the applying of conventional production or investment functions to obtain a new system state. More precisely, the model consists in, for each time step:

(1) update fishing costs: they depend on new values of stock abundances;

(2) update parameters according to scenario

(3) compute a resulting economic equilibrium using a variational inequality approach: compute matrices of the linear function related to the equilibrium of a supply chain (section 8.4), solve the linear variational inequality, get resulting flows and prices;

(4) use the values of flows given by the economic equilibrium to compute new values of fish stocks, of fleets profit, of fishing capacity, of canning industries profit, of new canning capacity.

14.2.5 Time Step and Length of Simulations

Based on the data and computing requirements, a time step corresponding to a trimester was chosen. Simulations were run over ten years.

14.2.6 Summary of Assumptions

The assumptions of this modeling attempt are given in box 14.1.

14.2.7 An Important Assumption about Tuna Migration

We decided not to consider migration between areas. We know that this is an unrealistic assumption, especially when one of our goals is to examine offshore protected areas. However, given the lack of knowledge about the determinant factors of tuna migration, we decided to examine a model where minimal assumptions are made about migration (and we have decided that here, minimal means no migration). Secondly, we examine the migration model for only one species (namely the Bluefin tuna). This model will be presented in a forthcoming chapter. If both analyzes provide significant

Entities	Tuna stocks, Fleets, Fresh traders, Canneries, Fresh markets, Can markets
Schedule of the model	Succession of network equilibrium and production (tuna stocks) or investment (fleets, canning industries) phases
Assumptions: equilibrium	There is inverse demand function on fresh and can markets A part of fishing cost inversely proportional to fish abundance Other costs are constant Profit of fleets and canning industries are related to limitations of fishing capacity and the subsequent shadow prices
Statics/Dynamics	At each step, the succession of an equilibration phase and a deterministic phase A production function for fish stocks (Schaeffer model) Investment is myopic; it is proportional to profit for fleets and canning industries
Calibration	Inverse modeling using databases Sardara and Fishstat (commodity)
Observations	The resulting dynamics for a 10 years period

Box 14.1: Design of the model of the global commodity chain for tuna

results, they could be the starting point for a further attempt to build a model of the global tuna supply chain that includes migration.

14.2.8 Approach

Modeling was carried out in four phases:

(1) Conjointly, analyze datasets, define the structure of the network, make scenarios explicit. This implies checking that equilibrium's computation is feasible; that the resulting structure of the network are illustrative; that necessary data are accessible; that solving the equilibrium allows analyzing scenarios.

(2) Using an inverse modeling approach, obtain values of necessary parameters.

(3) Implement the model, the equilibrium algorithm, the coupling of algorithms, the post processing (graphics and statistics) programs;

(4) Run and analyze the scenarios.

14.3 Details

14.3.1 Basic Entities

Basic entities used to define nodes are given in table 14.2. Geographical areas are shown in figure 14.1 together with corresponding catches. Figure 14.2 shows the production of selected commodities in most important countries.

Table 14.2: The global tuna supply chain. Entities of the model

Geographical areas	There are 33 oceanic areas. Resolution is greater in Indian Ocean.
Tuna species	Albacore, Bigeye, Bluefin, Yellow fin, Skipjack;
Countries	Brazil: BRA, Canada: CAN, Columbia: COL, Ecuador: ECU, Spain: ESP, France: FRA, Germany: GER, Indonesia: IDN, Italy: ITA, Japan: JPN, South Korea: KOR, Mexico: MEX, Morocco: MAR, Netherlands: NET, Philippines: PHL, Seychelles: SYC, Thailand: THA, Taiwan: TWN, UK, USA, Venezuela: VEN. Netherlands, Philippines, Seychelles, Thailand, Taiwan, United Kingdom, United States, Venezuela;
Fresh or frozen commodities	Albacore fresh: ALB-FRE, Albacore frozen: ALB-FRO, Blue eye tuna frozen: BET-FRE, Skipjack frozen: SKJ-FRO, Bluefin tuna frozen: BFT-FRO, Yellow fin tuna frozen: YFT-FRO, Tuna fresh: TUN-FRE, Tuna frozen: TUN-FRO;
Prepared commodities (can)	Albacore prepared: ALB-PRE, Skipjack prepared: SKJ-PRE, Tuna prepared: TUN-PRE.

14.3.2 Nodes

There are in the networks six kinds of nodes. They are as follows:

(1) Fish stocks, denoted s, correspond to an oceanic area and a species. They are characterized by a biomass θ_s carrying capacity ω_s, intrinsic growth rate ρ_s and catchability κ_s.

(2) Fleets, denoted b, correspond to a targeted species and a country. They are characterized by a fishing capacity v_b, depreciation rate η_b, and investment rate σ_b. Fleets b have a limitation on their inflow, related to their fishing capacity. They will be called constrained nodes.

(3) Trade and production systems of a fresh or frozen commodity, denoted n, correspond to a country and a fresh or frozen commodity.

(4) Trade and production systems of a can commodity, denoted t, correspond to a country and a can commodity. They are characterized by a canning capacity u_t, depreciation rate η_t, and investment rate σ_t. They have a limitation on their inflow, related to their canning capacity. They will also be called constrained nodes.

(5) Markets of a fresh or frozen commodity, denoted d, correspond to a country and a fresh commodity. They are characterized by an inverse demand function relating prices to quantities: $p_d = a_d - b_d q_d$.

(6) Markets of a can commodity correspond, denoted c, to a country and a can commodity. They are characterized by an inverse demand function relating prices to quantities: $p_c = a_c - b_c q_c$.

14.3.3 Links

There are in the network seven kinds of links.

(1) The catch of a fish stock s by a fishing fleet b is denoted x_{sb}. These links are characterized by exerted effort v_{sb} and fishing costs c_{sb}, representing the cost of fishing one ton of tuna. The conventional relationship between catch, stock, effort and catchability is:

$$x_{sb} = v_{sb}\theta_s \kappa_s$$

Effort is limited according to the fishing capacity of fleets: $\sum_s v_{sb} \leq v_b$. This results in constraints on flows (catches). According to previous relationship between catch, stock and effort, they are given by:

$$\sum_s x_{sb}/\kappa_s\theta_s \leq v_b$$

Fishing costs $c_{sb} = c_{sb}^E + c_{sb}^X$ have a component related to effort c_{sb}^E and a component related to catch. The value of c_{sb}^E is inversely proportional to fish abundance θ_s. We express total costs are $c_{sb}^X x_{sb} + c_{sb}^E v_{sb}$. Then, unitary costs are:

$$\begin{aligned} c_{sb} &= (c_{sb}^X x_{sb} + c_{sb}^E v_{sb})/x_{sb} \\ &= c_{sb}^X + c_{sb}^E v_{sb}/(x_{sb}) \\ &= c_{sb}^X + c_{sb}^E/(\kappa_s\theta_s) \end{aligned}$$

(2) Transformation flows of a caught fish b into a fish commodity n are denoted x_{bn} and are characterized by transformation costs c_{bn}. These costs will be set to zero in this study.

(3) Transformation flows of a fish commodity n into a can commodity t are denoted x_{nt}; they are characterized by a transformation cost c_{nt} (including cannery costs) and a transformation ratio: $\rho \simeq 2$. These flows are limited by a physical constraint related to canning capacity:

$$\sum_n x_{nt} \leq u_t$$

(4) Trade flows of fresh, frozen commodities from production system n to production system n' are denoted $x_{nn'}$ and are characterized by trading costs $c_{nn'}$ (including transport, taxes).

(5) Trade flows of can commodities from production system t to production system t' are denoted $x_{tt'}$ and are characterized by trading costs $c_{tt'}$ (including transport, taxes).

(6) Sales of a fresh or frozen commodity from a production system n to a market d are denoted $x_{nd} = q_d$ and are characterized by a selling cost c_{nd}.

(7) Sales of a can commodities from a production system t to a market c are denoted $x_{tc} = q_c$ and are characterized by a selling cost c_{tc}.

14.3.4 Equilibrium Equations

Let us make explicit in this context the complementarity relationships that are characteristic of supply chains (section 8.4). At equilibrium, on intermediate nodes: (1) either ingoing flow is greater than outgoing flow and there is no price; or (2) ingoing flow equals outgoing flow and there is a price. That is:

$$\begin{aligned}
p_n \perp f_n &= (\sum_b x_{bn} + \sum_{n'} x_{n'n}) \\
&\quad -(\sum_{n'} x_{nn'} + \sum_t x_{nt} + x_{nd}) \\
p_t \perp f_t &= (\sum_n x_{nt} + \sum_{t'} x_{t't}) - (\sum_{t'} x_{tt'} + x_{tc}) \\
p_b \perp f_b &= (\sum_s x_{sb} - \sum_n x_{bn})
\end{aligned} \tag{14.3-1}$$

At equilibrium, on links that do not end in a constrained node: (1) either price at origin plus transportation costs are greater than price at destination and there is no flow; or (2) price at origin plus transportation costs equal price at destination and there is a flow. That is:

$$\begin{aligned}
x_{nn'} &\perp f_{nn'} = (p_n + c_{nn'} - p_{n'}) \\
x_{tt'} &\perp f_{tt'} = (p_t + c_{tt'} - p_{t'}) \\
x_{bn} &\perp f_{bn} = (p_b + c_{bn} - p_n) \\
x_{nd} &\perp f_{nd} = (p_n + c_n - p_d) \\
x_{tc} &\perp f_{tc} = (p_t + c_t - p_c)
\end{aligned} \tag{14.3-2}$$

There are inflow constraints for fleets and canning industries:

$$\sum_s x_{sb}/\kappa_s \theta_s \leq v_b$$

$$\sum_f x_{nt} \leq u_t$$

These constraints create shadow prices λ_b and λ_t that are non-negative and positive only when fishing or canning capacity is fully used. At equilibrium, on links that end in a constrained node: either (1) price at origin plus transportation cost plus shadow cost is greater than observed price at destination and there is no flow; or (2) price at origin plus transportation costs plus shadow cost equals price at destination and there is a flow. That is:

$$\begin{aligned} \lambda_b \quad &\perp \quad \mu_b = (v_b - \sum_s x_{sb}/\kappa_s\theta_s) \\ x_{sb} \quad &\perp \quad f_{sb} = (c_{sb} + \lambda_b - p_b) \end{aligned} \tag{14.3-3}$$

and

$$\begin{aligned} \lambda_t \quad &\perp \quad \mu_t = (u_t - \sum_n x_{nt}) \\ x_{nt} \quad &\perp \quad f_{nt} = (p_n + c_{nt} + \lambda_t - p_t) \end{aligned} \tag{14.3-4}$$

14.3.5 Complementarity Problem

We now make explicit the theorems that were described in detail in section 8.4. A state of the system is made up of: (1) flows on all links; (2) prices on intermediate nodes (in other words excluding fish stocks and markets); and (3) shadow prices on constrained nodes (fleets and canning industries):

$$X = (x_{sb}, x_{bn}, x_{nt}, x_{nn}, x_{tt'}, x_{nd}, x_{tc}, p_b, p_n, p_t, \lambda_b, \lambda_t)$$

We consider the function: $X \to F(X)$, where

$$F(X) = (f_{sb}, f_{bn}, f_{nt}, f_{nn}, f_{tt'}, f_{nd}, f_c, f_b, f_n, f_t, \mu_b, \mu_t)$$

and the values of $f_{sb}, f_{bn}, \ldots, \mu_t$ are those given in the equations of section 14.3.5.

We know that a state X is an equilibrium state if and only if it is a solution of the complementarity problem $\mathcal{CP}(F)$. Recall that we denote the set of links $L = \{sb, bn, nt, nn, tt', nd, tc\}$, the set of intermediate nodes $N = \{b, n, t\}$, the set of constrained nodes $Q = \{b, t\}$.

Expressing $f_{sb}, f_{bn}, \ldots, \mu_t$, it appears that F is a linear function $F(X) = \mathbf{M} \cdot X + N$, the matrix $\mathbf{M}$ and the vector N being given by:

$$\mathbf{M} = \begin{pmatrix} \mathbf{M}_{LL} & \mathbf{M}_{LN} & \mathbf{M}_{LQ} \\ \mathbf{M}_{NL} & 0 & 0 \\ \mathbf{M}_{QL} & 0 & 0 \end{pmatrix}, \; N = \begin{pmatrix} N_L \\ 0 \\ N_Q \end{pmatrix}$$

Where the sub-vectors of N are:

- N_L that is such that: $N_{sb} = \theta_s + c_{sb}$ for $l \in L_{sb}$, $N_{nd} = -a_d + c_{nd}$, $N_{tc} = -a_c + c_{tc}$, $N_l = c_l$ elsewhere.
- N_Q that is such that $N_b = v_b$, $N_t = u_t$.

Where the sub-matrices of $\mathbf{M}$ are:

- $\mathbf{M}_{LL}$ is given by $M_{tc,tc} = b_c$, $M_{nd,nd} = b_d$, $M_{ll'} = 0$ elsewhere.
- $\mathbf{M}_{LN}$ is given by $M_{sb,b} = M_{bn,n} = M_{nt,t} = M_{nn',n'} = M_{tt',t'} = 1$, $M_{bn,b} = M_{nt,n} = M_{nn',n} = M_{tt',t} = M_{nd,n} = M_{tc,t} = -1$.
- $\mathbf{M}_{NL} = -^T\mathbf{M}_{LN}$.
- $\mathbf{M}_{LQ}$ is given by $M_{sb,s} = 1$, $M_{nt,n} = 1$.
- $\mathbf{M}_{QL}$ is given by $M_{s,sb} = -1/(\kappa_s\theta_s)$, $M_{n,nt} = -1$.

14.3.6 Equilibrium Computation

Figure 14.4 shows matrix $\mathbf{M}$. This is a large, sparse, square matrix. In the present implementation of the model, it has 865 lines. We solve the linear complementarity problem with the interior point method given in section 6.2.8. Even if we cannot be sure that $\mathbf{M}$ is positive, the algorithm quickly converges.

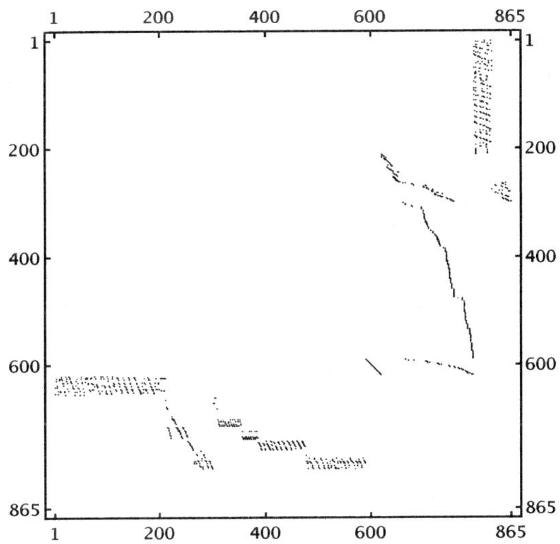

Figure 14.4: The matrix M. Non zero entries are represented by a dot.

14.3.7 Dynamics

Here we describe in detail the dynamics of the model.

Post-equilibrium. Once an equilibrium has been computed, we know

(1) the flows on all links:

$$(x_{sb}, x_{bn}, x_{nt}, x_{nn}, x_{tt'}, x_{nd}, x_c)$$

(2) the values of prices on intermediate nodes (in other words excluding fish stocks and markets):

$$(p_b, p_n, p_t)$$

(3) the values of shadow prices on constrained intermediate nodes:

$$(\lambda_b, \lambda_t)$$

Step 1: Computing profits. From flows, using inverse demand functions, we can compute the prices on markets c and d.

From flows and prices, we can compute profits of fleets and canning industries. Profit of fleet b is given by

$$i_b = \text{sales} \quad - \quad \text{selling costs} \quad - \quad \text{fishing costs}$$

The profit of fleets is related to shadow prices. We have:

Theorem 14.3.1

$$i_b = \lambda_b \sum_s x_{sb}$$

Proof. According to previous definition, we have

$$i_b = \sum_n x_{bn}(p_n - c_{bn}) - \sum_s x_{sb}c_{sb}$$

From equations (14.3-2) and (14.3-3), we extract: $x_{bn} \perp (p_b - c_{bn} - p_n)$ and $x_{sb} \perp (c_{sb} + \lambda_b - p_b)$; we get:

$$\begin{aligned} i_b &= \sum_n x_{bn}p_b - \sum_s x_{sb}(p_b - \lambda_b) \\ &= p_b(\sum_n x_{bn}p_b - \sum_s x_{sb}) + \sum_s x_{sb}\lambda_b \end{aligned}$$

Using equation (14.3-1): $p_b \perp (\sum_n x_{bn}p_b - \sum_s x_{sb})$, we get:

$$i_b = \lambda_b \sum_s x_{sb}$$

□

Profit of a cannery t is given by

$$\begin{aligned} i_t \quad = \quad & \text{sales} \ - (\text{purchase of fresh} \ + \text{production costs}) \\ & + (\text{export} - \text{import}) \end{aligned}$$

Using previous complementary relationships, we have in a similar way:

Theorem 14.3.2

$$i_t = \lambda_t \sum_n x_{nt}$$

Proof. By definition, we have:

$$i_t = p_t x_{tc} - \sum_n x_{nt}(p_n + c_{nt}) + \sum_{t'} x_{tt'} p_{t'} \\ - \sum_{t'} x_{t't}(p_t + c_{t't})$$

From complementarity equations (14.3-4) and (14.3-2), we extract: $x_{nt} \perp (p_n + c_{nt} + \lambda_t - p_t)$ and $x_{tt'} \perp (p_t + c_{tt'} - p_{t'})$; we get:

$$\begin{aligned} i_t &= p_t x_{tc} - \sum_n x_{nt}(p_t - \lambda_t) + \sum_{t'} x_{tt'}(p_t + c_{tt'}) \\ &\quad - \sum_{t'} x_{t't}(p_t + c_{t't}) \\ &= p_t \left(x_{tc} - \sum_n x_{nt} + \sum_{t'} x_{tt'} - \sum_{t'} x_{t't} \right) + \lambda_t \sum_n x_{nt} \end{aligned}$$

Using the previous complementarity equation (14.3-1):
$p_t \perp ((\sum_n x_{nt} + \sum_{t'} x_{t't}) - (\sum_{t'} x_{tt'} + x_{tc}))$, we get

$$i_t = \lambda_t \sum_n x_{nt}$$

□

Step 2: Dynamics of stocks. Stocks evolve according to the following production function:

$$\theta_s \to \theta_s + \rho_s \theta_s (1 - \theta_s / \omega_s) - \sum_b x_{sb}$$

Step 3: Dynamics of fleets and canning industries. Fishing capacity and canning capacity evolve according to the investment function given below. Using the above values of profit i_b, i_t, with depreciation rates η_b, η_t, investment rates σ_b, σ_t, we have:

$$\begin{aligned} v_b &\to v_b - \eta_b v_b + \sigma_b i_b \\ u_t &\to u_t - \eta_t u_t + \sigma_t i_t \end{aligned}$$

End of dynamic step. At the end of the dynamic step, we know:
(1) Production costs

$$(c_{sb}, c_{bn}, c_{nt})$$

Some of these costs (fishing costs (c_{sb}) depend on the state of fish stocks at this time-step;
(2) Transportation costs

$$(c_{nn'}, c_{tt'}, c_n, c_t)$$

They are assumed to be constant, or to depend on a scenario;
(3) Fishing and canning capacities

$$(v_b, u_t)$$

(4) The inverse demand function on markets:

$$p_d = a_d - b_d x_{nd}$$
$$p_c = a_c - b_c x_c$$

Coefficients a_c and b_c are assumed to be constant, or to depend on a scenario. These are the necessary quantities for computing the next equilibrium step.

14.3.8 Input and Initialization

We detail in an annex to this chapter (14.7) how we obtained from existing databases, initial parameters values (input) and the initial state of the system (initialization). Recall that, in this chapter, we illustrate how to apply network economics to a global supply chain. We have chosen, for every parameter, the simplest solution. Making parameters values more realistic would improve the model. However, it implies making a more thorough analysis and assessment of elementary processes, while maintaining the system structure.

14.3.9 Building Scenarios

The principle for building scenarios lies in: (1) associating variation in a parameter to a scenario; and (2) establishing (before running the model), what will be observed in terms of system dynamics.

In all the simulations, we observe the resulting dynamics for stocks, catches, trade, prices, consumption, profit, and capacities.

Petrol price. Petrol prices have an effect mainly on the effort part of fishing costs. We define a sensitivity parameter τ and consider the results of simulation when changing c^E_{sb} with τc^E_{sb} for all stocks s and all fleets sb.

Marine protected areas. To represent the protection of a marine area, we assume that fishing costs in this area are becoming too high to allow any fishing. We consider the results of simulation when changing c^Y_{sb} with ∞ for stocks in the protected area s and all fleets sb.

Fishing efficiency changes. To represent fishing efficiency changes, we assume that they are encapsulated in the catchability coefficient of fleets κ_b. We define a sensitivity parameter τ and consider the results of simulation when changing κ_b with $\tau\kappa_b$ for all fleets sb.

Globalization. To represent the effects of globalization, we consider the effects of changing trade costs. We define a sensitivity parameter τ and consider the results of simulation when changing $c_{tt'}$ with $\tau c_{tt'}$ for all trading countries t and t'.

Demand changes for fresh-frozen commodities. To represent changes in demand, we transform the inverse demand function. We define a sensitivity parameter τ and consider the results of simulation when changing the intercept parameter of the demand function a_c with τa_c for all fresh markets c or a_d with τa_d for all can markets d.

Biological productivity changes. To represent the effects of climate change of the fish production, we assume that it affects the carrying capacity or the intrinsic growth rate of fish stocks. In the forthcoming experiments, we have chosen to analyze the effects of changing carrying capacity. We define a sensitivity parameter τ and consider the results of simulation when changing the carrying capacity ω_s with $\tau\omega_s$ for all fish stocks s.

14.4 Results: Existence and Unicity of Solutions

During the equilibration step of the model, we have to solve a linear complementarity problem $\mathcal{LCP}(\mathbf{M}, N)$. Even if the matrix M appears no to be different of a monotone one, we have no mathematical proof of that property. This is why we have tackled the issue of the existence and the unicity of solutions with computing experiences.

14.4.1 Existence

It has been the case that the algorithm we have used (the interior point algorithm for $\mathcal{LCP}$ seen in section 6.2) has given a solution in most of cases we have tested. However, in simulations using high values for demand parameters the algorithm does not converge immediately. This may indicate a highly stressed system that has no equilibrium.

14.4.2 Unicity

Is the solution unique? To check, we ran the equilibration algorithm 20 times with randomly chosen values for the initial interior point. The obtained solutions were different.

14.4.3 Variability of Solutions

To quantify the variability of these solutions, we computed the ratios of means and standard deviation of resulting values (flows and prices) for all links and nodes of the network. Table 14.3, shows the most significant values. Variation coefficients are multiplied by 10^6. It only shows network entities with variation greater than 0.0005, that is a very low value.

It appears that only few variations levels are significant. The order of magnitude of the greatest ones is 0.1. Corresponding entities may be interpreted as being less constrained in the network structure.

Table 14.3: Variation coefficient for main variables characterizing nodes of the network). The last column corresponds to: 10^6 Standard Deviation/Mean

Type	Node	Mean	Variation
Trade Fresh	IDN	6.93378	133621
Trade Fresh	KOR	18.4621	107365
Profit Can	USA	0.938755	12979
Profit Can	SYC	2.21517	12821
Trade Can	NET	295.745	7403
Trade Fresh	ESP	144 723.	6820
Trade Fresh	USA	7 368.76	5948
Trade Can	BRA	2 733.49	5602
Trade Fresh	MEX	15 023.1	5295
Trade Can	COL	3 291.37	4159
Trade Fresh	VEN	18 420.9	4130
Profit Fresh	KOR	1 588.85	3228
Trade Fresh	FRA	304 087.	2902
Profit Fresh	FRA	8 841.05	2362
Profit Fresh	ESP	9 121.52	1439
Trade Fresh	ECU	123 387.	1285
Catches	Indian	895 029.	1280
Profit Fresh	VEN	2 566.32	1125
Catches	Non Indian	1235690	968
Trade Fresh	COL	72 250.3	948
Profit Fresh	COL	8 347.3	899
Profit Fresh	PHL	8 000.62	888
Trade Fresh	JPN	136 823.	814
Profit Fresh	ECU	7114.5	803
Prices Fresh	Non Indian	1.86118	646
Profit Fresh	JPN	11 049.9	622
Profit Fresh	MEX	10 314.6	587
Trade Can	SYC	155 88.3	531

14.5 Results: Scenarios

14.5.1 Reference Simulation

Figure 14.5 shows the dynamics of the system with reference values for parameters. Like forthcoming figures showing the results of scenarios, it presents estimates of tuna stock, catches, trade of canned tuna, trade of fresh and frozen tuna, profit of fleets, profit of canning industries, and canning capacity.

The relative stability of almost all series is worth noting. This is not surprising: most parameters were estimated assuming a stationary network.

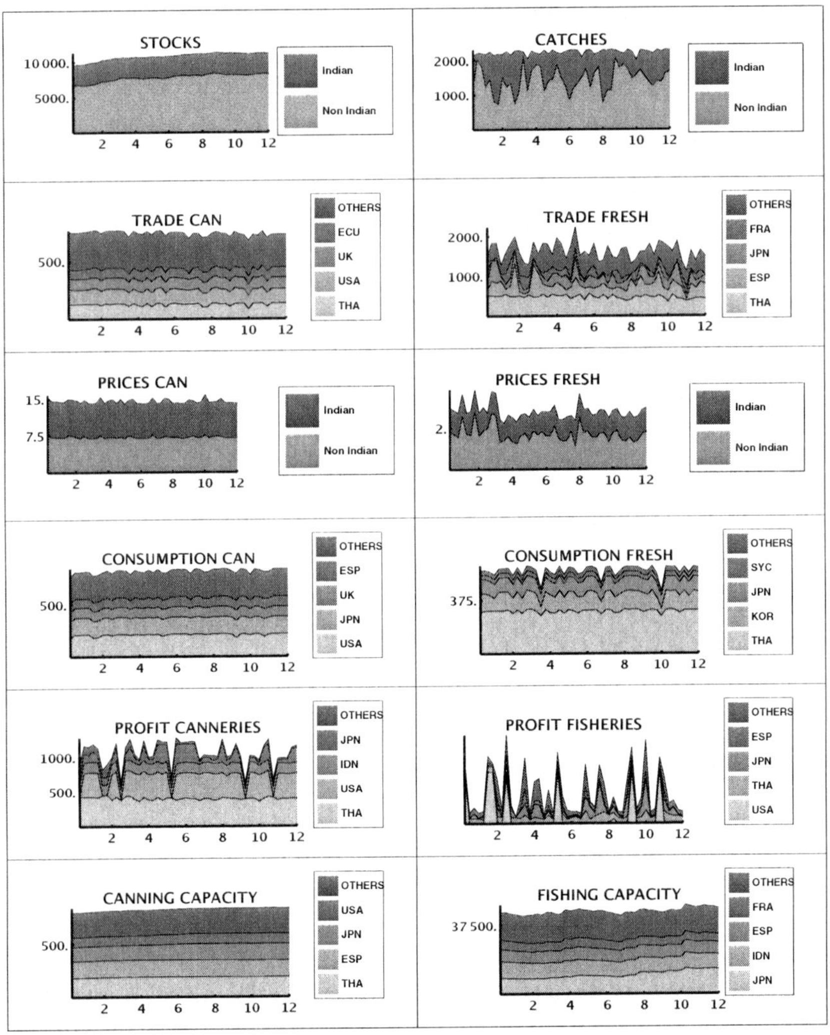

Figure 14.5: Reference scenario. Resulting series.

The series corresponding to the profit of fleets is the most variable and indicates where is instability in the model. Profit variability is an important feature of the economic of fisheries (Kasperski and Holland, 2013), and the approach seems appropriate to reproduce it. However, one must have in mind the possibility of an artifact related to the non-unicity of the solution of the equilibrium equations. Recall how the model computes profits: they are related to shadow prices at economic equilibrium, and thus to constraints, both for fleets (fishing capacity) and canning industries (canning capacity).

Emphasizing the variability of profits has consequences in terms of the management of a renewable resources on a global scale. There is a big difference in the assumption that fisheries must cope with intrinsic profit variability rather than an intrinsic natural variability.

We must also pay attention to the level of decision-making. At the level of fishermen, there are ways to cope with natural variability and the resulting catches. However, there may be no way to cope with price variability. How can this dilemma be addressed in the case of a national fleet?

On the other hand, it could be argued that these results question the assumption of myopic investment behavior. To what extent is this a valid argument?

14.5.2 Increasing Demand

We explore the consequences of a significant increase in demand for fresh, frozen, and canned tuna. This increase is set at 10% per year in all consuming countries. Technically it consists in increasing the intercept coefficients a_c and a_d in the inverse demand function. All other parameters retain their reference values.

The story we are told is depicted in figure 14.6, in which we observe a classical process of a fishery's collapse:

(1) increase in prices,
(2) increase in profit,
(3) increase in capacity,
(4) overexploitation and depletion of stocks.

We remark that after some years, profit increases. The increase of prices compensates the decrease of catches. One may ask if the flexibility of prices to quantities is well calibrated in the model? Let us recall how this flexibility has been estimated. On a given market, we have series for prices and quantities. We take the mean values of prices $\overline{p}$ and quantities $\overline{x}$, and derive a linear relationship $p = a - bx$ that (1) fits the mean values: we have $\overline{p} = a - b\overline{x}$ and (2) has flexibility -1 at this point. It is obvious that $a = 2\overline{p}$ and $b = \overline{p}/\overline{x}$. With a linear relationship $p = a - bx$, revenue px is maximum for $x = \overline{x}$.

These elements question the assumption of independent markets, whose functioning is characterized by their own linear inverse demand function. An alternative is constant flexibility functions of type $p = ax^{-b}$, or functions where prices depend on a market, or depend on the quantities sold on several inter-related markets.

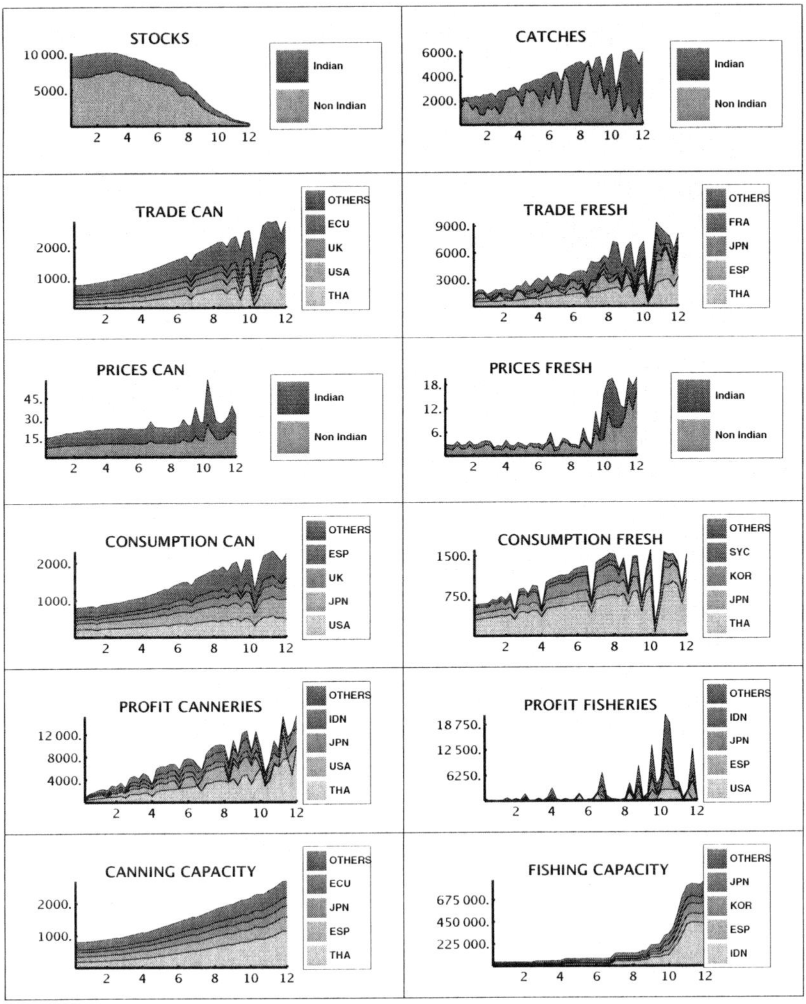

Figure 14.6: Increase of demand. Resulting series.

14.5.3 Stop Fishing in Indian Ocean

In this scenario we explore the closure of a large area (such as the Indian Ocean) to fishing. This is an unrealistic hypothesis, even in the event of the implementation of a global system of offshore protected areas (for example in response to high levels of piracy).

The story we are told is depicted in figure 14.7, in which we can observe a process whose main lines are:

(1) report of effort on the other oceans,
(2) an increase of prices related to a diminishing supply,
(3) the recovering of Indian ocean stock,
(4) an increase in profit of fleets,
(5) an increase in fishing capacity,
(6) the depletion of other stocks related to over-fishing,
(7) the collapse of the system.

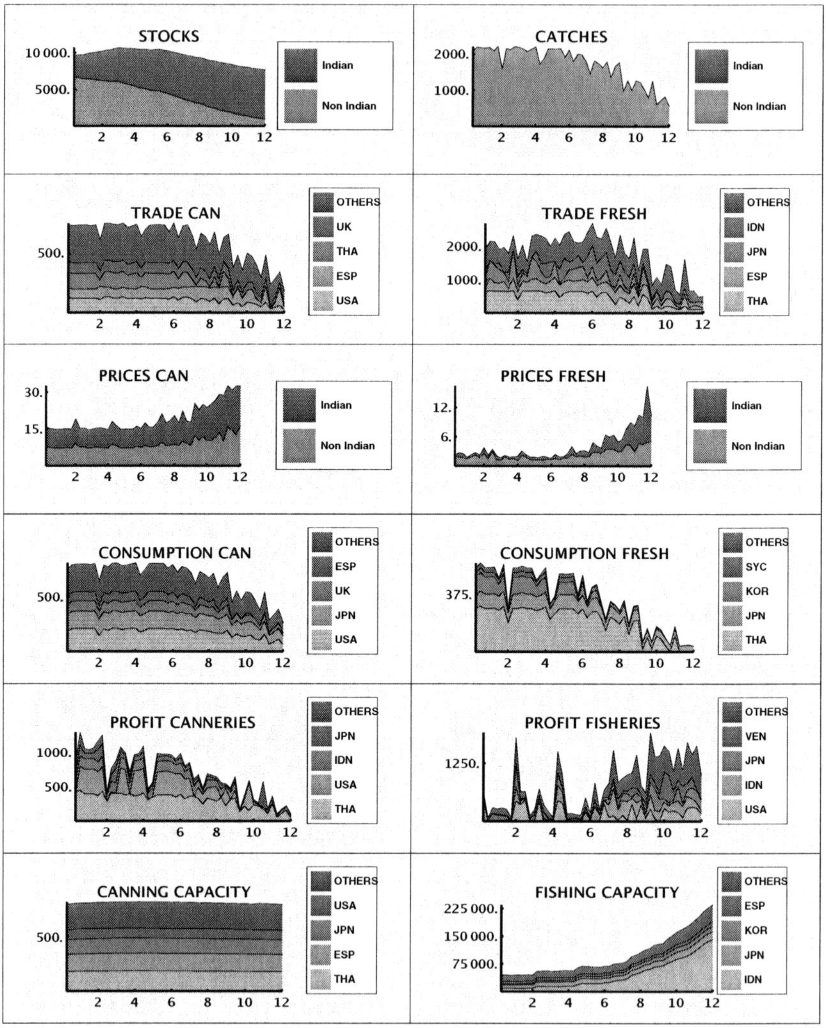

Figure 14.7: Ban on Indian Ocean fishing. Resulting series.

This is consistent with the economic theory of fisheries collapses, except that here the scale is global and the basic entities are national fleets.

Figure 14.7 shows results at the scale of countries, for example the inertia of Thai canning industries. Thailand appears as the central node in the network. Other canning industries in troubled times have more variable profits and their situation resembles that of fleets. The inertia of Thailand in the network may be related to a network property. In the network as it is designed, Thai canning industries are linked to several fleets. Some canned tuna markets are linked only to the Thai canning industry. Other countries having their own canning industries, whose canned tuna markets are supplied by the Thai canning industry.

Let us indicate that the consumption of fresh tuna in Thailand is an modeling artifact. It is due to the fact that we have chose a common transformation coefficient (around 2) for all fleets, and that it is too small for Thailand; by difference, it creates a non consumed volume of fresh fish and its destination is consumption.

14.5.4 Increase of Oil Price

In this scenario we explore the consequences of a significant increase in oil prices affecting both fishing and shipment costs. Technically, it consists of increasing the part of fishing costs due to effort by 20% per year. This can be seen as the exploration of a network process: these changes modify the connectivity of a part of the network.

The story we are told is depicted in figure 14.8, in which we observe a process whose main lines are:

(1) a decrease in effort, at the beginning of period, especially in Indian ocean, where fishing is less beneficial,
(2) the recovering of Indian ocean stock,
(3) a decrease in global supply,
(4) a resulting increase in profit.

As expected, an increase in oil price is good for fish stocks. It has a negative effect on consumption of fresh and frozen fish. Surprisingly, it is also good for fishermen but not for canning industries. These results indicate a kind of ordering. Recall that profit is related to shadow prices, and thus to production constraints. The underlying mechanism could be as follows:

(1) because of the increase of oil prices, some fleets decrease their catches;
(2) because of the global decrease of supply, prices for fresh frozen commodities (globalized) increase; consequently, the average global profit of fleets remains high;

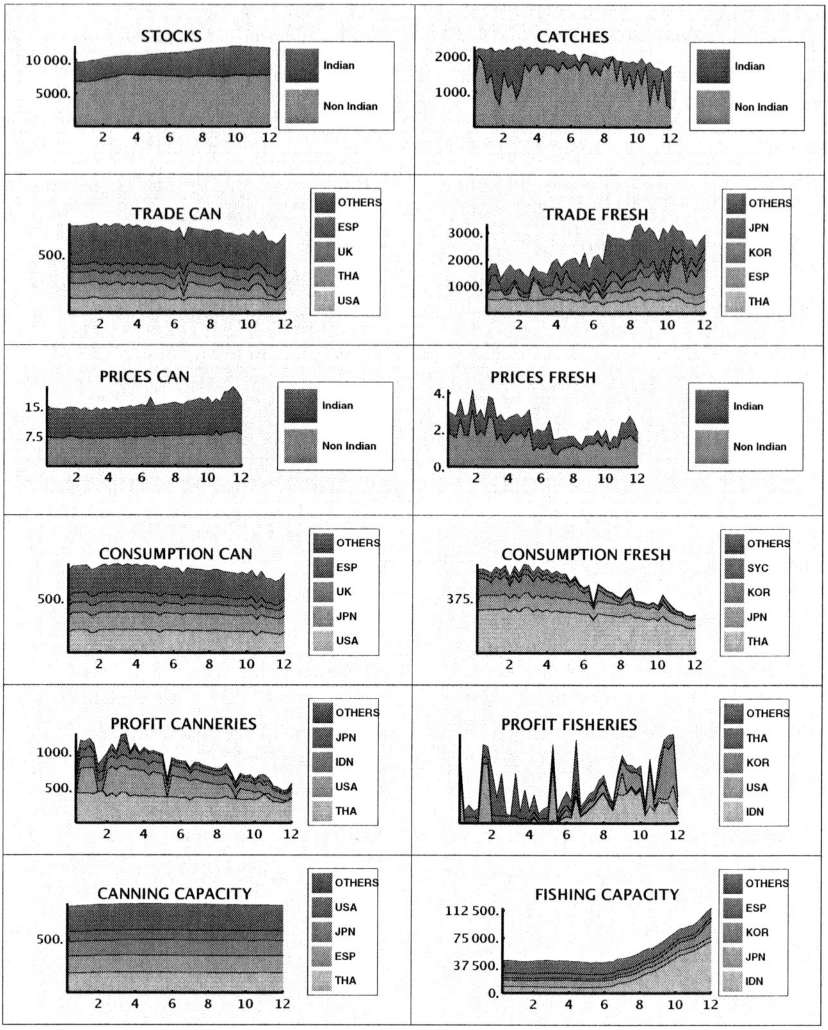

Figure 14.8: Increase of oil price. Resulting series.

(3) because of the transmission of oil prices along the supply chain, prices for canned tuna commodities increase, sales decrease, shadow prices for canning industries decrease, and the average global profit of canning industries decreases.

We note that at the end of simulation, fishing capacity has significantly increased because of myopic investment behavior. This means that there is overcapacity and a crisis should occur.

Of course, this mechanism is due to competition between fleets and canning industries. Although overall the situation is good for fisheries, it is clearly bad for some of them.

14.5.5 Sensitivity to Investment

The previous simulations underline the importance of the assumption of a myopic investment behavior for fleets and canning industries. All simulations showed high variability in the profit of fleets and lower variability in the profit of canning industries. This demonstrates the role of the investment rate in ensuring the flexibility of the system. It seems useful to systematically study the effects of this coefficient on the behavior of the system.

To that purpose, we define a sensitivity parameter $\tau =$ and examine the results of simulation when changing the investment rates of fleets σ_b with $\tau\sigma_b$, the investment rates of canning industries σ_t with $\tau\sigma_t$. We then observe aggregated values of the main variables.

Results are shown in figure 14.9. As expected, the global profit of fleets and canning industries depends on investment rates. Increasing reactivity decreases their profit and increases their wealth. There are few effects on other characteristics of the network.

We highlight two aspects of this sensitivity analysis. (1) It is based on uniform changes in the investment rate. It is obvious that different countries will have different investment rates depending on their level of economic development. (2) The simulation indicates good economic conditions, where demand ensures that there is sufficient profit for investment. We need to go further and study the model using more difficult conditions.

14.6 Concluding Remark

14.6.1 Features

In this chapter we built a network model of the global tuna supply chain. This system has the particularity of ending in two types of commodity: fresh or frozen tuna, and canned tuna. The size of the model is intermediate: it involves several hundreds of nodes and links. The aim was to illustrate several contrasting scenarios. It is a dynamic model: scenarios concern the trajectories of entities in future years. It was calibrated using inverse modeling and recent datasets of catches and trade. The availability of data was an important element in determining the structure of the model (entities, nodes and links).

The dynamics of this system are represented as a succession of steps consisting of an equilibration phase and a deterministic phase. In most cases, the solution of the equilibration phase exists, but is not unique. This is usually a delicate issue when simulating natural systems. Nevertheless, simulations showed that the observed patterns of variability were significant, especially those concerning the variability of profits.

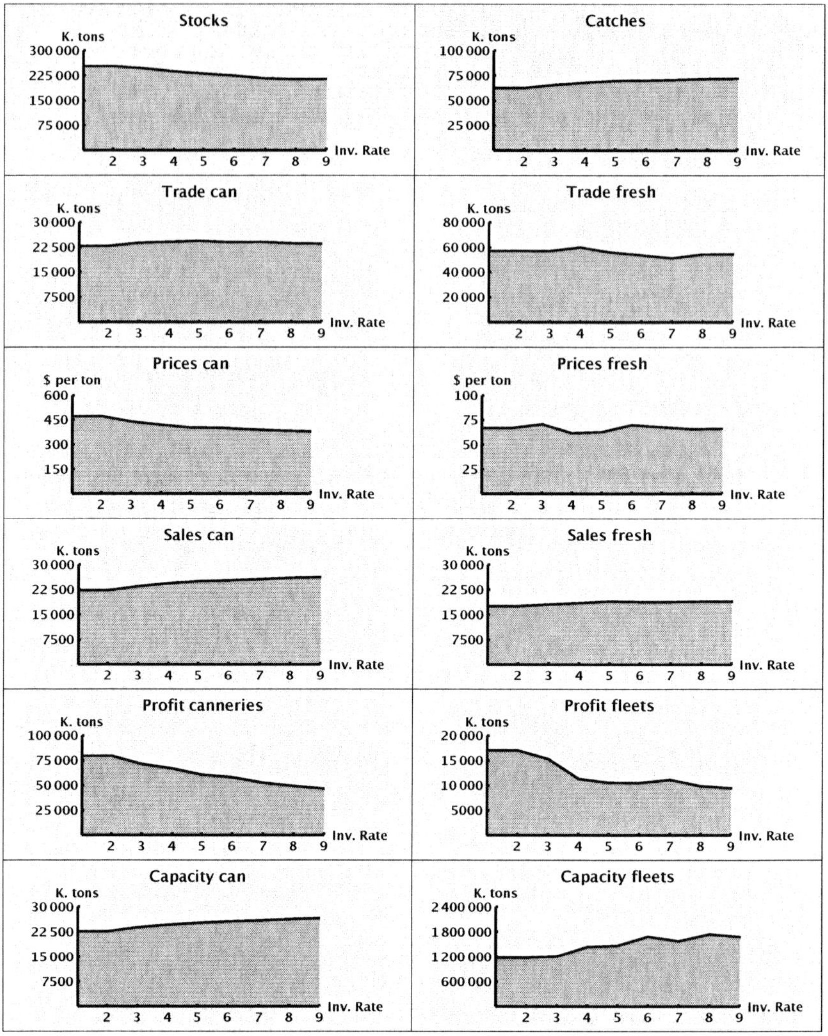

Figure 14.9: Sensitivity to investment rate.

The analysis of contrasting scenarios confirmed some features, and in other cases, identified surprising facts. Examples are:

(1) The profit of fleets as the adjustment variable and how it could be used to interpret the variability of corresponding series.

(2) Confirmation of the role inside the network of key players such as Thailand.

(3) A network effect: Is access to fish stocks too easy? Is the connectivity of the network at this level too high?

(4) Demand as an important factor of change in the commodity chain rather than supply (stock productivity) that is influenced by climate change.

14.6.2 Teachings

The approach has proven to be feasible in terms of mathematical formulation, model calibration and computation. Network entities were defined as a trade-off between: (1) data availability, the structure and contents of existing databases, and (2) the requirement for projected scenarios. This led to the definition of a state vector of dimension 900. The interior point algorithm for solving the associated linear complementarity problem proved to be convergent in a reasonable computation time.

Network economics made it possible to establish a bridge between knowledge of an exploited marine resource in economic (trade) terms and biological (catches) terms. It led to a synthesis of the dynamics resulting from different hypotheses (scenarios). This modeling experiment made it possible to look inside an intermediate-size, bio-economic system. The synthesized dynamics resulting from alternative scenarios (such as those shown in figures 14.6 and 14.7, etc.,) can be used in the perspective of global management. They define a scale and a resolution, make it possible to think in terms of networks, and illustrate network effects such as the importance of connectivity (represented by the opening or closing of links, or changing transportation costs). This network thinking can become an element of global governance as decision-makers are able to take account of the indirect effects of their decisions. The decisions that they take concerning one part of the network may have an impact on something happening in another part.

14.6.3 Going Further

The simulations and sensitivity analysis lead to questions about the choices and hypothesis of the model. They suggest scenarios based on differentiation between entities, e.g. different increases in demand, or different investment rates. They suggest that it is time to carry out new experiments in partnership with other stakeholders involved in global governance. In this work, models are not a magic predictive tool, but only a means for sharing points of view and we have made clear its speculative character.

The assumption that tuna do not migrate would be the first to be relaxed. Following the computations carried out in this first phase, there are reasons to be optimistic about resolving the computing issues this implies. It is more difficult to see how to represent migratory behavior. In a subsequent chapter we present one way to deal with it.

14.7 Annex: Input and Initialization

This section is more technical. Initial parameterization is an important step in any modeling attempt. It is especially important in the context of global studies that combine data from heterogeneous datasets. Here we present a simplification of our method (see Fortilus (2012) for all details). We use network equilibrium principles, together with inverse modeling techniques.

14.7.1 Parameters to Quantify

Recall that parameters that characterize entities and that we have to quantify concern:

(1) fish stocks: $\theta_s, \rho_s, \omega_s$,
(2) fleets: v_b, σ_b, η_b,
(3) canning industries: u_t, σ_t, η_t,
(4) costs (fishing, producing, trading, selling): c_{sb}, c_{bn}, c_{nn}, c_{nt}, c_n, $c_{tt'}, c_t$,
(5) demand (the inverse demand function): a_d, b_d, a_c, b_c.

14.7.2 Stocks

Renewal rates ρ_s are computed from life span of species v according to $\rho_s = 2/v$ given in table 14.4.

Table 14.4: Tuna species. Renewal rates

Species	v (year)	$\widetilde{\rho_s}$
SKJ	3	0.67
YFT	7	0.29
BET	7	0.29
ALB	5	0.4
BFT	20	0.1

For each stock s we have, using FAO reports, an estimate of the health of the stock h_s. It is supposed to represent the ration between catches and carrying capacity: $h_s = \theta_s/\omega_s$.

If we assume biological equilibrium, in other words that catches equal renewal of the stock, we have the following relationships:

$$\sum_b x_{sb} = \rho_s \theta_s (1 - \theta_s/\omega_s)$$

Thus we may assume that the stock is:

$$\theta_s = \sum_b x_{sb}/\rho_s(1 - h_s)$$

We have:

$$\sum_b x_{sb} = \kappa_s \theta_s \sum_p b_{sb}$$

We get an estimate of the carrying capacity:

$$\kappa_s = \sum_b x_{sb} / (\theta_s \sum_b v_{sb})$$

14.7.3 Fleets and Canning Industries

We estimate fishing capacities v_b and canning capacities u_t as the mean production these last years given by FAO Sardara base.

14.7.4 Inverse Modeling

To compute other characteristics of the system (parameters and reference state), we apply the principles of inverse modeling enunciated in section 6.5. We use the complementarity relationship:

$$X \perp \mathbf{M} \cdot X + N$$

We have to express sub-matrices and sub-vectors of $\mathbf{M}$ and N as in formula (as in section 14.3.5). We know:

(1) b_n, thus $\overline{\mathbf{M}_{LL}}$,
(2) the topology of the network (incidence matrices), thus $\overline{\mathbf{M}_{LN}}$ and $\overline{\mathbf{M}_{LN}}$;
(3) q_e, p_e, thus $\overline{\mathbf{M}_{LQ}}$ and $\overline{\mathbf{M}_{QL}}$,
(4) $\overline{X_L}$.

We have to determine:

(1) X_N, prices,
(2) X_Q, shadow prices,
(3) N_L, transport costs,
(4) N_Q, maximum fishing capacity of canning capacity.

We assume that known quantities are representative of an equilibrium state. We have thus:

$$\overline{\mathbf{M}_{LL}} \cdot \overline{x_L} + \overline{\mathbf{M}_{LN}} \cdot x_N + \overline{\mathbf{M}_{LL}} \cdot x_Q + N_L \perp \overline{x_L}$$
$$\overline{\mathbf{M}_{NL}} \cdot \overline{x_L} \perp x_N$$
$$\overline{\mathbf{M}_{QL}} \cdot \overline{x_L} + N_Q \perp x_Q$$

We assume that $\overline{x_L} > 0$ and we look for solutions such that $x_N > 0$ and $x_Q > 0$. This means that $\overline{\mathbf{M}_{NL}} \cdot \overline{x_L} = 0$. We obtain:

$$\overline{\mathbf{M}_{LL}} \cdot \overline{x_L} + \overline{\mathbf{M}_{LN}} \cdot x_N + \overline{\mathbf{M}_{LL}} \cdot x_Q + N_L = 0$$
$$\overline{\mathbf{M}_{QL}} \cdot \overline{x_L} + N_Q = 0$$

Thus N_Q can be computed as $N_Q = -\overline{\mathbf{M}_{QL}} \cdot \overline{x_L}$. In the first equation, there are n_L equations. Recall that intercepts of inverse price functions a_m are known and that we have:

$$\begin{aligned} N_l &= c_l + c_{o(l)} && \textit{for a link from a fishing area} \\ N_l &= c_l - a_{d(l)} && \textit{for a link to a market} \\ N_l &= c_l && \textit{for other links} \end{aligned}$$

There are:

(1) n_N unknowns related to prices: x_N,
(2) n_Q unknowns related to shadow prices x_Q,
(3) n_L unknowns related to transport costs c_L,
(4) n_Z unknowns related to fishing costs c_z.

There are thus more unknowns than equations. But these unknowns must be positive. We express N_L as a linear function $F(c_L, c_N)$. We put

$$S(x_N, x_Q, c_L, c_N) = \overline{\mathbf{M}_{LL}} \cdot \overline{x_L} + \overline{\mathbf{M}_{LN}} \cdot x_N + \overline{\mathbf{M}_{LL}} \cdot x_Q + F(c_L, c_N)$$

We have two alternatives.

(1) in general, solve the quadratic program: Find the values of parameters (x_N, x_Q, c_L, c_N) that minimize:

$$(x_N, x_Q, c_L, c_N) \cdot (x_N, x_Q, c_L, c_N)$$

subject to:

$$(x_N, x_Q, c_L, c_N) \geq 0 // S(x_N, x_Q, c_L, c_N) = 0$$

(2) when the equation $S(x_N, x_Q, c_L, c_N) = 0$ is too drastic, we solve instead the quadratic program: Find the values of parameters (x_N, x_Q, c_L, c_N) that minimize:

$$\begin{aligned} &(x_N, x_Q, c_L, c_N) \cdot (x_N, x_Q, c_L, c_N) \\ &+ S(x_N, x_Q, c_L, c_N) \cdot S(x_N, x_Q, c_L, c_N) \end{aligned}$$

subject to:

$$(x_N, x_Q, c_L, c_N) \geq 0$$

We chose the second term for the modeling experiment as it appears to give more robust results.

14.7.5 Fleets and Canning Industries

For a fleet b, according to previous estimates, we know the fishing capacity v_b, and profit i_b. Then we select values of investment rate i_b, depreciation rate ρ_b, price of one unit of fishing capacity π_b that verify the balance:

$$i_b r_b = \rho_b v_b \pi_b$$

This is the same for a cannery: we know canning capacity and U_t and profit i_t; we select values of parameters, investment rate i_t, depreciation rate ρ_t, price of one unit of fishing capacity π_t that verify the balance:

$$i_t r_t = \rho_t v_t \pi_t$$

Chapter 15

Modeling the Worldwide Fish Supply Chain

15.1 Overview

15.1.1 Questioning

The importance of a global approach to fisheries has been underlined (Pauly et al., 2002). This is the aim of *The state of World fisheries*, published by FAO (FAO, 2012) and forms the program of the *Sea Around Us* project (Pauly, 2007).

There is a need to create scenarios for future global fisheries systems (Pauly et al., 2003; Pereira et al., 2010; Merino et al., 2010). Such scenarios should help to formulate future global governance and answer questions such as: What is at stake? Who are the stakeholders? How can they be encouraged to share their points of view?

15.1.2 Modeling

One way to create scenarios is to build a model, although it is not the only one (Van Notten et al., 2003).

In this chapter we present a large-scale, low resolution model of the worldwide supply chain for fish. It represents oceanic primary production, the dynamics of fish stocks, fisheries, aquaculture and international trade. It therefore integrates ecological and economic processes. It aims to provide a general perspective on large global commodity chains, demonstrating what is at stake for the global governance of marine systems.

This network economics model (despite huge simplifications) is based on an intermediate-size network structure. There are several hundreds of nodes and links. We show how the network equilibrium of the system depends on the values of parameters associated with scenarios. In this short presentation, we analyze the results of four very different scenarios:

(1) changes of fishing costs,

(2) changes of demand for all fish,

(3) changes of productivity (due to climate change),

(4) increase of trade costs (reversing globalization).

Building a model implies making choices about processes and entities (for example). We first had to decide if it was even worthwhile to try to build a model that could be considered relevant from a global management perspective. We were aware it was a risky decision and this incertitude formed the backdrop for our efforts.

15.2 Design

15.2.1 Goals

Our aim was to analyze how network economics contribute to the identification of the stakes in the development of a worldwide system for fisheries and aquaculture.

15.2.2 Rationale

The model was explicitly designed in order to provide sensitivity analysis of the main factors of change (climate change or economic globalization) impacting the global supply chain for fish. It also sought to establish the effects of the network structure on this complex system.

The principles behind the sensitivity analyses were:

(1) Define a reference network that corresponds to the present observed state of production and exchanges,

(2) Assess parameters values making it a stationary state,

(3) Change the values of these parameters related to global changes,

(4) Observe the resulting network equilibrium.

15.2.3 Choices

There are many important choices that must be made in order to carry out such a project. In particular:

(1) Entities. The range of possibilities is huge. There are hundreds of possible species, tens of fishing areas, tens of fishing pavilions, thousands of fish commodities, and thousands of trading links between counties. There cannot be too many entities for computational reasons. However, there must not be too few entities if we are to be able to use the model in a governance perspective, which implies several categories of actors interested in the representation of their own function in the system.

(2) Processes. What kind of processes are represented? At which level of detail? For example, do we take into account the differences between fleets? Between fishing techniques?

(3) Time. How is time represented in the model?

(4) The organization of the model. What is an input? What is an output? These choices have important consequences in terms of governance; they implicitly define the functions of actors and the means of governing the system.

15.2.4 Network Structure

The structure of the network is given in figure 15.1. It represents the following nodes: primary production areas, fish stocks, fleets, and commodity and consumption markets, and the following flows: paths from a primary production area to a fish species (grazing), paths between fish species (predation), paths from a fish species to a fleet (catches), paths from a fleet to a commodity market (transformation), paths from a fish farm to a commodity market, paths from a fishmeal or fish oil market to a fish farm, paths between commodity markets (trade), and paths from a commodity market to a consumption market (supply).

The bottom of figure 15.1 shows at the level of a group of countries, the characteristics of nodes (prices) and links (flows) that are the subject of our approach.

15.2.5 A Static Model

Unlike the analysis of the global tuna commodity chain, sensitivity analyzes are performed according to a resulting network equilibrium, and not through a dynamic process composed of a succession of equilibration and evolution steps. In previous chapter, we defined fish production and investment functions. These relate the status of the fish stock or fishing capacity to

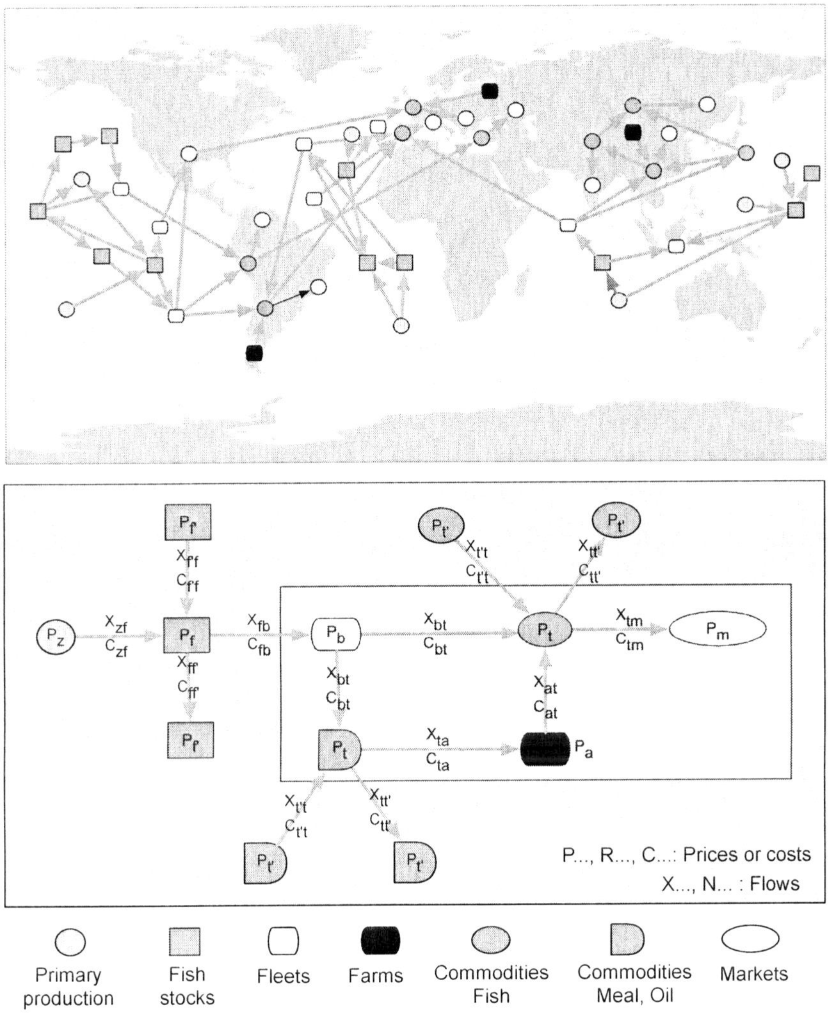

Figure 15.1: The supply chain associated to the worldwide supply chain for fish. Geographical and functional representations of the network. At the top, we emphasize the geographical structure of the network. Top: emphasizes the geographical structure of the network. Bottom: the functional structure for a geographical entity (set of countries).

previous years, fish catches or fleets profit that occurred previous years. We then examined, for each year, the resulting dynamics of: (1) the economic network equilibrium; and (2) the application of these functions. In this chapter, we compute the equilibrium between biological factors (relationships between production and fish biomass through the idea of accessibility) and economic factors (relationships between catches, stock abundance and demand for fish). Once the values of input parameters associated with the sensitivity analysis have been set, we compute the resulting equilibrium, considered to be the state towards which the system should tend.

15.2.6 Assumptions

Assumptions resulting of the preliminary analysis are given in box 15.1.

Entities	Plants, Fish stocks, Fleets, Fish transformers and traders, Fish markets
Assumptions	Assimilation in fish stocks is proportional to inflow Losses in fish stocks are proportional to biomasses A mass balance principle is operating in fish stocks, whence a relationship between biomass and trophic flows There is inverse demand function on fish markets A part of fishing cost inversely proportional to fish abundance Grazing costs are positively related to grazer abundance and negatively to plant abundance Predation costs are positively related to predator abundance and negatively to prey abundance Other costs are constant Profit of fleets are related to limitations of fishing capacity and the subsequent shadow prices There is a network equilibrium
Observations	The sensitivity of the resulting network structure to global coefficients such changes in primary production, fishing costs, catch ability, productivity
Calibration	Inverse modeling using databases Fishstat and Comtrade

Box 15.1: Design of the model of the world fish supply chain

15.2.7 Approach

Like the attempt to model the global tuna supply chain, this attempt was carried out in four phases:

(1) Analyze datasets, define the network structure, and make scenarios explicit. Solving the equilibrium must be feasible. The resulting network structure must be illustrative. The necessary data must be available. Solving the equilibrium must make it possible to analyze scenarios.

(2) Using an inverse modeling approach, obtain values for necessary parameters.

(3) Implement the model, the equilibrium and coupling algorithms, and post-processing (graphics and statistics) programs.

(4) Run and analyze scenarios.

15.3 Details

15.3.1 Basic Entities

Based on constraints related to the existence of data, the objective of building scenarios, and limits in computing, the following network entities were selected.

The basic network entities are given in table 15.4. There are several different types: geographic areas, plankton colonies, groups of fish species, groups of countries, and groups of fish commodities. They have been selected taking account the different nomenclature used in databases, the groupings they enable and the need to keep the network to a reasonable size.

Table 15.1: Basic entities of the world fisheries model. Geographical areas correspond to FAO oceanic areas. Groups of fish species and fish commodities are issue from to ISCAAP classification

Geogra-phical areas	Pacific, Southeast, Pacific, Northwest, Atlantic, Northeast, Pacific, Western Central, Indian Ocean, Western, Atlantic, Eastern Central, Pacific, Northeast, Indian Ocean, Eastern, Pacific, Eastern Central, Atlantic, Southeast, Atlantic, Southwest, Mediterranean and Black Sea
Production	One generic plankton species
Grouped species	Herrings, sardines, anchovies, Miscellaneous pelagic fishes, Cods, hakes, haddocks, Miscellaneous coastal fishes, Tunas, bonitos, billfishes, Miscellaneous demersal fishes, Flounders, halibuts, soles, Salmons, trouts, smelts.
Grouped countries	Eastern Asia, Northern Europe, South America, Northern America, South-Eastern Asia, Eastern Europe, Southern Europe, Western Europe, Southern Africa, Central America, Southern Asia, Northern Africa, Australia and New Zealand.
Grouped commodities	Marine fishes not identified (Frozen), Tunas, bonitos, billfishes (Frozen), Cods, hakes, haddocks (Frozen), Miscellaneous pelagic fishes (Frozen), Herrings, sardines, anchovies (Frozen), Salmons, trouts, smelts (Frozen), Herrings, sardines, anchovies (Meal/Oil), Miscellaneous coastal fishes (Frozen), Flounders, halibuts, soles (Frozen), Miscellaneous demersal fishes (Frozen), Miscellaneous freshwater fishes (Frozen), Miscellaneous pelagic fishes (Meal/Oil), Sharks, rays, chimaeras (Frozen).

15.3.2 Nodes

Nodes consist in pairs of above basic entities. They are: plankton stocks, fish stocks, fleets, farms, trade markets, consumption markets. Their types are given in table 15.2.

Table 15.2: Nodes of the world fisheries model. Columns correspond to the type of node, the corresponding basic entity, the number of nodes of that type, notation and characteristics

Node type	Pairs	#		Characteristics
Production area	FAO area × autotroph species	19	z	biomass b_z
Fish stocks	FAO area × grouped species	52	f	biomass b_f
Fleets	grouped countries × grouped species	50	b	departure price p_b
Farming systems	grouped countries × grouped species	30	a	departure price p_a
Trade markets	grouped countries × grouped commodities	95	t	trade price p_t
Consumption markets	grouped countries × grouped commodities	84	m	consumption price p_m

15.3.3 Links

Links concern grazing, predation, farming, catches, trade, transformation, and sales. They relate two nodes that meet coherency constraints such as: grazing is only possible when there are primary production and fish stocks in the same area; or fishing flow from a stock (area, fish species) to a fleet (a group of pavilions, targeted fish species) is only possible if the species are the same. These are given in table 15.3. Compatibility conditions between the origin and end of link are as follows. Grazing occurs within the same area. Predation occurs within the same area and follows biological knowledge of who eats whom. Fishing concerns the same group of species. Transformation of wild fish occurs within the same group of countries; species and commodities are compatible. Aquaculture consumption of fish meal and oil occurs within the same group of countries. Aquaculture production occurs within the same group of countries; species and commodities are compatible. Trade concerns the same group of commodities. Sales occur within the same group of countries, and concern the same commodity.

15.3.4 Notations and Basic Equations

Here we define how prices and costs are attributed to network nodes and links.

Autotrophic and fish stocks. As usual, trophic flows are assumed to determine the biomass of autotrophic stocks according to:

$$b_z = (\gamma_z e_z - \sum_f x_{zf})/\mu_z$$

In this equation γ_z is a trophic efficiency coefficient, μ_z a trophic maintenance coefficient and e_z corresponds to energy input. In the same way, trophic flows

Table 15.3: Links of the world fisheries model. Columns correspond to: the type of link, the number of links of that type, the direction, their origin, their endpoint, flow notation, transportation cost notation and characteristics

Link	#		Origin × End	*Fl.*	*Co.*
Grazing	45	$z \to f$	production area × fish stock	x_{zf}	c_{zf}
Predation	92	$f \to f'$	fish stock × fish stock	$x_{ff'}$	$c_{ff'}$
Fishing	86	$f \to b$	fish stock × fleet	x_{fb}	c_{fb}
Transformation of wild fish	101	$b \to t$	fleet × trade market	x_{bt}	c_{bt}
Meal, oil aquaculture consumption	13	$t \to a$	trade market × farms	x_{ta}	c_{ta}
Aquaculture production	41	$a \to t$	farms × trade market	x_{at}	c_{at}
Trade	191	$t \to t'$	trade market × trade market	$x_{tt'}$	$c_{tt'}$
Sales	84	$t \to m$	trade market × market	x_{tm}	c_{tm}

and catches determine biomass of fish stocks according to:

$$b_f = (\gamma_f(\sum_z x_{zf} + \sum_{f'} x_{f'f}) - (\sum_{f'} x_{ff'} + \sum_p x_{fb}))/\mu_f$$

Grazing. As in section 7.5, costs of grazing have a constant part and a part proportional to grazer biomass; they are given by $d_{zf} + s_f b_f$, s_z being an intra specific competition coefficient. Advantage of grazing is proportional to plankton biomass; it is denoted $s_z b_z$. In the same way, costs of predation of a fish species f' on another species f are given by $d_{ff'} + s_{f'} b_{f'}$. The benefits of predation are denoted by $s_f b_f$.

Fishing. Fishing costs c_{fb} are given by $c_{fb} = u_{fb} + e_{fb}/b_f$ with a constant part u_{fb} and a part that is inversely proportional to biomass: e_{fb}/b_f.

Transformation. Transformation rate η_t from fish to fishmeal or fish oil commodities t is set to 5; for other commodities, η_t is set to 1. There are two types of fish farms, those that grow herbivore fish and others that grow carnivore fish. The first type has a limited production capacity v_a. The production of the second type is limited by their meal and oil consumption. The transformation rate from fishmeal to farmed fish is denoted θ_a. We have $\theta_a = 2$ for carnivores' farms, $\theta_a = 0$ for other farms.

Markets. On a consumption market m, price p_m depends on traded quantity x_m according to a linear inverse demand function:

$$p_m = a_m - b_m x_{tm}$$

15.3.5 Equilibrium Equations

Changes in the global fish supply chain are examined using the network economics approach to supply chains seen in section 8.4. Recall that in this framework a node is characterized by a price, and a link is characterized by a cost. Complementarity equations express: (1) there is price on a node if there is no excess of inflow or outflow, and (2) there is a flow on a path if origin price plus transportation costs are not lower than destination price. Complementarity relationships are given in table 15.4.

Table 15.4: Complementarity equations. There is a complementarity relationship between quantities in the second and the third columns: both are positive, and only one is non-negative.

Node, link	Flow, cost	Function	$\perp$
z	p_z	$f_z = b_z$	$p_z \perp f_z$
f	p_f	$f_f = b_f$	$p_f \perp f_f$
b	p_b	$f_b = \sum_f x_{fb} - \sum_t \eta_t x_{bt}$	$p_b \perp f_b$
a	p_a	$f_a = v_a + \theta_a \sum_t x_{ta} - \sum_t x_{at}$	$p_a \perp f_a$
t	p_t	$f_t = \sum_b x_{bt} + \sum_b x_{at} + \sum_{t'} x_{t't}$ $- \sum_a x_{ta} - \sum_{t'} x_{tt'} - \sum_m x_{tm}$	$p_t \perp f_t$
$z \to f$	x_{zf}	$f_{zf} = d_{zf} + s_f b_f - s_z b_z$	$x_{zf} \perp f_{zf}$
$f \to f'$	$x_{ff'}$	$f_{ff'} = d_{ff'} + s_{f'} b_{f'} - s_f b_f$	$x_{ff'} \perp f_{ff'}$
$f \to b$	x_{fb}	$f_{fb} = e_{fb}/b_f + y_{fb} - p_b$, or $f_{fb} = e_{fb} + b_f(y_{fb} - p_b)$	$x_{fb} \perp f_{fb}$
$b \to t$	x_{bt}	$f_{bt} = \eta_t(p_b + c_{bt}) - p_t$	$x_{bt} \perp f_{bt}$
$t \to a$	x_{ta}	$f_{ta} = p_t + c_{ta} - \theta_a p_a$	$x_{ta} \perp f_{ta}$
$a \to t$	x_{at}	$f_{at} = p_a + c_{at} - p_t$	$x_{at} \perp f_{at}$
$t \to t'$	$x_{tt'}$	$f_{tt'} = p_t + c_{tt'} - p_{t'}$	$x_{tt'} \perp f_{tt'}$
$t \to m$	x_{tm}	$f_{tm} = p_t + c_{tm} - p_m$	$x_{tm} \perp f_{tm}$

The state of the system is described by the vector:

$$X = (p_z, p_f, p_t, p_a, p_b, x_{zf}, x_{ff'}, x_{fb}, x_{bt}, x_{ta}, x_{at}, x_{tt'}, x_{tm})$$

The functional is $X \to F(X)$ where:

$$F(X) = (f_z, f_f, f_t, f_a, f_b, f_{zf}, f_{ff'}, f_{fb}, f_{bt}, f_{ta}, f_{at}, f_{tt'}, f_{tm})$$

Then X^* is a vector equilibrium if and only if it is a solution of $\mathcal{CP}(F)$.

15.3.6 Equilibrium Computation

We use the interior point algorithm for the general complementarity problem given in section 6.2. This is a highly dimensional problem: vector dimension is 980. Non-linearity comes from the expression of fishing costs as inversely proportional to biomass, and the fact that they themselves depend linearly on flows that are components of the state vector. It has been the case that, for all the sensitivity analyzes we have performed, the algorithm (once suitably parameterized) has converged to equilibrium. Although the computer code has not been optimized, and computation can take a while, it is reasonable and several sensitivity analyzes can be run in a day.

15.3.7 Input

A set of parameters are used for sensitivity analyzes. All these parameters have an effect on competition between predators or competition between fleets. They concern: (1) climate, (2) biological production, (3) fishing, (4) trading, (5) demand and the relationships between them. Later we will focus on the effects of warming on plankton and fish productivity, and the effects of circulation changes on accessibility or catchability. The model is run with different values e_z, or γ_f and we observe how the other characteristics of the system are affected. We are also interested in the effects of the demand for fish on fishing, shipment and capital costs, and on other system characteristics. The model is then run with different values for e_{fb}, or a_m.

15.3.8 Output

In all sensitivity analyzes, we observe the spatial structure of the network according to different values of a related coefficient and compare global values of stocks, catches, trade, sales and prices.

15.4 Results

15.4.1 Reference State

After data structuring, calibration, and equilibration with the appropriate parameters, we obtain a reference state for the system. This is mapped in figure 15.2. Overall, dataset flows and flows after equilibration are very similar, although there are differences for some components, such as those related to fishmeal or fish oil trade. This indicates the need to improve the calibration step.

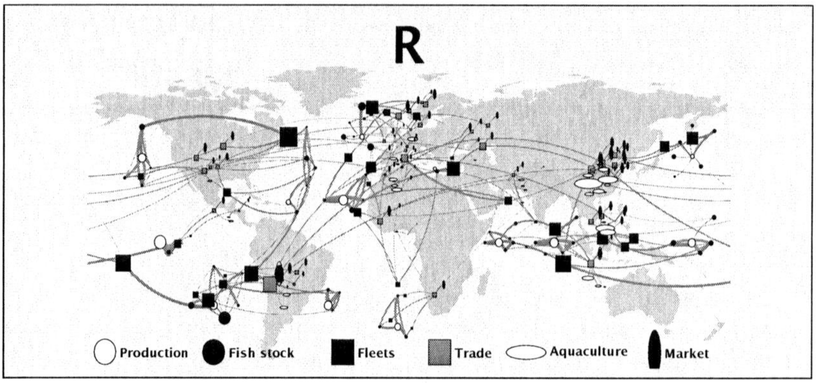

Figure 15.2: The system of world fisheries. A map of the network corresponding to the reference state. In this figure, as well as in the following ones, geographic positions are approximate. This is due to the necessities of a cartographic representation.

This map makes it possible to identify several geographical centers in world fish production, trade and sales. These are Eastern and Southern Asia, Europe and North Africa, and South America.

15.4.2 Sensitivity Analyses

Sensitivity analyzes make it possible to relate the effects of systematically changing given characteristics of network entities (nodes or links) on its equilibrium states. An equilibrium state can be considered as a stationary state that the network would reach in an indefinite time, all else being equal. Recall that it is characterized by the existence of flows according to the value of prices, and the existence of prices according to the value of flows, as shown in section 8.4.

15.4.3 Sensitivity to Fishing Costs

In this sensitivity analysis, the value of the fishing costs parameters e_{fb} are multiplied successively by 0.5, 0.75, 1, 2, 5, 10.

Figure 15.3 maps the network structure resulting from two contrasting scenarios. In the first case, the network resembles the reference state. In the second case, as expected, most of the fishing links disappear. Those that remain concern trade in fishmeal and fish oil from small pelagic fisheries to aquaculture systems, and sales of farmed fish.

The results of a sensitivity analysis shown in figure 15.4 confirm this analysis. There is a threshold at which aquaculture production suddenly replaces fisheries in order to satisfy demand. For values below this threshold, there are no drastic changes in catches, only an increase of biomass stocks

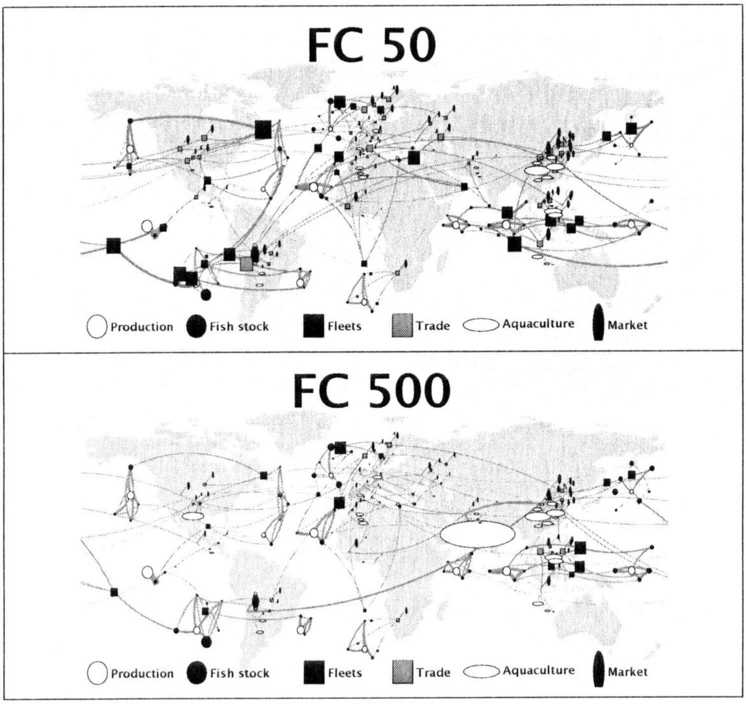

Figure 15.3: The system of world fisheries: sensitivity to fishing costs. Maps correspond to extreme scenarios. Top: a small decrease ($\times 0.5$) of fishing costs. Bottom: a high increase ($\times 5$).

(except in the South East Pacific, which supplies the aquaculture industry with forage fish).

15.4.4 Sensitivity to Demand

In this second sensitivity analysis, the value of the demand parameters a_c and a_d are multiplied successively by 0.5, 0.7, 1.1, 1.5, 1.8, 2.

We highlight again that we have seen convergence problems when demand is high (high values of above coefficients). It appears that the structure of the system is not able to support the stress of a high demand.

Figure 15.5 maps the network structures corresponding to a decrease and an increase in demand. Figure 15.6 shows the results of the sensitivity analysis.

These two figures show how dangerous an increase in demand is for fish stocks. They suggest that the collapse of world fisheries is not far away. The model as it is parameterized here suggests that fishing continues until fish stocks are exhausted, then aquaculture systems are developed to supply markets.

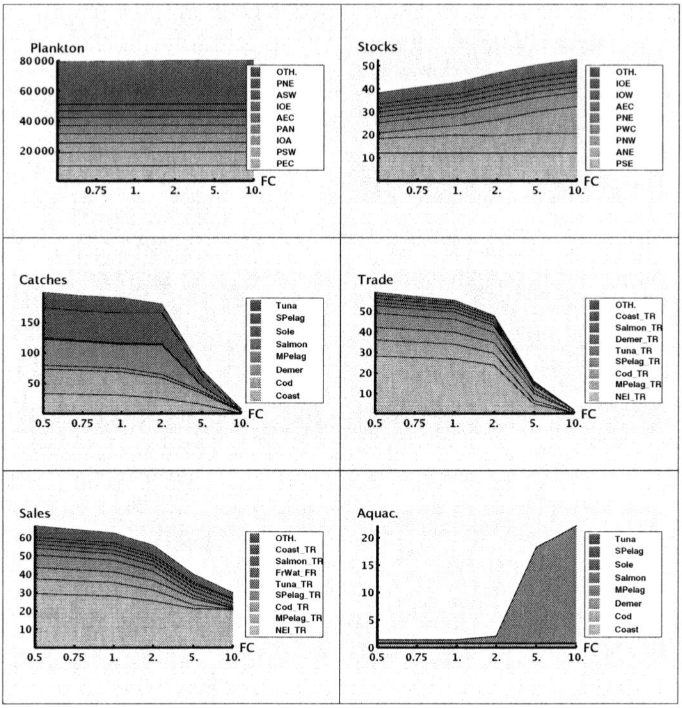

Figure 15.4: The system of world fisheries: sensitivity to fishing costs.

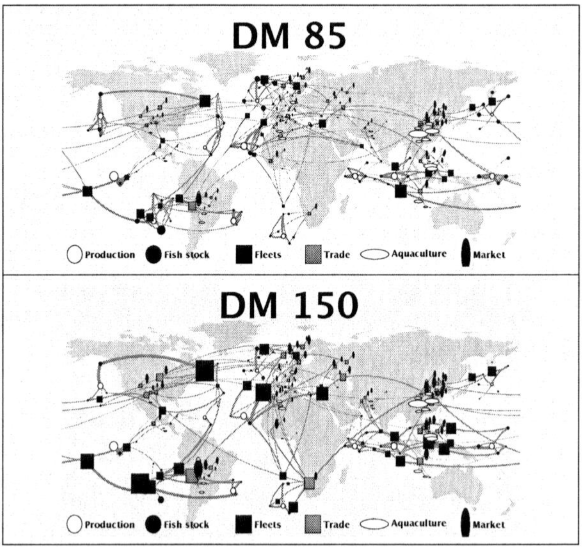

Figure 15.5: The system of world fisheries: sensitivity to demand. Maps correspond to extreme scenarios. Top: a low demand. Bottom: a high demand.

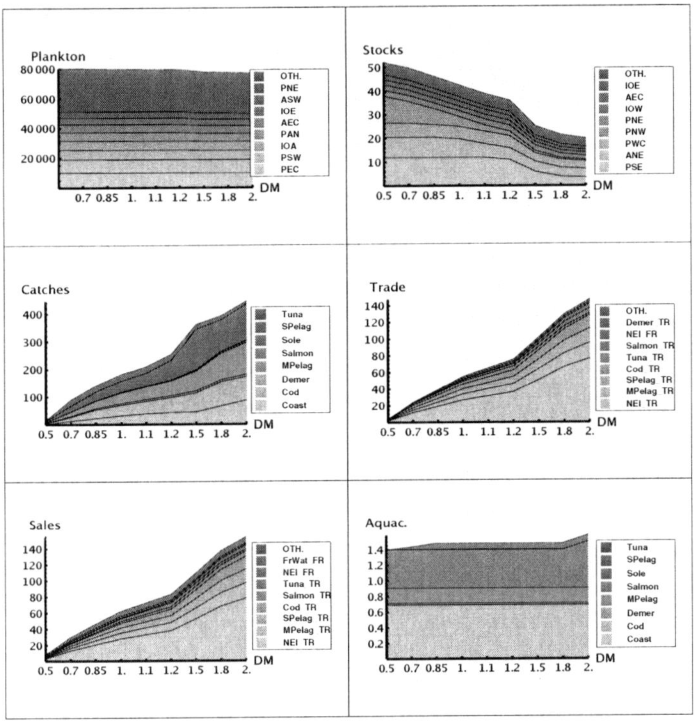

Figure 15.6: The system of world fisheries: sensitivity to demand.

15.4.5 Sensitivity to Primary Production

This sensitivity analysis represents the exposure of the global fish supply chain to climate change. The values of energetic input parameters e_z have been successively multiplied by 0.5, 1, 2, 4. Results are shown in figures 15.7 and 15.8.

Figure 15.7 shows that the connectivity of the network increases with primary production. A low level of primary production favors the development of aquaculture. If in a given oceanic area initial fish stocks are not replaced by more opportunistic ones (as is the case in this analysis) the trophic efficiency of the system is saturated. At high levels of primary production the ocean turns green as it is full of plankton, with no corresponding increase in fish stocks and prices. This suggests a further analysis of what happens with the arrival of new species in an oceanic area.

15.4.6 Sensitivity to Trade Costs

This sensitivity analysis represents the exposure of the global fish supply chain to economic globalization. The values of trade cost parameters $c_{tt'}$ have been successively multiplied by 0.5, 1, 2, 4. Results are in figures 15.9 and 15.10.

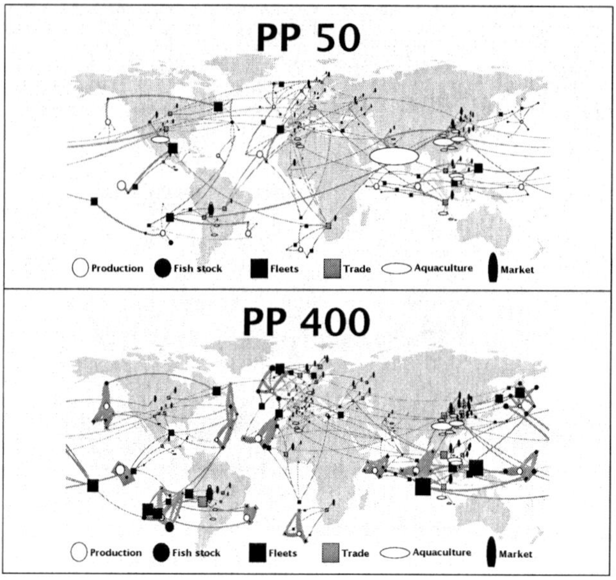

Figure 15.7: The system of world fisheries: sensitivity to primary production. Maps correspond to extreme scenarios. Top: a low primary production. Bottom: a high primary production.

In the model as it is parameterized, trade costs have the expected impact on the network structure. Although connectivity decreases with trade costs (figure 15.9) they do not have a significant impact on global system characteristics. This may be due to how the model is parameterized. It is possible that trade costs are underestimated.

There is a slight negative effect on fish stocks that is not easy to explain. This could indicate that, at equilibrium, local consumption first pushes the system towards more profitable fleets, then towards higher levels of exploitation and lower fish stocks. This point should be analyzed further.

Here again, we underline the fact that in this sensitivity analysis, we only consider uniform changes in trade costs, although local changes in trade costs (e.g. through customs unions) is an important issue.

15.5 Concluding Remark

15.5.1 Features

In this chapter we have attempted to model the system of world fisheries. We developed a bio-economic model that includes both biological (primary production, predation) and economic processes (transformation, trading, sales). The following is a summary of its main features.

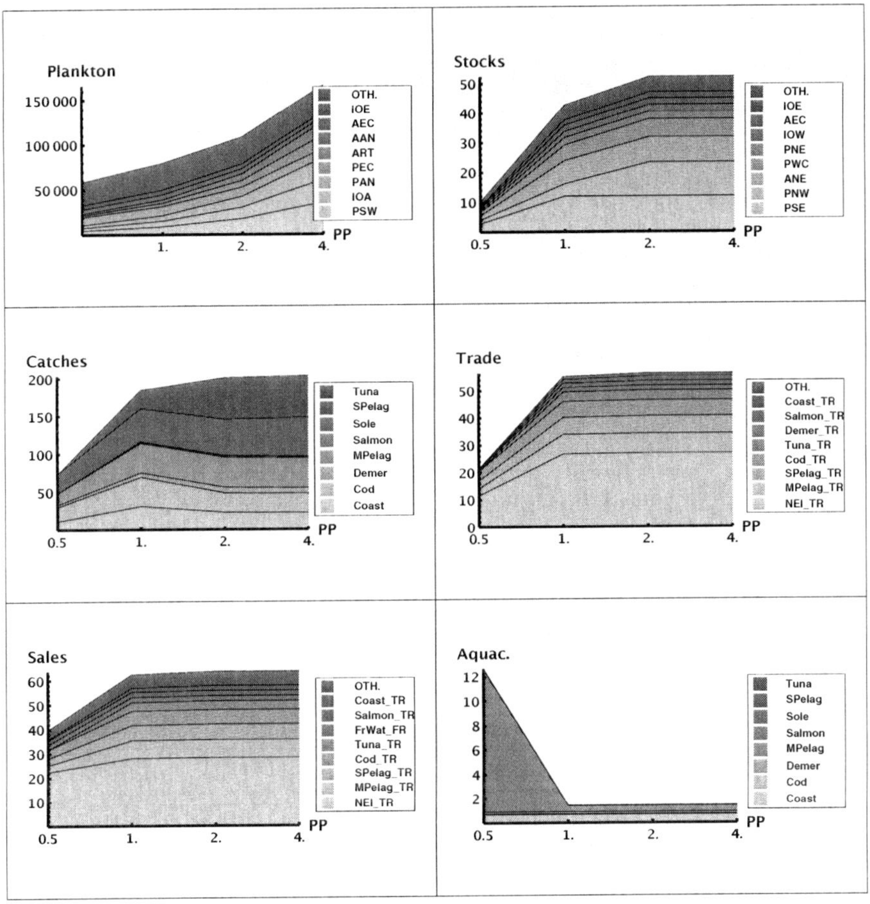

Figure 15.8: The system of world fisheries: sensitivity to primary production.

(1) It is designed using the framework of network economics with the objective of representing a large-scale system subject to double exposure.

(2) It is fairly realistic. The network structure, entities, and processes are defined according to existing datasets.

(3) It has an intermediate level of complexity; it is a based on a network with hundreds of nodes and links.

(4) It has been designed in order to provide graphical representations that illustrate the network effects generated by the system;

(5) It has been designed to enable sensitivity analyzes and the development of scenarios.

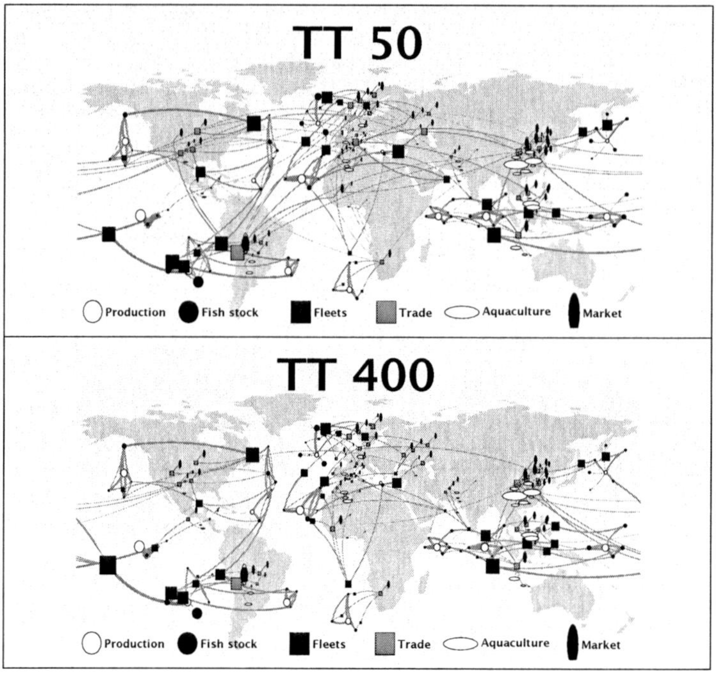

Figure 15.9: The system of world fisheries: sensitivity to trade costs. Maps correspond to extreme scenarios. Top: low trade costs. Bottom: high trade costs.

The functioning of the model has been investigated in many other sensitivity analyzes that are in the process of publication.

15.5.2 Teachings

The approach has proven to be feasible in terms of mathematical formulation, calibration and computation. The design of this model using the framework of network economics has required the definition of a state vector (made up of flows, prices, and shadow prices), and the definition of the complementarity function. Although the mathematical formulation is long (because of the different types of nodes and links) it is not complicated: all formulae are as simple as possible. The computer implementation of the model required the development and testing of an algorithm to solve a non-linear complementarity problem of intermediate size (the dimension of the vector state is around 1000). The implementation of a standard interior point algorithm has provided solutions in a reasonable computing time, even without code optimization.

As a result of our experience, we suggest the following phases in modeling the global supply chain:

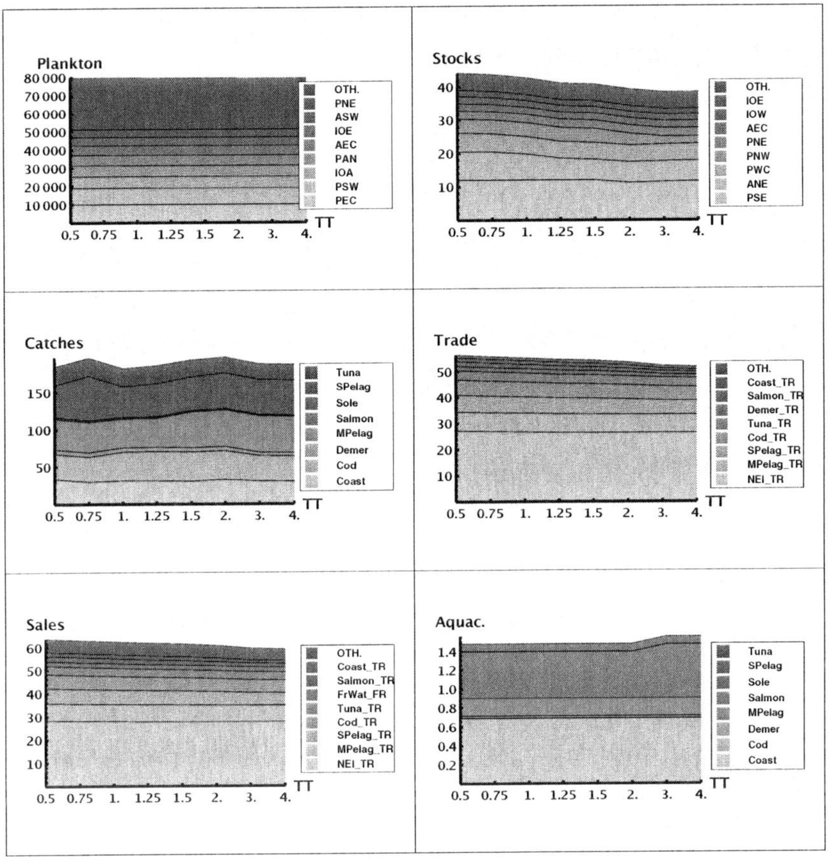

Figure 15.10: The system of world fisheries: sensitivity to trade costs.

(1) Create the structure of the model according to datasets.
(2) Calibrate the model using inverse modeling.
(3) Solve the equilibrium for numerous systems.
(4) Build scenarios.

This approach was taken to modeling the global tuna supply chain. It appears therefore that the methodology can be generalized and applied to new models of marine systems subject to double exposure.

15.5.3 Going Further

This is the time to assess this modeling attempt by communicating its results in an interdisciplinary context and analyzing answers to questions such as:

(1) Are biologists aware of the importance of the economic investment rate from a long-term perspective?

(2) Are economists aware of the importance of the access that fish have to food and that this could be the main effect of climate change?

(3) Have both economists and ecologists noticed the analogy between ecological and economic processes when analyzed in terms of a network?

(4) Are managers and other stakeholders aware of the arbitrariness in the choice of entities in modeling attempts, particularly when the model is supposed to provide a framework for common governance?

It was difficult for us, not only to obtain a graphical representation of a network with a high number of entities, but also to share our results in a management context. Our work to date must be considered as only a first step.

A further improvement is, of course, to develop less aggregated models that focus on particular parts of the system.

15.6 Annex: Input and Initialization

15.6.1 Datasets

We have defined network entities using the nomenclature of the most voluminous datasets concerning ocean productivity, fisheries, aquaculture and commodity trade. The following statistics were used.

(1) Primary production has been extracted from colour satellite images. We have used the method that is described in Strömberg et al. (2009). Data are issued from Dunne et al. (2012).

(2) Catches, farm production and commodity production have been extracted from the FAO Fishstat database[1].

(3) Trade flows (importer, exported volumes, values) have been extracted from the UN Comtrade database[2].

Fortunately, Comtrade and Fishstat nomenclature are compatible and based on the International Statistical Standard Classification of Aquatic Animals and Plants (ISCAAP) for fish species.

[1]Fishstat is made of a software and datasets achieved and made available by FAO (Food and Agriculture Organization). Datasets concern catches (production), farming (aquaculture) and fish commodities (production, importation, exportation). Data are provided by reporting counties. Fishstat (software and datasets) is available on the web site (FAO, 2013).

[2]Comtrade is the acronym of United Nations Commodity Trade Statistics Database. It groups together statistics about exchanges reported by 170 countries. It contains about 1.7 billion data records for 45 years. Data are available on the web site (Nations, 2013).

15.6.2 Initialization and Parameterization

In this section we show how, from existing datasets concerning catches, prices, and fishing capacities and with a few simple assumptions, we were able to obtain a rough estimate of parameters and an initial state used for all simulations. Parameters that are necessary to run the model are given in table 15.5.

Table 15.5: Parameters of the world fisheries models. Notations and meaning

γ_z	assimilation (autotroph)	γ_f	assimilation (fish species)
e_z	energetic input	μ_z	maintenance (autotroph)
μ_f	maintenance (fish species)	s_z	grazing costs coefficient;
s_f	predation costs coefficient;	d_{zf}	grazing advantage
$d_{ff'}$	predation advantage;	a_m	demand function (intercept);
b_m	demand function (slope);	v_a	farming capacity;
u_{fb}	fishing cost;	e_{fb}	effort cost;
c_{bt}	transformation cost	c_{at}	transformation cost
c_{ta}	transformation cost	$c_{tt'}$	trade cost
c_{tm}	sale cost		

We have proceeded as follows:

Step 1: setting the values of some biological parameters. For autotrophic stocks, for each FAO area, energetic input is computed as follows: the primary production density, pp, expressed in g of $C/m^3/day$, is converted into a secondary production biomass (in million $tons/km^2$), using the formula $sp = pp * 0.16/0.3$ (from Strömberg et al. (2009)); then coefficient e_z is obtained by multiplying sp by the surface of the area; then we use the following values: assimilation efficiency $\gamma_z = 0.1$, somatic maintenance: $\mu_z = 0.1$. Biomass is computed from energetic input obtained from production: $b_z = \gamma_z e_z / \mu_z$; accessibility parameter s_z is set to b_z.

For fish stocks, we used: assimilation efficiency $\gamma_f = 0.1$, somatic maintenance $\mu_f = 0.1$. Biomass is computed as follows. We start from catches x_{fb} from Fishstat; we use l_f the exploitation level of stock (from FAO); then we put $b_f = l_f \sum_b x_{fb}$; accessibility parameter s_f is set to b_f and grazing advantage or predation advantages are set to 1.

Step 2: getting a coherent set of trophic flows. Grazing and predation must be coherent with connectivity matrix given in table 15.7 at the end of this chapter.

From previous step, we know biomasses b_z, b_f and catches x_{fb}. The problem consists in finding the trophic flows x_{zf} and x_{ff} that (1) are non negative, (2) are coherent with above trophic connectivity matrix, and (3) minimize the differences between known values of biomasses and catches and computed values according to previous equations, i.e., the sum:

$$\begin{gathered}\sum_z(b_z - \tfrac{1}{\mu_z}(\gamma_z e_z - \sum_f x_{zf}))^2 \\ + \sum_f(b_f - \tfrac{1}{\mu_f}(\gamma_f(\sum_z x_{zf} + \sum_{f'} x_{f'f}) - (\sum_{f'} x_{ff'} + \sum_p x_{fb})))^2\end{gathered}$$

It is obvious that minimization can be done on a per-area basis.

Step 3: getting a coherent set of transformation flows. Transformation of a fish into a fish commodity must be coherent with the connectivity matrix given in table 15.8.

We know trade flows $x_{tt'}$, catches x_{fb}, farm production x_a. Our problem consists in finding the transformation flows x_{at}, x_{ta}, x_{bt}, x_{tm} such that (1) x_{at}, x_{ta}, x_{bt}, x_{tm} are non negative, (2) x_{bt} and x_{at} are coherent with transformation connectivity matrix, and (3) minimize

$$\begin{gathered}\sum_a(\theta_a x_a - \sum_t x_{ta})^2 \\ + \sum_a(x_a - \sum_t x_{at})^2 \\ + \sum_b(\sum_f x_{fb} - \sum_t \eta_t x_{bt})^2 \\ + \sum_t \left(\sum_b x_{bt} + \sum_a x_{at} + \sum_{t'} x_{t't} - \sum_{t'} x_{tt'} - \sum_a x_{at} - \sum_m x_{tm}\right)^2\end{gathered}$$

Step 4: getting coherent trade prices and costs. We define observed trade prices:

$$r_t = \frac{\sum_{t'} y_{tt'} + \sum_{t'} y_{t't}}{\sum_{t'} x_{tt'} + \sum_{t'} x_{t't}}$$

The high level of aggregation means that groups of fish commodities include high- and low-price commodities, and commodities that are proper to a given group of countries. Datasets include biological incoherencies such as the importation of a species in a given commodity group, despite the fact that the mean price of this commodity group is lower locally. We suggest the following method to reduce this kind of incoherency. Knowing observed trade prices r_t, trade flows $x_{tt'}$, find prices p_t that (1) are such that $x_{tt'} > x_{t't} \Rightarrow p_t < p_{t'}$ and (2) minimize

$$\sum_t (p_t - r_t)^2$$

Other prices are and coefficient are defined in an elementary way: market prices $p_m = 1.5\, p_t$, selling costs $c_{pm} = 0.5 p_t$, inverse demand coefficients $a_m = 2\, p_m$, $b_m = p_m / q_m$.

Table 15.6: Parameters estimates of the world fisheries model. Calibration

Biological coefficients	$\gamma_z = \gamma_f = 0.1$, $\mu_z = \mu_f = 0.1$
Plankton biomass	b_z according to MODIS estimates
Energetic input	$e_z = 2\,b_z$
Fish biomass	b_f according to FAO estimates
Accessibility	$s_z = b_z$, $s_f = b_f$, $d_{ff'} = 1$, $d_{zf} = 1$
Aquaculture transformation	$\theta_a = 5$, $\eta_t = 2$
Trophic flows	x_{zf}, x_{ff}: inverse modeling detailed in step 2
Trade flows	x_{at}, x_{ta}, x_{bt}, x_{tm}: inverse modeling detailed in step 3
Trade prices	p_t: inverse modeling detailed in step 4
Sales	$x_m = \sum_t x_{tm}$
Aquaculture production	$x_a = \sum_t x_{at}$
Selling process and selling costs	$p_m = 1.5\,p_t$, $c_{tm} = 0.5\,p_t$
Inverse demand function	$a_m = 2\,p_m$, $b_m = p_m/q_m$
Aquaculture limit (carnivore)	$v_a = 0$
Aquaculture limit (herbivore)	$v_a = x_a$
Fishing capacity	$u_{fb} = p_b/2$
Fleet price	$p_b = (\sum_t x_{bt}p_t/\sum_t x_{bt})/2$
Farm price	$p_a = (\sum_t x_{at}p_t/\sum_t x_{at})/2$
Fishing cost	$e_{fb} = (p_b/s_z)/2$
Transformation costs	$c_{bt} = p_t - \eta_t p_b$,
Farming costs	$c_{at} = p_t - p_a$, $c_{ta} = p_a - \theta_a p_t$,
Trade costs	$c_{bt} = p_t - \eta_t p_b$, $c_{tt'} = p_t - t_{t'}$

Summary of the calibration process. Now, we complete the procedure shown in table 15.6. It is clear that these are rough estimates. For our purposes, the focus is mainly on the building of the network and the network processes that operate in the system. Obtaining more accurate parameter values could be a second step, after the analysis of the results of the first step and identification of its weak points. For example, one objective could be to reduce incompatibilities between biological and economic datasets.

15.6.3 Connectivity Tables

We used the following connectivity tables that constrain network links. First table (15.7) is about trophic relationships: who eats whom?

Table 15.7: Trophic connectivity. Columns correspond to prey. Rows correspond to predators. X indicates the possibility of a trophic flow

	CST	COD	DEM	MPE	PP	RAY	SAL.	SPE	SOL
COAST	X				X			X	
COD	X	X	X	X	X			X	X
DEME	X	X	X	X	X			X	X
MPEL				X	X				
PP					X				
RAYS	X	X	X	X	X	X		X	X
SALM.	X	X		X	X		X	X	
SMPEL	X				X			X	
SOLES	X	X	X		X			X	X
TUNA		X	X	X				X	

The second table (15.8) shows transformation between caught species and fish commodities. Particular note should be taken of the NEI (non-identified) category. This is widely used in the data provided by several important countries. Removing this category would distort the data; however keeping it creates an artificial connectivity. We have chosen to keep the NEI category.

Table 15.8: Connectivity between fish species and fish commodities. Columns correspond to groups of fish species. Columns abbreviations are as follows. CARP (1), COAST (2), COD (3), DEME (4), DIADR (5), EEL (6), MFRESH (7), MNEI (8), MPEL (9), RAYS (10), SALMON (11), SMPEL (12), SOLES (13), STURGEON (14), TILAPIA (15), TUNA (16). Rows correspond to commodities

	1	2	3	4	5	6	7	8	9	10	11	12	13	14	15	16
CARP FR	X															
COAST FR		X														
COAST TR		X						X								
COD FR			X													
COD TR			X					X								
DEME FR				X												
DEME TR				X				X								
DIADR FR					X											
EEL FR						X										
EEL TR						X		X								
MFRESH FR							X									
MNEI FR		X	X	X				X	X		X	X	X			X
MNEI TR		X	X	X		X		X	X	X	X	X	X		X	X
MPEL FR									X							
MPEL TR								X	X							
RAYS FR										X						
RAYS TR								X		X						
SALMON FR											X					
SALMON TR								X			X					
SMPEL FR												X				
SMPEL TR								X				X				
SOLES FR													X			
SOLES TR								X					X			
STURGEON FR														X		
TILAPIA FR															X	
TILAPIA TR															X	
TUNA FR																X
TUNA TR								X								X

Chapter 16

Fishing Under Finance

16.1 Overview

16.1.1 Questioning

For several decades fisheries have been managed according to a principle that is mainly due to due to due to Beverton and Holt (1957) and that can be summarized as follows: (1) fish populations are highly reproductive and fisheries must be careful to respect the reproduction phase (Hjort, 1926); (2) there is, in the long term, a relationship between constant fishing pressure (effort) and the resulting stock level (Schaefer, 1957), and (3) there exists, in the long term, an optimal harvesting strategy such that stocks are preserved and yield is maximized (Gordon, 1954; Schaefer, 1957). Managing a fishery consists in being careful about short-term optimization behaviors that may endanger the long-term stability of the system (Botsford et al., 1997; Beddington et al., 2007).

What is implicit in this approach is that the behavior of fisheries is related to the long-term optimization of either their yield or their profit. The question we put forward here is whether there are other motivations for the behavior of fisheries? We are particularly skeptical about the way the capitalistic behavior of fishermen has been represented.

Rising the issue of the financing fisheries is becoming more and more current. It relates to questions of excess or overcapacity (Ward et al., 2005), subsidies (Steenblik and Munro, 1998; Sumaila et al., 2010), and national decommissioning schemes, i.e., boat buy-back (Hannesson, 2007). See Nostbakken et al al (2011) for a recent review.

These are really issues of finance, i.e., managing the relationship between profit and fishing capacity. Let us recall that, in a financially organized world, production increases until the ratio between earnings and invested capital reaches a given threshold; there is a minimal rate of return. Moreover, the goal of agents is supposed to be the maximization of their wealth rather than their profit. Most of the first chapter of the renowned treatise on finance by Bodie and Merton (1999) is dedicated to this point. However, for a fishery, maximizing its wealth can sometimes resemble traditional precautionary behaviors in a situation of uncertainty, and consists in maximizing its fishing capacity (Charles, 1983). This demonstrates the potential for a mismatch, and we express this dilemma using the complementarity framework. This approach could lead to an explanation for the difficulty in managing fishing capacity (Cunningham and Greboval, 2001; Pauly et al., 2002).

16.1.2 Modeling

This chapter addresses the question of financing fisheries, in a very elementary way, using the complementarity principles that have been described in detail in previous chapters.

16.2 Design

16.2.1 Goals

We provide a simple mathematical model, based on complementarity principles that represent the consequences of different capitalistic behaviors by fishermen under financial constraints.

16.2.2 Rationale

We create an explicit bio-economic system where such principles operate. We focus on the effects of changing the value of the rate of return required by financiers. In the first example, we study a system with one fishery and its own way of operating (costs, efficiency and rate of return). We perform a sensitivity analysis of the effect of the value of the rate of return on other system characteristics (stock, yield, and profit). In a second example, we study a system with two fisheries each of which behaves differently (costs, efficiency and rate of return). We perform a sensitivity analysis of the joint effects of both rates of return on the other characteristics of the system. This second case illustrates the differences between industrial fisheries and small-scale fisheries.

We express complementarity equations and observe the consequences of strategies that expect a minimal rate of return. As usual, this is based on the study of the long-term equilibrium according to a selected set of parameters. In this case, these are the value of the rate of return required by the financial sector.

16.2.3 Network Structure

We examine two very simple supply chains in the network economics framework (figure 16.1). They represent one or two fisheries and define their strategy according to the abundance of fish stocks and demand in fish markets.

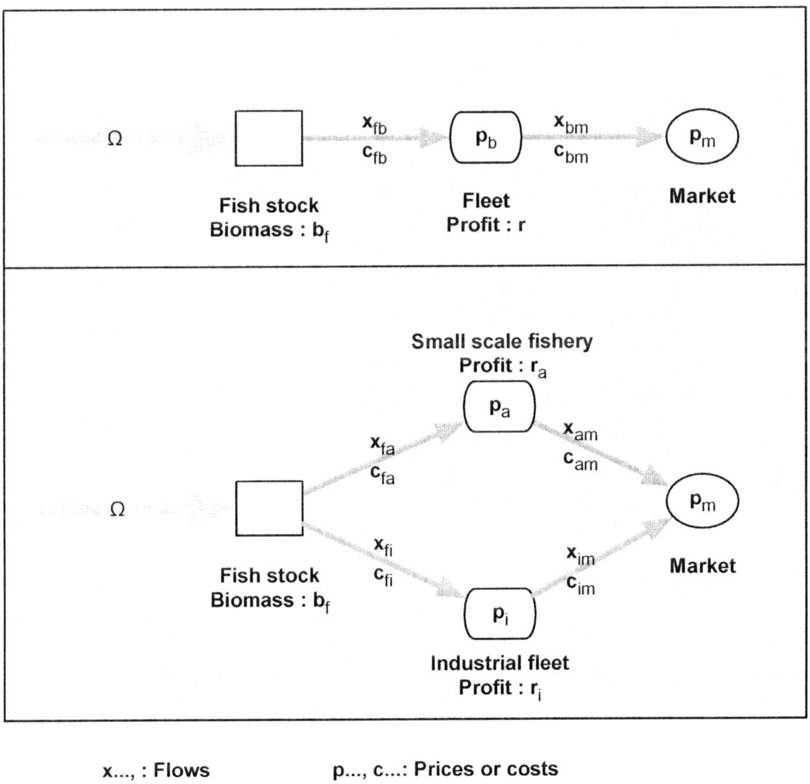

Figure 16.1: Fishing under finance. Top: a system with one fishery. Bottom: a system with two competing fisheries.

16.2.4 Summary of Assumptions

Main assumptions of this modeling attempts are given in box 16.1.

Entities	One fish stock, one or two fleets, one market
Assumptions	Assimilation of fish stock is proportional to a given energetic inflow Losses in fish stocks are proportional to biomass There is a mass balance principle, whence a relationship between biomass and trophic flows (grazing and fishing) for fish stock Costs of fishing are negatively to fish biomass There is a complementarity relationship between fishing costs and fishing flows There is a relationship between yield, fishing effort and fish stocks Fishing effort is limited by fishing capacity, whence the profit of fleet Fleets musts satisfy a minimal return rate (ratio between profit and capital) There is a network equilibrium
Statics/Dynamics	Equilibrium analysis
Calibration	There is no need for calibration. Parameters are given standard values
Observations	The relationship, according to different values of the return rate, of the relationship between effort and profit, or capital The same in the case of two competing fleets

Box 16.1: Design of a model for analyzing productivity differences between upwelling ecosystems

16.3 Details: One Fishery, Formulation and Analysis

16.3.1 Mathematical Notations and Basic Equations

We consider a system (figure 16.1, top) where entities are: f a fish stock, b a fishery, m a fish market. These entities are characterized as follows.

There is constant inflow Ω_f from plankton to fish. Biomass b_f of fish stock depends on catches x_{ef}; as usual in this book we have a relationship between inflow, outflow and biomass: $b_f = (\gamma_f \Omega_f - x_{ef})/\mu_f$.

There is a linear inverse demand function on fish market: $p_m = \alpha_m - \beta_m x_{bm}$; α_m is the intercept, β_m the slope of the inverse demand function.

There is a conventional relationship between yield, fishing effort and fish stocks: $x_{fb} = q_f b_f e_b$, where q_f is the catchability.

The fishery is able to define a departure price $p_b > 0$ when demand equals supply. Its fishing effort e_b is limited by fishing capacity v_b. A part of fishing costs c_{fb} is constant; an other part, the one related to effort, is inversely proportional to fish abundance: $c_{fb} = y_{fb} + e_{fb}/(q_f b_f)$. Profit of fishery is computed as $r_b = x_{bm}(p_m - c_{bm}) - x_{fb}c_{fb}$, where c_{bm} corresponds to selling costs.

There is a constraint of a minimal rate of return κ_b; let k_b the invested capital of the fishery: $k_b = \pi_b v_b$, where π_b is the value of one unit of fishing capacity; we must have $r_b \geq \kappa_b k_b$.

16.3.2 Elementary Mathematical Analysis

Let the state of the system be the vector $X = (x_{fb}, x_{bm}, p_b, v_b)$. We must have:

$$\begin{aligned} b_f &= (\gamma_f \Omega_f - x_{fb})/\mu_f \geq 0 \\ v_b &\geq 0 \\ r_b &\geq \kappa_b \pi_b v_b \end{aligned}$$

Associated to the constraint of a limitation of effort by fishing capacity, there exists a shadow price λ_b giving the complementarity relationships:

$$\begin{aligned} \lambda_b &\perp (v_b - x_{fb}/(q_f b_f)) \\ p_b &\perp f_b = (x_{fb} - x_{bm}) \\ x_{fb} &\perp f_{fb} = (\lambda_b + c_{fb} - p_b) \\ x_{bm} &\perp f_{bm} = (p_b + c_{bm} - p_m) \end{aligned}$$

Of course, there are two types of solutions, ones with fishing and those that do not. Suppose that $x_{fb} > 0$, $p_b > 0$. Then we successively have

(1) $x_{fb} = x_{bm}$,
(2) $f_{fb} = f_{bm} = 0$,
(3) $r_b = (p_m - c_{fb} - c_{bm})x_{fb} = \lambda_b x_{fb}$,
(4) $\lambda_b > 0$, (5) $v_b = x_{fb}/(q_f b_f)$.

Using the above definitions of b_f, r_b, c_{fb}, p_m and expanding inequality $\kappa_b \pi_b v_b \leq r_b$, we find that x_{fb} must be such that $0 \leq Ax_{bf}^2 + Bx_{bf} + C$ where coefficients A, B, C depend on the parameters of the system:

$$\begin{aligned} A &= b_m q_f \\ B &= q_f(a_m - y_{fb} - c_{bm} - q_f) \\ C &= -\mu_f(\kappa_b \pi_i + \eta_{fb} q_f) + (a_m - y_{fb} - c_{bm}) q_f (\gamma_f \Omega_f) \end{aligned}$$

In this equation A is positive; this implies that x_{bm} must be also outside the roots of the second degree polynomial. Indeed, it is easy to show that it must be smaller than the smallest of theses roots (this is related to the inequality $x_{fb} \leq \gamma_f \Omega_f$. Results of this analysis are depicted in figures 16.2 and 16.3. To every value of the rate of return, the fishery can choose a yield between the x-axis and the upper curve (plot at the top, left); it will determine a value for the other characteristics: the possible ranges for these values appear on the other plots.

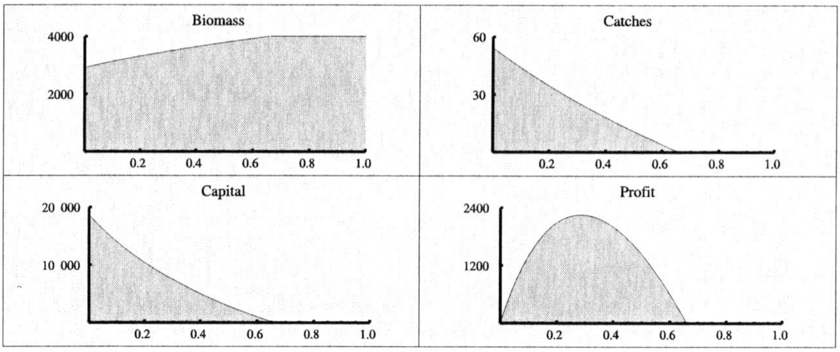

Figure 16.2: Changing rate of return. Rate of return and other characteristics of equilibrium states. x-axis corresponds to rate of return between 0 and 1. Parameters values corresponding to this system are: $\gamma_f = 0.1$; $\mu_f = 0.05$; $\omega_f = 2000$; $q_f = 1$; $y_{fb} = 20$; $c_{bm} = 10$; $a_m = 200$; $e_{fb} = 20000$; $b_m = 3$; $\pi_b = 1000000$; and $q_f = 1$.

Figure 16.2, shows that at rates of return less than 70 percent the fishery is viable. It may choose a yield resulting in a sustainable stock and an profit insuring a better rate of return. Its possibilities are greater for low values of minimal rate of return: for a minimal rate of return of 29 percent, it may choose to fish between 0 and 4 millions tons; for a minimal rate of 60 percent, it may only choose to fish between 0-1 million tons; a minimal rate of return equal to 25 percent allows the fisheries to optimize it profit.

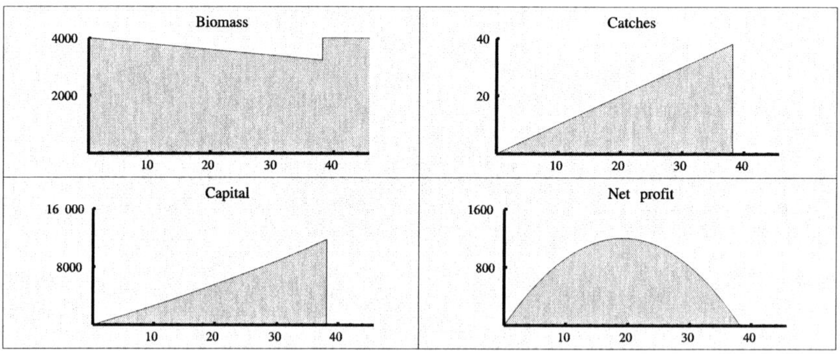

Figure 16.3: Results of strategies that are feasible according to a given rate of return equal to 20 percent. How does the characteristics of a sate depend on yield. Values of rate of return 0.16.

In figure 16.3, we have represented the effects of the possible choices of the fishery for a given value of rate of return (20 percent). It may decide to fish between 0 and 33.7 millions tons. Outside this range, the fishery is not enough profitable. We observe the effects of the level of yield on net profit $(r_b - \kappa_b \pi_b v_b)$ or on fishing capacity v_b.

This figure highlights an issue we would like to emphasize. There are two possible goals for the fishery. It may decide either to optimize its net profit, resulting in an optimal sustainable yield, or maximize its own capital in terms of fishing capacity. As the figure shows, the first strategy protects stock and the profit of fisheries, while the second strategy entails a high level of fishing, low stock and low levels of net profit, just at the level of economic sustainability.

Is the second strategy unrealistic? It supposes (1) a finance based reasoning (optimizing wealth rather than profit) and (2) that fisheries have few possibilities to diversify their investment.

16.4 Details: Two Fisheries, Formulation and Analysis

16.4.1 Mathematical Notations and Basic Equations

To represent competition in a fishing system under finance, let us consider a system where there are two kinds of fisheries: for example a small scale fishery a, and an industrial fishery i that have different production costs and different mandatory rate of returns. Notations and basic formulas are immediate generalization of previous ones. We give them in a telegraphic style.

State variables. They are:

(1) Catches from fish stocks to fishery x_{fa}, x_{fi},
(2) Sales of fishery x_{am}, x_{im},
(3) Fish prices at the level of fishery (departure price) p_a, p_i,
(4) Fishing capacity v_a, v_i,
(5) Shadow price due to fishing capacity limitation λ_a, λ_i.

Intermediate variables. They are:

(1) Fish stock $b_f = (\gamma_f \Omega_f - x_{fi} - x_{fa})/\mu_f$,
(2) Observed prices $p_m = \alpha_m - \beta_m(x_{am} + x_{im})$,
(3) Fishing costs $c_{fa} = y_{fa} + \eta_{fa}/b_f$, $c_{fi} = y_{fi} + \eta_{fi}/b_f$,
(4) Selling costs $c_{am} = \rho_{am}$, $c_{im} = \rho_{im}$,
(5) Income of fishery $r_a = x_{am}(p_m - c_{am}) - x_{fa}c_{fa}$, $r_i = x_{im}(p_m - c_{im}) - x_{fi}c_{fi}$,
(6) Capital of fishery $k_a = \pi_a v_a$, $k_i = \pi_i v_i$,

(7) Overcapacity $g_a = v_a - e_{fa}$, $g_i = v_i - e_{fi}$,
(8) Fishing effort $e_{fa} = x_{fa}/(q_f b_f)$, $e_{fi} = x_{fi}/(q_f b_f)$,
(9) Net profit of fishery $w_a = r_a - \kappa_a k_a$, $w_i = r_i - \kappa_i k_i$,
(10) Balance for fishing $f_{fa} = (\lambda_a + c_{fa} - p_a)$, $f_{fi} = (\lambda_i + c_{fi} - p_i)$,
(11) Balance for selling $f_{am} = (p_a + c_{am} - p_m)$, $f_{im} = (p_i + c_{im} - p_m)$,
(12) Balance for fishery $f_a = x_{fa} - x_{am}$, $f_i = x_{fa} - x_{im}$.

Constraints. They concern:

(1) Biomass: $b_f \geq 0$,
(2) Overcapacity: $g_a \geq 0$, $g_i \geq 0$.
(3) A minimal rate of return: $w_a \geq 0$, $w_i \geq 0$.

Complementarity equations. They are:

(1) Concerning fishing $x_{fa} \perp f_{fa}$, $x_{fi} \perp f_{fi}$,
(2) Concerning selling $x_{am} \perp f_{am}$, $x_{im} \perp f_{im}$,
(3) Concerning fishery $p_a \perp x_{fa} - x_{am}$, $p_i \perp x_{fi} - x_{im}$,
(4) Concerning fishing capacity $\lambda_a \perp g_a$, $\lambda_i \perp g_i$.

16.4.2 Elementary Mathematic Analysis

Let us examine what happens to fishery a. Suppose that $x_{fa} > 0$, $p_a > 0$. Just as before, we find that yield x_{fa} must satisfy an equality:

$$0 \leq A_a x_{fa}^2 + B_a x_{fa} + C_a$$

with

$$\begin{aligned} A_a &= b_m q_f \\ B_a &= q_f(a_m - b_m x_{fi} - y_{fa} - c_{am} - q_f) \\ C_a &= -\mu_f(\kappa_a \pi_a + \eta_{fa} q_f) \\ &\quad + (a_m - b_m x_{fi} - y_{fa} - c_{am}) q_f(\gamma_f \Omega_f - x_{fi}) \end{aligned}$$

This shows that, for a, the range of possible yields x_{fa} depends on the strategy x_{fi} chosen by i. In the save way we get an inequality for the possible values of x_{fi} and this inequality depends on x_{fa}. This is the framework for the Generalized Nash Equilibrium problem that was been presented in section 6.4. Another way to express previous result is: both fleets may fish together if the sum of their yield $x = x_{fa} + x_{fi}$ is such that, altogether we have inequalities:

$$\begin{aligned} 0 &\leq A_a x^2 + B_a x + C_a \\ 0 &\leq A_i x^2 + B_i x + C_i \end{aligned}$$

with

$$
\begin{aligned}
A_a = A_i &= b_m q_f \\
B_a &= q_f(a_m - b_m - y_{fa} - c_{am} - q_f) \\
B_i &= q_f(a_m - b_m - y_{fi} - c_{im} - q_f) \\
C_a &= -\mu_f(\kappa_a \pi_a + \eta_{fa} q_f) + (a_m - y_{fa} - c_{am}) q_f (\gamma_f \Omega_f) \\
C_i &= -\mu_f(\kappa_i \pi_i + \eta_{fi} q_f) + (a_m - y_{fi} - c_{im}) q_f (\gamma_f \Omega_f)
\end{aligned}
$$

So the game is now clear in terms of strategies. Fleets a and i determine their strategy in order that previous conditions are satisfied and to optimize something that remains to be defined. Several cases can be observed according to the values of solutions $t_a < t'_a$, $t_i < t'_i$ of equations

$$
\begin{aligned}
0 &= Ax^2 + B_a x + C_a \\
0 &= Ax^2 + B_i x + C_i
\end{aligned}
$$

Possible situations are as follows. If $t'_a < 0$ and $t'_i < 0$, no fishing is possible for both fleets: $x_{fa} = x_{fi} 0$. If $t'_a < 0$ and $t_i > 0$, a does not fish and $x_{fa} = 0$ and i may choose a strategy $x_{fi} \in [Max(0, t'_i), t_i]$. If $t'_i < 0$ and $t_a > 0$, i does not fish $x_{fi} = 0$ and a may choose a strategy $x_{fa} \in [Max(0, t'_a), t_a]$. If $t_a > 0$ and $t_i > 0$, a may choose a strategy $x_{fa} \in [Max(0, t_a), t'_a]$ and in the same way i may choose a strategy $x_{fi} \in [Max(0, t'_i), t_i]$, but these strategies must be such that $x = x_{fa} + x_{fi}$ satisfies the inequalities:

$$
\begin{aligned}
0 &\leq Ax^2 + B_a x + C_a \\
0 &\leq Ax^2 + B_i x + C_i
\end{aligned}
$$

It should be noted that in previous formulas the total of both yields $x_{fa} + x_{fi}$ determines the level of stock and the market price.

Following figures illustrate shows what happens with a given pair of rates of return: $rA = 12$ percent, $rI = 20$ percent. Both fisheries decide their own yield and observe the result of this competition on their capital (figure 16.4) and net profit (figure 16.5). High values of rate of return prevent over-fishing. There are obviously more possibilities for fishery a: it is more competitive. Let us try to identify Nash equilibrium of this system. If fisheries are optimizing their net profit, we find a equilibrium for a couple of yields around $x_{fa} = 2.57$, $x_{fi} = 3.41$. If fisheries are optimizing their capital (fishing capacity), we find an equilibrium around $x_{fa} = 5, x_{fi} = 0$.

As expected, competition is fiercer in the case of capital maximization.

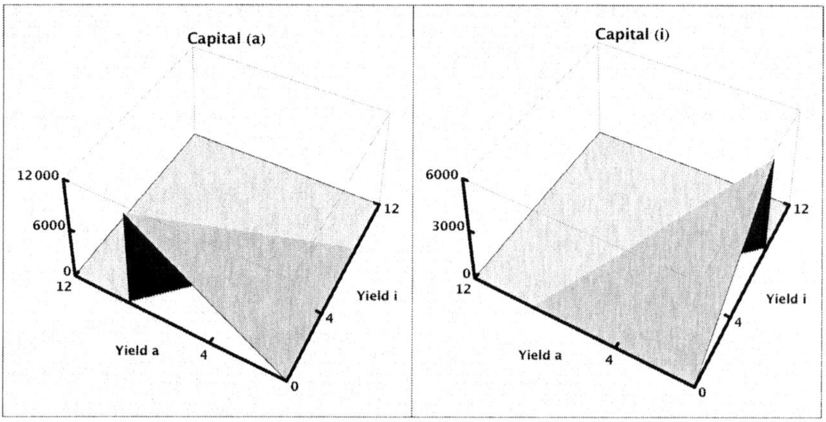

Figure 16.4: Results of pairs of strategies given a pair of feasible strategies. How does capital depends yields of a and i. Parameters values are: $\gamma_f = 0.1$; $\mu_f = 0.05$; $\Omega_f = 2000$; $a_m = 200$; $b_m = 3$; $q_f = 1$; $c_{am} = 8$; $y_{fa} = 7$; $e_{fa} = 10000$; $\pi_a = 5000000$; $c_{im} = 12$; $y_{fi} = 25$; $e_{fi} = 14000$; $\pi_i = 2500000$.

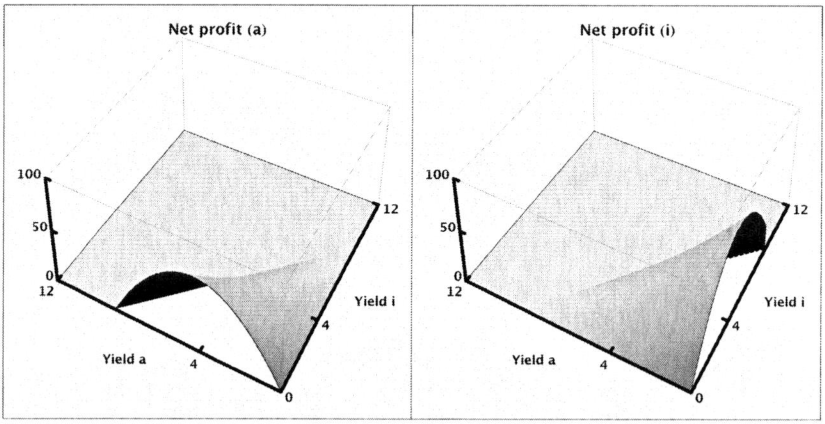

Figure 16.5: Results of pairs of strategies given a pair of feasible strategies. How does net profit depends on both yields.

16.5 Details: Several Fisheries, Several Stocks and Several Markets, Formulation

In the previous sections we defined a simple framework and observed that, as expected, a given rate of return defines a range of possible yields. Can we go further and illustrate these questions in the network economics framework by studying a general fish commodity chain? In this section, we generalize the principles of previous sections. We take the example of several fisheries exploiting several stocks and selling on several fish markets. We use the example to investigate the behavior of fisheries. In particular, we look at two behaviors from the set of possible strategies defined by a rate of return. Do fisheries choose the strategy that maximizes their net profit or the one that maximizes their capital?

The purpose of this exercise is to illustrate the mathematical issues that arise.

16.5.1 Mathematical Notations and Basic Formulas

Using the same notations as before, we compute:

$$\begin{aligned}
b_f &= \frac{\gamma_f \Omega_f - \sum_b x_{fb}}{\mu_f} \\
c_{fb} &= y_{fb} + \frac{e_{fb}}{q_f b_f} \\
p_m &= \alpha_m - \beta_m \sum_m x_{bm} \\
r_b &= \sum_m x_{bm}(p_m - c_{bm}) - \sum_f x_{fb} c_{fb}
\end{aligned}$$

We will make two hypothesis. According to them, the objective function of a fishery b is either $o_b = v_b$ or $o_b = r_b - \kappa_b \pi_b v_b$. Strategy of fishery b consists in determining $(v_b, (x_{fb}), (x_{bm}))$ that maximize its objective o_b subject to:

$$\begin{aligned}
f_b &= \sum_f x_{fb} - \sum_m x_{bm} \geq 0 \\
g_b &= v_b - \sum_f \frac{x_{fb}}{q_f b_f} \geq 0 \\
h_b &= r_b - \kappa_b \pi_b v_b \geq 0
\end{aligned}$$

Let us remark that condition $g_b = v_b - \sum_f x_{fb}/(q_f b_f) \geq 0$ is not linear in v_b and x_{fb}, and, even, does not define a convex set. We will use the full formulation of the Karush-Kuhn-Tucker (see subsection 5.6.6):

$$\begin{aligned}
v_b \perp k_b &= -\frac{\partial o_b}{\partial v_b} - \lambda_b \frac{\partial f_b}{\partial v_b} - \eta_b \frac{\partial g_b}{\partial v_b} - \theta_b \frac{\partial h_b}{\partial v_b} \\
x_{fb} \perp k_{fb} &= -\frac{\partial o_b}{\partial x_{fb}} - \lambda_b \frac{\partial f_b}{\partial x_{fb}} - \eta_b \frac{\partial g_b}{\partial x_{fb}} - \theta_b \frac{\partial h_b}{\partial x_{fb}} \\
x_{bm} \perp k_{bm} &= -\frac{\partial o_b}{\partial x_{bm}} - \lambda_b \frac{\partial f_b}{\partial x_{bm}} - \eta_b \frac{\partial g_b}{\partial x_{bm}} - \theta_b \frac{\partial h_b}{\partial x_{bm}} \\
\lambda_b &\perp f_b \\
\eta_b &\perp g_b \\
\theta_b &\perp h_b
\end{aligned}$$

The computing of derivatives is straightforward. We get a linear complementarity problem $\mathcal{CP}(F)$ with state:

$$X = (v_b, x_{fb}, x_{bm}, \lambda_b, \eta_b, \theta_b)$$

And function F given by

$$F(X) = (k_b, k_{fb}, k_{bm}, f_b, g_b, h_b)$$

This complementarity problem is not a linear one. It is indeed far from being one. In most of cases, the solution, if it exists, is difficult to compute. The problem is quite general. It could be the starting point of an original research.

16.6 Concluding Remark

16.6.1 Features

This chapter was mainly theoretical. We addressed the question of the financing of fisheries and illustrated the dilemma between optimizing profit and optimizing wealth in the case of one or two fisheries. Then we formulated the general case.

16.6.2 Teachings

Network economics makes it possible to represent, mathematically, the various consequences of alternative capitalistic behaviors of fisheries.

16.6.3 Going Further

It is obviously necessary to look for, in the economic behavior of existing fisheries, evidence of specific financing behaviors (based on the maximization of fishing capacity rather than the maximization of net profit), and evidence of a dependency on finance, its emergence, and its predictable consequences.

Then the taking into account of the capitalistic behavior of fisheries could become an important issue in forthcoming fisheries policies. We can only imagine the reactions to such an abrupt formulation.

Chapter 17

The Coexistence of Small and Big Fish

17.1 Overview

17.1.1 Questioning

Imagine a world in which small fish prefer living on the surface and escape big fish, and in which big fish prefer living at the bottom of the ocean and pursue small fish (see figure 17.1). This is not a totally unrealistic world (Giske et al., 1998). It defines a pursuit evasion game, which is a part of mathematical game theory (Nahin, 2007). Nor is it anecdotal ecological theory; the possibility of an ecology of fear has already been examined (Brown et al., 1999). A game theoretical approach can be applied to habitat selection by predators and prey (Hugie and Dill, 1994), the relationship between fish foraging and habitat choice, (Mittlebach, 2002), and a re-examination of the ideal free distribution theory under predation risk (Moody et al., 1996).

17.1.2 Modeling

We present a very simple model of this situation based on network economics principles.

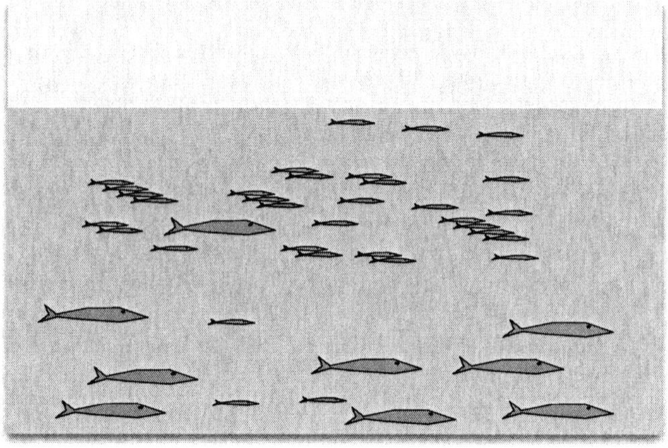

Figure 17.1: A vertical section of a marine ecosystem. Small fish stay near the surface and big fish remain in deep water.

17.2 Design

17.2.1 Goals

Our goal is speculative. The model is designed to: (1) illustrate competition for space in ecology; (2) provide an alternative point of view that does not use differential equations (ordinary or partial derivatives), and (3) prove feasibility using schematic processes with realistic orders of magnitude.

17.2.2 Rationale

We represent the mutual trade-offs between two populations with habitat preferences and attraction/repulsion behaviors. There is no predation in the model in its present state. We observe the movements of the two populations in a limited number of time steps (approximately 10). We define a pursuit escape game and observe the emergence of a stationary distribution and the relationships between this stationary distribution and preference parameters.

17.2.3 Network Structure

The underlying network is defined by a water column (figure 17.2). Nodes correspond to layers and links correspond to moves between layers. The method proceeds as follows. At any time step, small and big fish populations that are distributed on the various layers move to areas they prefer, based on transportation costs and anticipating the behavior of the other population. This creates a new distribution for the next step. We represent

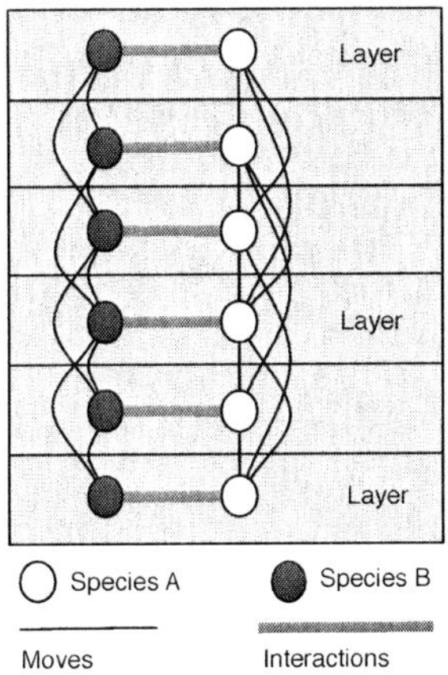

Figure 17.2: Water column. Layers.

the water column as a succession of layers and split fish populations according to their size.

17.2.4 Summary of Assumptions

Main assumptions are summarized in box 17.1.

Entities	Small fish and big fish populations in the layers of a water column
Assumptions	Migration costs proportional to distance between layers Advantage of living in a layer depending on depth (positively for big fish, negatively for small fish) and to the presence of the other population (attraction of big fish for small fish, repulsion of small fish for big fish) At any time, a network equilibrium defines the effective migrations
Statics/Dynamics	At each step, an equilibration phase
Observations	Sensitivity analysis of the spatial distribution dynamics according to ecological parameters (density dependence of the living costs in an area, dying parameters in the Leslie matrix)

Box 17.1: Design of a migration model of the attraction/repulsion game between small and big fish

17.2.5 Model Entities

The proposed model is an implementation of the migration model that has been presented in chapter 9. Entities are as follows. Spatial entities are nl layers; they are denoted: i, j, k, l. At the beginning of a time step, there are

$\overline{p}_i^S$ small fish and $\overline{p}_i^B$ big fish in layer i. Both populations may move. We denote: x_{ij}^S the number of small fish and x_{ij}^B the number of big fish moving from layer i to layer j. After moves, at the end of a time step, populations are:

$$p_i^S = \overline{p}_i^S - \sum_j x_{ij}^S + \sum_j x_{ji}^S$$
$$p_i^B = \overline{p}_i^B - \sum_j x_{ij}^B + \sum_j x_{ji}^B$$

Moves are costly. Moving costs are denoted c_{ij}^S for small fish and c_{ij}^B for big fish: both are increasing with $|i-j|$. For a small fish, living costs in layer i are denoted u_i^S; they are increasing with i (small fish prefer the surface); they are increasing with p_i^B (small fish try to avoid big fish). For a big fish, they are denoted u_i^B; they are decreasing with i (big fish prefer the depth); they are increasing with p_i^S (big fish prefer to be with small fish). Living costs are given by $u_i^S = \alpha^S i - \beta^S p_i^B$, $u_i^B = -\alpha^B i + \beta^B p_i^S$. Migration costs are given by: $c_{ij}^S = \theta^S|i-j|$, $c_{ij}^B = \theta^B|i-j|$. Constraints on moves are:

$$x_{ij}^S \geq 0, \quad x_{ij}^B \geq 0$$
$$\sum_j x_{ij}^S \leq \overline{p}_i^S, \quad \sum_j x_{ij}^B \leq \overline{p}_i^B$$

17.2.6 Model Equilibrium

We consider a vector of moves $X = (x_{ij}^S, x_{ij}^B)$ and the constrained set K of vectors X satisfying above constraints. We define the function $X = (x_{ij}^S, x_{ij}^B) \rightarrow F(X) = (f_{ij}^S, f_{ij}^B)$ with:

$$f_{ij}^S = u_i^S - u_{ij}^S + c_j^S$$
$$f_{ij}^B = u_j^B - u_{ij}^B + c_j^B$$

A migration equilibrium is a moves vector X such that for all layer i, there exists $\lambda_i^S \geq 0$ and $\lambda_i^B \geq 0$ such that for all j

$$\lambda_i^S = 0 \perp (\overline{p}_i^S - \sum_j x_{ij}^S)$$
$$x_{ij}^S \perp (f_{ij}^S + \lambda_i^S)$$
$$\lambda_i^B = 0 \perp (\overline{p}_i^B - \sum_j x_{ij}^B)$$
$$x_{ij}^B \perp (f_{ij}^B + \lambda_i^B)$$

Then, as in the general case presented in chapter 9, we have:

Theorem 17.2.1 *$X^* \in K$ is a migration equilibrium if and only if X^* is a solution of the variational inequality $\mathcal{VI}(F, K)$.*

We also have in this case:

Theorem 17.2.2 *F is an affine function*

Proof. We have:

$$f_{ij}^S = n_{ij}^S + \sum_{kl} M_{ij,kl}^{SB} x_{kl}^B$$
$$f_{ij}^B = n_{ij}^B + \sum_{kl} M_{ij,kl}^{BS} x_{kl}^S$$

where: $n_{ij}^S = \alpha^S(j-i) - \beta^S(\overline{p}_j^B - \overline{p}_i^B) + \theta^S|i-j|$, $M_{ij,kl}^{SB} = \beta^S(\delta_{jk} - \delta_{jl} - \delta_{ik} + \delta_{il})$, $n_{ij}^B = -\alpha^B(j-i) + \beta^B(\overline{p}_j^S - \overline{p}_i^S) + \theta^B|i-j|$, $M_{ij,kl}^{BS} = -\beta^B(\delta_{jk} - \delta_{jl} - \delta_{ik} + \delta_{il})$. □

Moreover, it can be shown that M is a non negative matrix (its eigenvalues are $0, \pm i\sqrt{\beta^S}, \pm i\sqrt{\beta^B}$) and that the variational inequality problem has at least a solution.

17.2.7 Model Dynamics

We represent the pursuit evasion game as iterations of the following process:

(1) Determine equilibrium moves according to initial populations $(\overline{p}_i^S, \overline{p}_i^B) \to (x_{ij}^S, x_{ij}^B)$.
(2) Compute new populations: $p_i^S = \overline{p}_i^S - \sum_j x_{ij}^S + \sum_j x_{ji}^S$, $p_i^B = \overline{p}_i^B - \sum_j x_{ij}^B + \sum_j x_{ji}^B$.
(3) Set initial populations for the next step: $\overline{p}_i^S = p_i^S, \overline{p}_i^B = p_i^B$.

It has been proven by (Pan and Nagurney, 1994) that this kind of iterative process leads to a stationary distribution.

17.2.8 Model Parameters

The model consists of 6350 small fish and 6350 big fish distributed in ten layers of a water column. We study the dynamics during 10 time steps. The model relies on a very small number of parameters. Their reference values are as follows: $\alpha^S = 0.22$, $\alpha^B = 0.2$, $\beta^S = 0.002834$, $\beta^B = 0.001889$, $\theta^S = 0.2$, $\theta^B = 0.36$.

17.2.9 Model Initialisation

Every simulation starts with the distribution given in table 17.1.

Table 17.1: Vertical migrations Initial distribution

	Small fish	Big fish
1	1796	631
2	1500	497
3	1203	484
4	907	456
5	611	437
6	314	397
7	18	388
8	0	910
9	0	1020
10	0	1130

17.3 Results

We performed sensitivity analyzes of all model parameters and observed how the population distribution evolved according to changes in the values of selected parameters.

17.3.1 Preference of Big Fish for Depth

This consists in changing the values of parameter α^B. The resulting distribution are given in figure 17.3. The bottom plot shows the effects of the preference of big fish for depth (on the right, with a high value of α^B, they all stay in the lowest layer; on the left with a low value of α^B, they distribute themselves more equally). The top of the plot shows the reaction of small fish escaping to big fish (on the right they stay at the surface in the upper layers; on the left their distribution is complementary to the distribution of big fish).

17.3.2 Preference of Small Fish for Surface

This consists in changing the values of parameter α^S. The resulting distribution is shown in figure 17.4. The top of the plot shows the effects of preference of small fish for surface (on the left, with a low value of α^S, small fish distribute themselves equally, on the right, with a high value of α^S they stay in the upper layers. The bottom plot shows the trade-off for big fish between their pursuit behavior and their preference for depth: when α^S is high, the big fish population splits in two: one part moves to the surface where they find small fish, the other stays in deeper water, where they prefer to be.

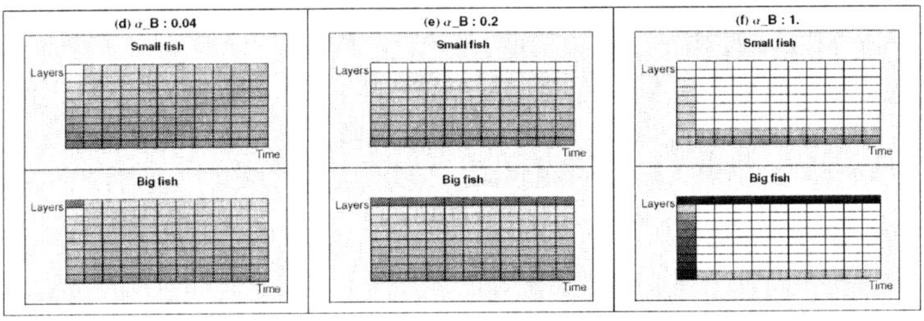

Figure 17.3: Sensitivity of big fish preference for depth. Left: low preference of big fish for depth. Right: high preference of big fish for depth. In this plot (as in those that follow), the x-axis corresponds to time steps and the y-axis corresponds to the distribution of the fish population in layers. For example, the absence of dark layers corresponds to a situation in which most fish remain in deep water; the absence of gray layers corresponds to a situation in which the fish population splits itself between deeper waters and the surface, etc. Left: low value sensitivity parameter. Middle: mean value. Right: high value. The middle plot corresponds to a stationary population distribution.

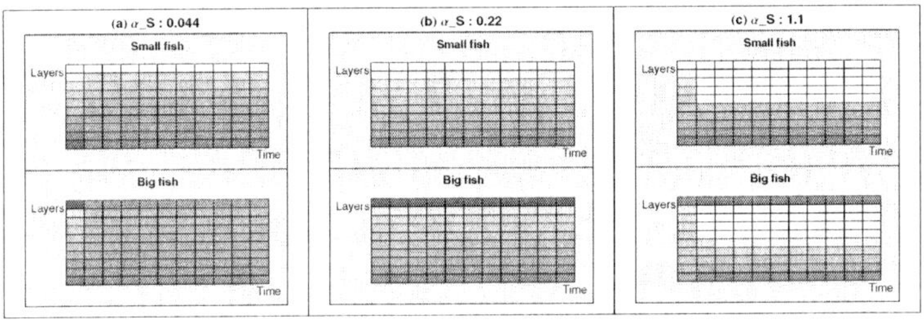

Figure 17.4: Preference of small fish for surface. Left: low preference of small fish for surface. Right: high preference of small fish for surface.

17.3.3 Escape Behavior of Small Fish

This consists in changing the values of parameter β^S. The resulting distribution is shown in figure 17.5. There is an indirect effect: with a high level of escape behavior (right-hand side) more small fish head for depths where there are usually more big fish.

17.3.4 Pursuit Behavior of Big Fish

This consists in changing the values of parameter β^B. The resulting distribution is shown in figure 17.6. At high levels of pursuit behavior (right-hand side) big fish head towards the surface, where they find small fish. At the same time, small fish head for deeper waters where there are fewer big fish. At low levels of pursuit behavior (left-hand side) big fish split themselves into two sub-populations; those that prefer to stay in deeper water and those that prefer to follow small fish.

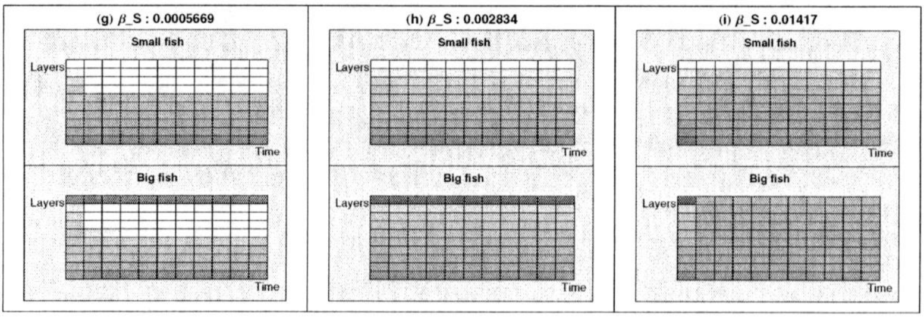

Figure 17.5: Escape behavior of small fish. Left: low level of escape behavior. Right: high level of escape behavior.

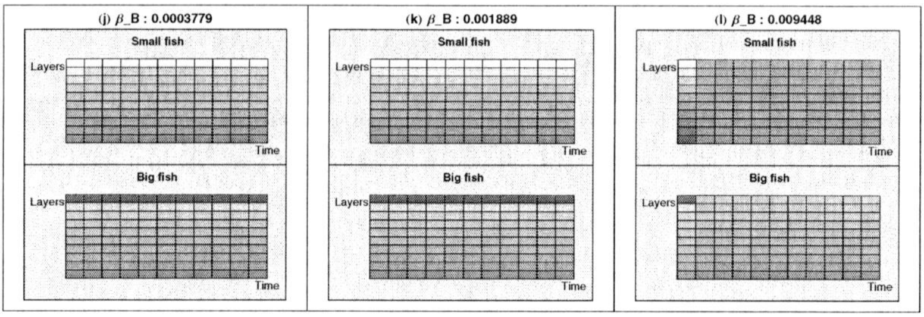

Figure 17.6: Pursuit behavior of big fish. Left: low level pursuit behavior. Right: high level of pursuit behavior.

17.4 Discussion

17.4.1 Features

This section has analyzed the movements of fish in a water column with principles and tools developed for human migration.

17.4.2 Teachings

The approach has proven to be feasible in terms of mathematical formulation and computation. The model involves a realistic number of layers and a non-negligible number of big and small fish. We were able to express the equations and compute the resulting equilibrium. The sensitivity of equilibrium to parameters values is important as it makes it possible to identify some indirect and a priori unexpected effects.

The network economics approach to migration provides a different perspective on space occupancy in ecology. It may offer improvements to the ideal free distribution theory that is based on similar principles.

17.4.3 Discussing Assumptions

This model aims at explaining animal moves in the framework of network economics. It represents an animal population in which several individuals migrate based on trade-offs between migration and living costs. We highlight that, as for human migration, this does not imply cooperative behavior of individuals in a given area. It simply means that these individuals have the same way of computing the benefits and costs of migration. By taking account of each individual decision, the overall migratory effect can be analyzed.

The most questionable assumption of this model is the representation of animal movements using economic models of human migration. There is clearly no housing market or rental market: what is at stake is competition for space, and the coordination of anticipatory behaviors.

We can try to define the biological meaning of such behavior. Is it adaptive? This would mean that through evolution, populations have acquired ways to recognize, at a given time, the global state of the competition for space and resources: Who is where? Who ought to go where? How will one part of the population move, knowing the movements of another part? Although it is a plausible hypothesis in the case of hourly moves of fish populations in a water column the question remains of the extent to which it is more generally applicable.

17.4.4 Going Further

Going further means using the approach with spatial ecosystems that are more general than a vertical water column. For example, the interplay between prey and predators in a complex spatial system such an archipelago with several islands, where the connectivity of habitats is a major feature (see figure 17.7).

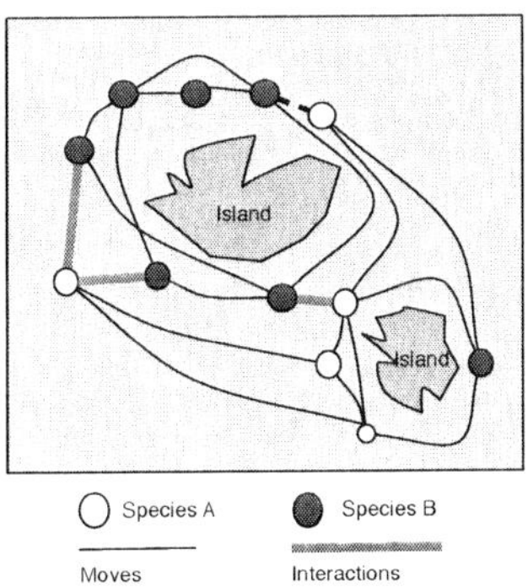

Figure 17.7: Connectivity. Islands.

Chapter 18

The Spatial Behavior of the Atlantic Bluefin Tuna

18.1 Overview

18.1.1 Questioning

This emblematic fish population has been described by various authors (Block et al., 2001; Ravier and Fromentin, 2001; Fromentin and Powers, 2005; Fromentin, 2003; Safina and Klinger, 2008). Presently, the Atlantic Bluefin tuna has a very high value on fish markets. There is over-fishing and risk of extinction.

It is known to reproduce in two spawning areas, the Gulf of Mexico, and the Mediterranean Sea. The tuna migrate to other areas ranging from Norway to the Brazilian coast for feeding, according to specific migration paths (see figure 18.1). Changes in migration paths have been observed during the last decades. The main change lies in the disuse of long-distance feeding areas such as Brazil and Norway.

18.1.2 Modeling

We apply the principles of network economics to study the migration of the Atlantic Bluefin tuna and to reproduce observed patterns.

The forthcoming simulations represent the migration of tuna species, rather than their exploitation. However, we will also examine a scenario involving over-fishing of adults in the Mediterranean Sea.

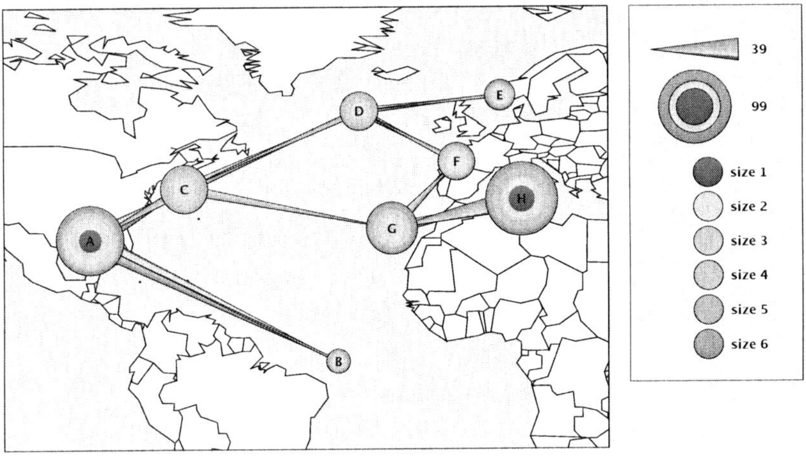

Figure 18.1: Spatial behavior of the Bluefin tuna in the Atlantic ocean. Habitats and migration paths.

18.2 Design

18.2.1 Goals

The goals of this model are: (1) to explore the feasibility of a network economics approach in a biologically important case study, and (2) to identify the kind of results it may provide; especially how it reproduces observed changes in migration paths.

18.2.2 Network Structure

The model uses a network representation of the movements of structured populations between several oceanic habitats. In the model, nodes represent a structured population in a habitat, and links are migrations between two nodes. Migration with several classes of species was mathematically formulated in chapter 9.

Nodes are distribution areas: Mexico Gulf, Brazil, Gulf of Maine, Mediterranean sea, Gulf of Biscaye, North Atlantic, Norway. Links correspond to observed migration paths between distribution areas.

18.2.3 Rationale

We provide a sensitivity analysis of the most significant parameters, and try to identify those that are associated with changes in migration patterns and others that should be taken into account in the case of over-exploitation of stocks in the Mediterranean Sea.

18.2.4 Summary of Assumptions

Main assumptions are summarized in box 18.1

Entities	Age classes of Bluefin tuna populations in their living areas
Assumptions: migration	Migration costs proportional to distance between living areas Advantage of living in a living area depending (1) on the primary production in this area, (2) on the total tuna population in this area At any time, a network equilibrium defines the effective migrations
Assumption: ecological processes	At any time steps, growing, spawning and dying in a living area according to a Leslie matrix; this matrix depends on the living area; mortality parameters are uniform; fertility parameters are density dependent.
Statics/Dynamics	At each step, the succession of an equilibration phase and a deterministic phase Leslie matrices are used to represent deterministic ecological processes
Calibration	Primary production according to observation. Other parameters by tuning.
Observations	Sensitivity analysis of the vertical distribution dynamics according to depth, preference or attraction/repulsion parameters

Box 18.1: Design of a migration model of the Atlantic Bluefin tuna

18.3 Details

18.3.1 Entities

The model is implemented with the entities, parameters values and migration paths shown in table 18.1. Two areas, Mediterranean sea, Gulf of Mexico, are spawning habitats.

Table 18.1: Atlantic Bluefin tuna habitats. κ carrying capacity; θ impact of temperature; σ seasonality

Habitat	Id.	κ	θ	σ
Gulf of Mexico	A	200	2	0.1
Brazil	B	100	2	0.2
Gulf of Main	C	500	2	0.5
North Atlantic	D	300	2	0.8
Norway	E	400	2	0.8
Gulf of Biscay	F	300	2	0.7
West Atlantic	G	300	2	0.2
Mediterranean sea	H	200	2	0.1

18.3.2 Mathematical Formulation

Migrations. Let us recall the mathematical notations and the principles of the application of network economics to the study of migrations with classes that have been described in chapter 9.

(1) p_i^k is the tuna population of class k in habitat i and u_i^k the cost of living in habitat i for individuals of age class k; u_i^k is opposite to the advantage of living: $u_i^k = -a_i^k$; x_{ij}^k the flow of tuna population of class k from habitat i to habitat j and c_{ij}^k the cost of migrating from i to j for class k. A state of the system is a set of effective migrations $X = (x_{ij}^k)$.

(2) Migrations are limited: $x_{ij}^k \geq 0$ and $\sum_j x_{ij}^k \leq \overline{p_i^k}$, where $\overline{p_i^k}$ is the initial population of class k in habitat i. This defines a constrained set is: $K = \{(x_{ij}^k)\}$.

(3) Let the function $X \to F(X)$ be defined by $F(X) = (f_{ij}^k)$ where $f_{ij}^k = u_i^l + c_{ij}^k - u_j^k$.

(4) A state $X = (x_{ij}^k) \in K$ is an equilibrium of the migration system if and only if is a solution the variational inequality $\mathcal{VI}(F, K)$. The characteristics of $\mathcal{VI}(F, K)$ makes it solvable.

We represent the dynamics of migrating fish populations, taking account of growth of individuals and consider weight classes k; let s^k the mean weight of individuals in class k; total weight of the population is $w_i = \sum_k s^k p_i^k$. We denote θ_i^k the sensitivity to temperature of individuals of class k in habitat i and κ_i the carrying capacity of habitat i. Both may depend on season. Then, the living advantage in area i, for class k is given by: $a_i^k = \theta_i^k(1 - w_i/\kappa_i)$ We denote d_{ij} the distance between habitats i and habitats j. The migration costs from area i to area j, at time t, for class k are given by: $c_{ij}^k = 0.001 \times d_{ij}$ if $k > 1$, $c_{ij}^1 = \infty$: size 1 individuals don't move.

Population dynamics. The dynamics of the system are represented through the coupling of migrations and other ecological processes: fertility, growth, aging, dying.

Let, at time t the population vector be $\overline{P_i} = (\overline{p_i^k})$. Let, at this time, the parameters be: θ_i^k, κ_i.

(1) Population parameters allow to specify function F and constrained set K that have been previously defined. We solve the variational inequality $\mathcal{VI}(F, K)$ and get the migration vector x_{ij}^k and the population after migration $P_i(t) = (p_i^k(t))$ with $p_i^k(t) = \overline{p_i^k} - \sum_j x_{ij}^k + \sum_j x_{ij}^k$.

(2) In each area, ecological processes are represented using according to a Leslie matrix L[1]. We obtain the population $\overline{P}_i(t+1) = L \cdot P_i(t)$ at the beginning of next time step.

[1]A general introduction to Leslie matrices is given in Caswell (2006).

We consider the following Leslie matrix:

$$L = \begin{pmatrix} q_1 & r_1 & r_2 & r_3 & r_4 & r_5 \\ g_2 & q_2 & 0 & 0 & 0 & 0 \\ 0 & g_3 & q_3 & 0 & 0 & 0 \\ 0 & 0 & g_4 & q_4 & 0 & 0 \\ 0 & 0 & 0 & g_5 & q_5 & 0 \\ 0 & 0 & 0 & 0 & g_6 & q_6 \end{pmatrix}$$

Parameters q correspond to survival, g to growth, r to reproduction. They may depend on time step t. They are not the same for all habitats.

18.3.3 Implementation

Let us give the formulation of the Leslie matrix that will be used in forthcoming simulations. In all habitats, we put $q_i = 0.6$, $g_i = 0.2$. In a spawning habitat i, let the scarcity be $\sigma_i = 1 - w_i/\kappa_i$, the spawning intensity be: ρ_i and $r_1 = r2 = 0$, $r_3 = \sigma_i \rho_i$, $r_4 = 2\sigma_i \rho_i$, $r_5 = \sigma_i \rho_i$.

In the simulations with an environmental effect, the living advantages in an habitat depend on season (through seasonal changes of the parameter θ_i given in table 18.1). This mainly concerns non-spawning habitats where the temperature is highly variable (Norway and the North Atlantic).

18.3.4 Sensitivity Analysis

We will observe the sensitivity of the previous model to the following parameters: (1) intensity of spawning ρ_i, (2) moving costs δ_{ij}. We aim to represent the simplest explanation to the disaffection for long distance feeding areas, namely Bluefin tuna do not go there because they can find enough food in areas that are closer.

18.4 Results

18.4.1 Favourable Situation

We first present the seasonal dynamics of age distributions obtained when spawning is high and migration costs are low (figure 18.2).

The dynamics of migrations shown in figure 18.2 for population distribution and in figure 18.3 for population migrations) appear to be seasonal and discontinuous. Migrations that occur only when there is an advantage happen very quickly, indicated by the peaks in distributions. We observe peak populations in the North Atlantic and the Gulf of Biscay that are a stopover point in migration towards feeding areas. These are explained

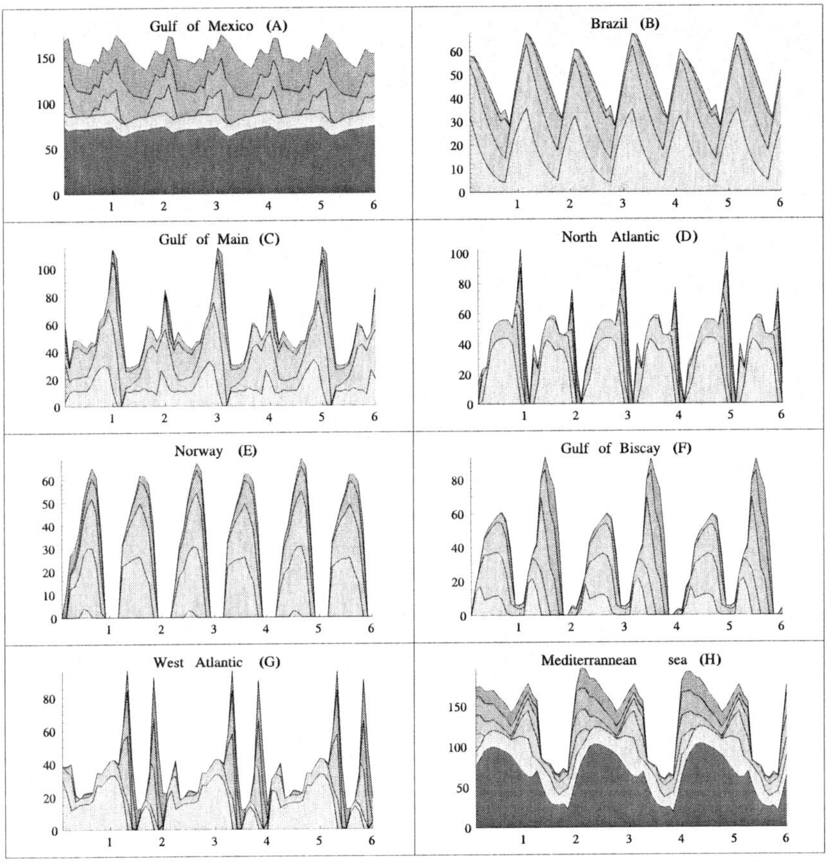

Figure 18.2: Migration of Atlantic Bluefin tuna in favorable conditions. Dynamics of population distribution in the eight areas. The x-axis corresponds to time: the simulation was run over the last six years. Gray levels correspond to size classes: dark gray represents juveniles, while larger fish are shown in a darker gray.

by a sudden change in the trade-off to migrate to high-seasonality feeding areas.

The plot scales in figure 18.3 show that similar numbers of fish migrate, whether they are abundant or not in the area of origin. Differences are due to a longer period of migration. It looks like a limitation of capacity on migration paths. At the present time, we have no explanation for this behavior and we do not know if it has been observed in the field.

In most areas there seems to be an annual cycle, although in the Mediterranean Sea, and to a lesser extent in the West Atlantic and the Gulf of Biscay, the dominant cycle seems to be two years. According to other simulations (not shown), there seems to be several similar regimes. In all cases, there is a cycle. In some cases, it is annual, in others it lasts

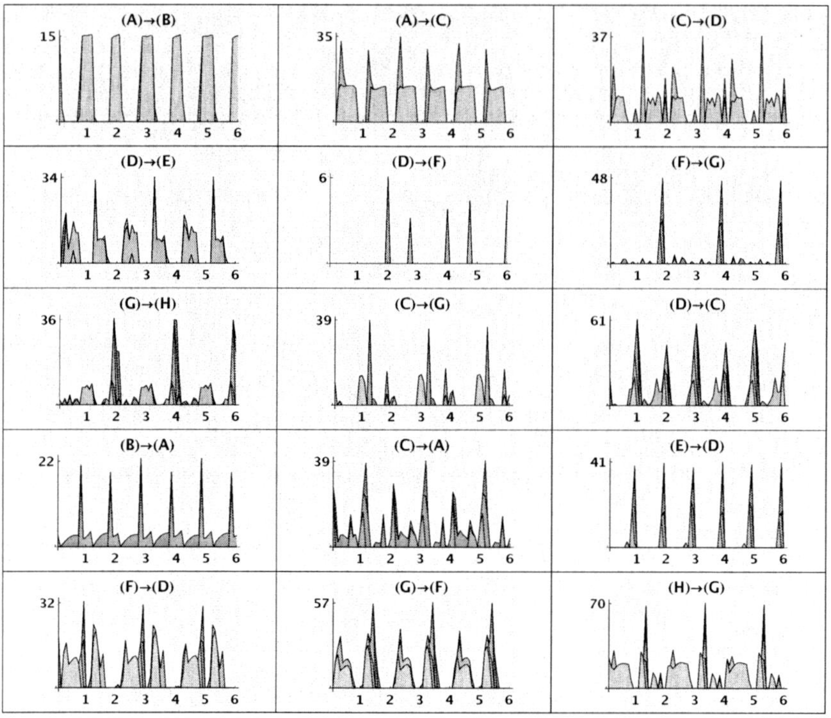

Figure 18.3: Migration of Atlantic Bluefin tuna: favourable situation. For the 15 possible migration paths, we represent the dynamics of migration. As in figure 18.2, x-axis correspond to time and grey levels correspond to size classes.

two years. This is obviously a network effect and suggests that more experimentation is needed.

18.4.2 Unfavourable Situation

We present now the dynamics of age distributions obtained when spawning is lower and migration costs are higher. We get dynamics depicted in figure 18.4. As expected, abundance is smaller; and there is almost no fish in feeding areas (Brazil, North Atlantic, Norway) that are distant from spawning areas.

This is a simple representation of the halt in migration to distant areas. It indicates the opportunistic migration behavior of Bluefin tuna.

18.4.3 Over-fishing in Mediterranean Sea

We present now the age distribution dynamics obtained when fishing mortality is high in the Mediterranean Sea. In this case, we adjust the mortality parameter in the Leslie matrix and consider intermediate values for spawning intensity and migration cost parameters.

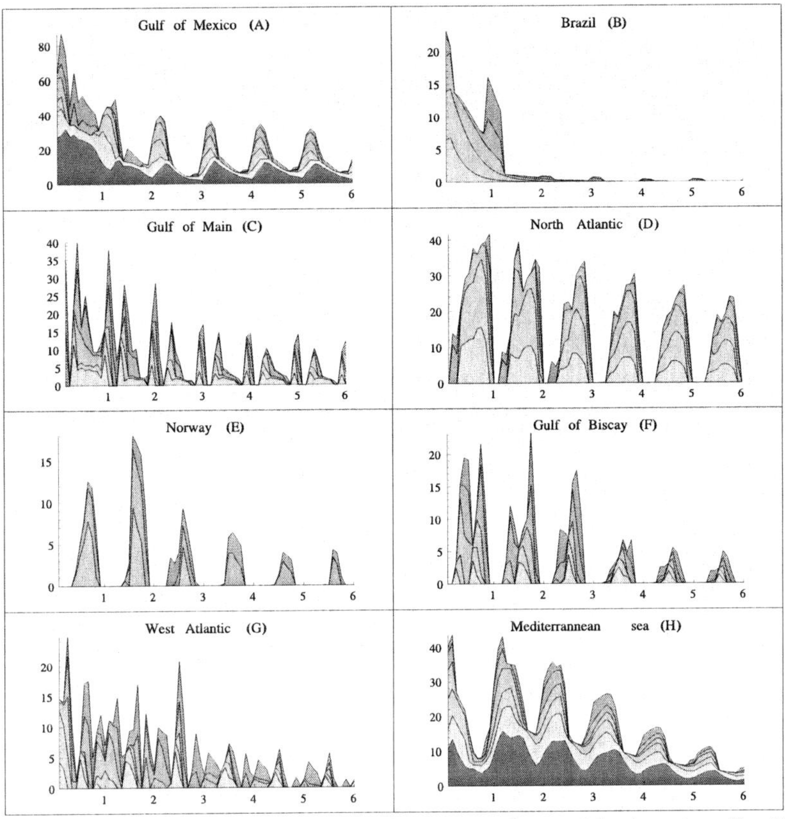

Figure 18.4: Migration of Atlantic Bluefin tuna, unfavourable situation. See figure 18.2.

First of all, we argue that large tuna are found in the Mediterranean Sea because of return migration. Although the general dynamics of the system do not seem to change (an annual cycle is observed in all areas), there are network effects. For example, increasing fishing mortality results in a decrease of biomass in the Gulf of Mexico.

18.4.4 Sensitivity Analysis

We now observe in figure 18.6 the sensitivity of the previous model to the following parameters: (1) intensity of spawning ρ_i, (2) moving costs δ_{ij}.

Increasing spawning intensity has the expected effects on population dynamics. Increasing migration costs seem to affect population dynamics to a certain extent. One and two-year cycles emerge, depending on the value of sensitivity parameters (mainly migration costs). We have no explanation for this difference and do not know if similar patterns have been observed in real data.

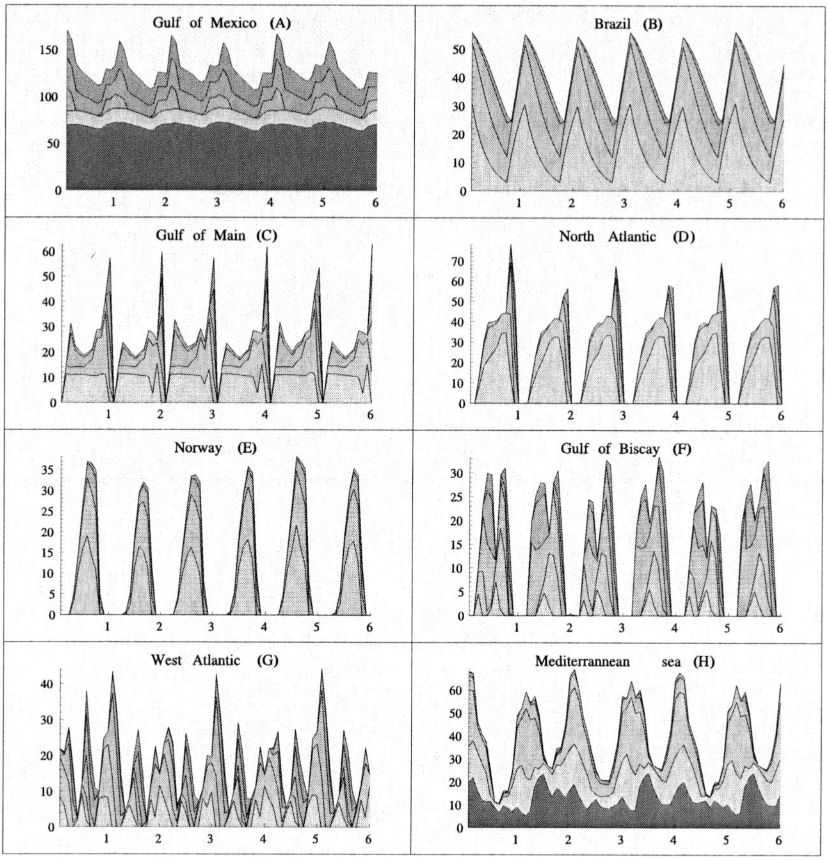

Figure 18.5: Migration of Atlantic Bluefin tuna. Over-fishing in Mediterranean Sea: size distribution.

Figure 18.6 also shows the effects of the coupling of both parameters and makes it possible to identify the emergence of a cycle.

18.5 Concluding Remark

18.5.1 Features

This simple implementation of a migration model has been developed in the context of a real population with known spatial distribution dynamics. The main principle is opportunism. The results of the sensitivity analysis are in line with expectations. They suggest simple explanations for the spatial dynamics of Bluefin tuna.

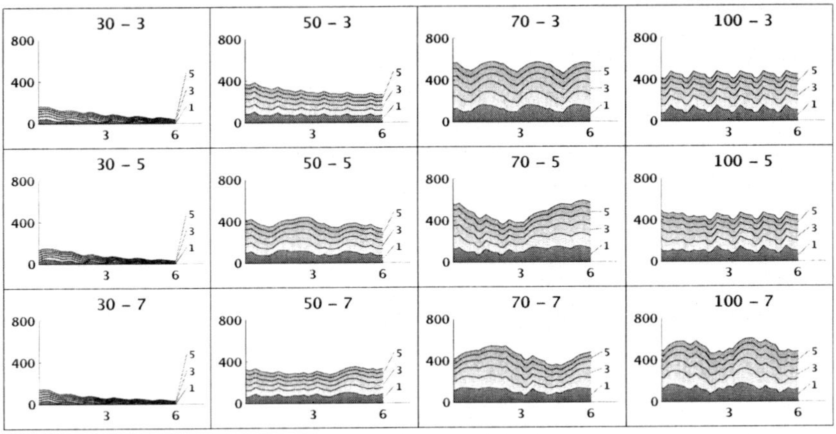

Figure 18.6: Sensitivity of migration of Bluefin tuna to the intensity of spawning and the energetic costs of migration. Spawning 30, 40, 50, energetic costs proportional to distance and to 1, 3, 5, 7. Gray levels correspond to size distributions. Simulation covers six years (x-axis).

18.5.2 Teachings

The approach has proven to be feasible in terms of mathematical formulation and computation. The mathematical formulation is simple and the computer implementation is easy. The model is not over-parameterized and appears to be robust. Computing times are such that it is possible to envisage other experiments with more habitats, and more age classes.

The approach makes it possible to explore theoretical issues, such as the reason for the cycles observed. Other sensitivity analyzes (not presented here) show that this seasonal behavior is firstly related to ecological processes, and secondly to the effect of environment. Without ecological processes (spawning, growth, and natural mortality) and with a yearly seasonal effect there is convergence towards a stationary distribution. When ecological processes are included, without any environmental seasonal effect there is a dynamic cycle, but it is not annual (it is about 14 months). It seems that it is the coupling of both effects (as shown in figure 18.2) that creates the annual cycle. This is a well-known feature of the use of Leslie matrices to represent ecological processes (Caswell, 2006).

18.5.3 Discussing Assumptions

There are difficult underlying assumptions when modeling animal migrations with network economics principles that were developed for human migration. They were discussed in the conclusion of the previous chapter.

18.5.4 Going Further

We cannot claim that the model presented here is a model of the Atlantic Bluefin tuna. It is simply a migration model suggested by the patterns observed in the spatial dynamics of the species. A closer analysis of assumptions could lead to the development of a more realistic model.

Part V

Conclusion

Chapter 19

Network Economics, Economic and Ecological Modelling

19.1 About Principles

Now we are ready to discuss the adequacy of network economics for the representation of the dynamics of marine ecosystems under the double exposure, climate change and economic globalization.

To summarize what has been presented before, the application of network economics to marine systems implies:

(1) Representing ecological or economic systems as networks. It aims at identying relationships between a structure and its dynamics (weakness, robustness).
(2) A focus on balance equations occurring on nodes.
(3) A focus on complementarity equations occurring on links.
(4) A focus on the resulting equilibrium.
(5) A discussion of the way to reach equilibrium: its effectiveness and the related dynamics.
(6) The production of network representations and scenarios to be shared by diverse stakeholders.

19.2 About Methodological Issues

Network economics has been applied to several case studies. Some were mainly theoretical (biological diversity and the financing of fisheries), while others were more applied (the global tuna supply chain). These case studies have made explicit the feasibility of network economics in terms of modeling (associating a model to a scientific question), computability (dealing with intermediate complexity systems), and communication (outlining what is at stake in public debates about the environment and natural resources, and underlying network structures).

19.2.1 A Specific Methodology

In the case studies, ideas from Pattern Oriented Modeling, and the Object-Design-Details protocol made it possible to adapt a methodology that is proper to a network economics approach by:

(1) identifying an adapted system description (how to define nodes and links at a specific level of complexity for research into ecology or economics consisting of tenths to hundreds of entities);
(2) parameterizing systems with inverse modeling principles; and
(3) using equilibrium principles to explore the consequences of alternative hypotheses (long-term equilibrium, or repeated equilibration process).

19.2.2 About the Purpose of Modeling

When developing case studies,

(1) We observed the performance of the resulting model (mainly using sensitivity analyzes).
(2) We tried to identify in the assumptions and the formulation of the model ways to explain the models performance.
(3) We conclude that the network economics approach is feasible: mathematical formulation, and computations are possible, results have meaning.
(4) In almost all cases we conclude that assumptions must be questioned, or new ones expressed in order to build a new model, make the model more realistic or prove the generality of observed performances with ad-hoc data analysis.
(5) No attempt has been made to hide the speculative nature of our attempts to model complex systems; we also highlight the need for iterative improvement.

19.2.3 About Calibration

Some case studies (the Chilean ecosystem, the worldwide supply chain for tuna, the worldwide supply chain for fish) involved an initial calibration phase. We have not concealed the arbitrariness of the choices that were made in this phase and we have not insisted on validation that would usually be necessary according to modeling precepts. This is due to the need to prioritize modeling choices. The most important choices in network economics concern network structure, the definition of nodes, links, and network connectivity. To analyze the effects of the network structure on the resulting equilibrium, we calibrated the system with sensible values and then carried out a sensitivity analysis using these values. This is not what is usually called model validation. However, we have shown that in several cases this makes it possible to go further, particularly in terms of stimulating discussion between actors interested in the future of marine systems.

19.2.4 An Orientation Toward Communication

Our approach has used models adapted to an interdisciplinary context defined by double exposure and future global management. Therefore, we must keep in mind that models represent the behavior of agents, and it must be possible to discuss the results of modeling with these agents. In this situation, to avoid misunderstanding or conflicts, it is always better if modelers are aware of a possible abuse of power related to the modeling capacity. In such a situation it is counterproductive for them to be seen as conjurers.

From the perspective of changes in global governance, it is important that the various actors involved in the management of a natural resource understand both the potential of the modeling approach and its imperfections and iterative nature. To emphasize the importance of communication between the different agents interested by the future of marine ecosystems and their exploitation, we took the following decisions:

(1) to not develop the most predictive or the most realistic model;

(2) to highlight the iterative character of the approach;

(3) with global governance as a backdrop, to initially build models that emphasize global effects, and take the time to analyze and discuss the lessons that can be learned from them with partners;

(4) to formulate models, having in mind the perspective of a second phase based on the taking into account of local effects, in such a way that, then, their computation should not be an issue.

19.2.5 Coupling Ecology and Economics

Network economics reveals deep connections between ecology and economics. For example, in the illustrations we have provided we have considered:

(1) Ecological systems in which the behavior of fish is determined by their well-being. They arbitrate in order to trade-off predation costs and benefits, and living and migration costs.
(2) Economic systems in which the behavior of agents is determined by the maximization of their wealth, independent of their well-being.

We note that this is contrary to the conventional assumptions in ecology and economics. In ecological systems, the principle is a trade-off between energy intake and energy expense (Odum's principle); in economic systems, the principle is that consumers make trade-offs between well-being and costs. This may appear as a contradiction: we apply to fish species a formalism we reject for economic agents. It is a contradiction and we have underlined the arbitrariness of modeling choices. Such contradictions are familiar to modeling practitioners and are well-illustrated by Chardin's painting of the young man building his house of cards shown earlier.

19.3 About Possible Developments

19.3.1 Statics and Dynamics

We discussed the questions of statics and dynamics in economics and ecology in section 3.3. In the interdisciplinary context of double exposure, there is a tendency for researchers from one discipline be mistaken about the value of the results from an other discipline, especially when it concern equilibration processes. For example, it may be that economists or fisheries managers have been too optimistic about the rate at which depleted fish stocks can be recovered. One the other hand biologists have been too optimistic about the disinvestment cycle of loss-making fisheries. This implies that we should use (and improve) tools dedicated to the analysis of dynamic systems in the network economics context. This is the goal of projected dynamical systems (Nagurney and Zhang, 1996). Their application in the context of double exposure is a very stimulating challenge, from both theoretical and practical points of view.

19.3.2 Develop the Network Approach

In the sensitivity analyzes we carried out, input related to connectivity parameters, and output concerned the identification of critical nodes. This was not done in a systematic way. Now that the feasibility of the approach has been verified, there is a need for a systematic investigation of network effects in the statics and the dynamics of marine ecosystems under the double exposure effect. We are ready to go further and characterize the ecological-economic connectivity of marine systems under the effect.

19.3.3 Local Scenarios

The case studies developed here have led us to conclude that there is a need to take into account local changes. This will be done in forthcoming experiments. It is relevant to ask, why only now? The answer is, everything in its own time! Here we have presented several models and for each of them we have underlined their weaknesses in terms of modeling (unrealistic or oversimplified assumptions, obvious artifacts). It is essential to take the time to reflect on progress so far, before building new models.

19.3.4 On A Theoretical Point of View

In our discussion of statics/dynamics issues, we mentioned the possibility of numeric ecological models without functional responses (this has been illustrated in the case studies).

In the discussion of the relationships between prices and utility functions, we mentioned the possibility of economic models without utility functions, or even hidden demand functions. In the example of supply chains between markets and avaricious constrained consumers, we exhibited such a model and discussed its singularity.

In the presentation of supply chains, we represented the equilibrium of an economic system with complementarity equations in their crude form: there is a flow between aggregates because conditions for the existence of a flow are satisfied. We did not mention optimizing behaviors of economic agents.

These examples demonstrate how network economics can be used for theoretical explorations and a renewed interdisciplinarity.

Bibliography

Abel, T., 2004. Systems diagrams for visualizing macroeconomics. Ecological Modelling 178 (1-2), 189–194.

Abramson, G., Kuperman, M., 2001. Social games in a social network. Physical Review E 63 (3), 030901.

Ahmed, A., Xing, E., 2009. Recovering time-varying networks of dependencies in social and biological studies. Proceedings of the National Academy of Sciences 106 (29), 11878–11883.

Allesina, S., Alonso, D., Pascual, M., 2008. A general model for food web structure. science 320 (5876), 658–661.

Allesina, S., Bodini, A., 2004. Who dominates whom in the ecosystem? energy flow bottlenecks and cascading extinctions. J. Theor. Biol. 230 (3), 351–358.

Allesina, S., Bodini, A., Bondavalli, C., 2006. Secondary extinctions in ecological networks: Bottlenecks unveiled. Ecol. Model. 194 (1-3), 150–161.

Allison, E., 2001. Big laws, small catches: global ocean governance and the fisheries crisis. Journal of International Development 13 (7), 933–950.

Amberg, M., 2007. A pattern-oriented approach to a methodical evaluation of modeling methods. Australasian Journal of Information Systems 4 (1).

Anderson, R. W., 1980. Some theory of inverse demand for applied demand analysis. European Economic Review 14, 281–290.

Andrasfai, B., 1989. Graph theory : flows, matrices. Adam Hilger, Bristol.

Arrow, K., Debreu, G., 1954. Existence of an equilibrium for a competitive economy. Econometrica: Journal of the Econometric Society, 265–290.

Asche, F., Salvanes, K. G., Steen, F., 1997. Market delineation and demand structure. American Journal of Agricultural Economics 79 (1), 139–150.

Aumann, R., Hart, S., 1994. Handbook of game theory with economic applications. Vol. 2. North Holland.

Avriel, M., Diewert, W., Schaible, S., Zang, I., 1988. Generalized Concavity. Plenum Press, New York and London.

Ayres, R., Ayres, L., Frankl, P., 1996. Industrial ecology: towards closing the materials cycle. Edward Elgar Cheltenham.

Bakun, A., 1996. Patterns in the Ocean: Ocean Processes and Marine Population Dynamics. California Sea Grant College System, Univ. California, La Jolla, USA.

Bakun, A., 2006. Wasp-waist populations and marine ecosystem dynamics: Navigating the "predator pit" topographies. Progress In Oceanography 68 (2-4), 271–288.

Balmford, A., Gravestock, P., Hockley, N., McClean, C., Roberts, C., 2004. The worldwide costs of marine protected areas. Proceedings of the National Academy of Sciences of the United States of America 101 (26), 9694.

Barabási, A., Albert, R., Jeong, H., 1999. Mean-field theory for scale-free random networks. Physica A: Statistical Mechanics and its Applications 272 (1), 173–187.

Barabsi, A., 2003. Linked. Plume.

Barange, M., Cheung, W., Merino, G., Perry, R., 2010. Modelling the potential impacts of climate change and human activities on the sustainability of marine resources. Current Opinion in Environmental Sustainability.

Bayliff, W., 2004. JI d. Leiva Moreno, and J. Majkowski. 2005. Management of tuna fishing capacity: conservation and socio-economics. In: Second Meeting of the Technical Advisory Committee of the FAO Project, Madrid Spain. pp. 15–18.

Beddington, J., Agnew, D., Clark, C., 2007. Current problems in the management of marine fisheries. science 316 (5832), 1713–1716.

Berkes, F., Colding, J., Folke, C., 2000. Rediscovery of traditional ecological knowledge as adaptive management. Ecological applications 10 (5), 1251–1262.

Bertrand, A., Segura, M., Gutierrez, M., Vasquez, L., 2004. From small-scale habitat loopholes to decadal cycles: a habitat-based hypothesis explaining fluctuation in pelagic fish populations off peru. Fish and Fisheries 5 (4), 296+.

Bestor, T., 2004. Tsukiji: the fish market at the center of the world. Vol. 11. Univ. of California Pr.

Beverton, R., Holt, S., 1957. On the dynamics of exploited fish populations. Vol. 11. Springer.

Blaug, M., 2002. Ugly currents in modern economics. Fact and fiction in economics, 35–56.

Blöchl, F., Theis, F., Vega-Redondo, F., Fisher, E., 2010. Which sectors of a modern economy are most central? CESifo, Center for Economic Studies & Ifo Institute for economic research.

Block, B., Dewar, H., Blackwell, S., Williams, T., Prince, E., Farwell, C., Boustany, A., Teo, S., Seitz, A., Walli, A., et al., 2001. Migratory movements, depth preferences, and thermal biology of atlantic bluefin tuna. Science 293 (5533), 1310–1314.

Boccaletti, S., Latora, V., Moreno, Y., Chavez, M., Hwang, D., 2006. Complex networks: Structure and dynamics. Physics reports 424 (4), 175–308.

Bodie, Z., Merton, R. C., 1999. Finance. Prentice-Hall.

Bollobs, B., 1998. Modern graph theory. Springer.

Boons, F., Baumann, H., Hall, J., 2012. Conceptualizing sustainable development and global supply chains. Ecological Economics.

Bopp, L., Monfray, P., Aumont, O., Dufresne, J., Le Treut, H., Madec, G., Terray, L., Orr, J., 2001. Potential impact of climate change on marine export production. Global Biogeochemical Cycles 15 (1), 81–100.

Border, K., 1989. Fixed point theorems with applications to economics and game theory. Cambridge Univ. Pr.

Borgatti, S., Halgin, D., 2011. On network theory. Organization Science 22 (5), 1168–1181.

Borgatti, S., Mehra, A., Brass, D., Labianca, G., 2009. Network analysis in the social sciences. Science 323 (5916), 892.

Bornholdt, S., 2005. Less is more in modeling large genetic networks. Science 310 (5747), 449.

Borwein, J., Lewis, A., 2006. Convex analysis and nonlinear optimization: Theory and examples. Vol. 3. Springer Verlag.

Botsford, L., Castilla, J., Peterson, C., 1997. The management of fisheries and marine ecosystems. Science 277 (5325), 509–515.

Boyd, S., Vandenberghe, L., 2004. Convex optimization. Cambridge Univ. Pr.

Braess, D., 1968. Uber ein paradoxon der verkehrsplanung. Unternehmensforschung 12, 258–268.

Brainard, W., Scarf, H., 2000. How to compute equilibrium prices in 1891. Cowles Foundation for Research in Economics.

Brander, K., 2007. Global fish production and climate change. Proceedings of the National Academy of Sciences 104 (50), 19709–19714.

Brandes, U., Kenis, P., Raab, J., 2006. Explanation through network visualization. Methodology: European Journal of Research Methods for the Behavioral and Social Sciences 2 (1), 16–23.

Breckling, B., Jopp, F., Reuter, H., 2011. Historical background of ecological modelling and its importance for modern ecology. Modelling Complex Ecological Dynamics, 29–40.

Brouwer, L., 1912. Uber abbildungen von mannigfaltigkeiten. Mathematische Annalen 71 (38), 97–115.

Brown, J., Laundré, J., Gurung, M., 1999. The ecology of fear: optimal foraging, game theory, and trophic interactions. Journal of Mammalogy 80 (2), 385–399.

Brown, M., October 2004. A picture is worth a thousand words: energy systems language and simulation. Ecol. Model. 178 (1-2), 83–100.

Bunn, D., 2004. Modelling Prices in Competitive Electricity Markets. Wiley.

Caddy, J., Cochrane, K., 2001. A review of fisheries management past and present and some future perspectives for the third millennium. Ocean & coastal management 44 (9), 653–682.

Camacho, J., Arenas, A., 2005. Food-web topology: Universal scaling in food-web structure? Nature 435 (7044), E3–E4.

Carpenter, S. e. a., 1994. Fitting predator-prey models to time series with observation errors. Ecology 75, 1254–1264.

Carrington, P., Scott, J., Wasserman, S., 2005. Models and methods in social network analysis. Cambridge Univ. Pr.

Caswell, H., 2006. Matrix population models. Wiley Online Library.

Catarci, C., 2005. The world tuna industry–an analysis of imports and prices, and of their combined impact on catches and tuna fishing capacity. In: Second meeting of the Technical Advisory Committee of the FAO project Management of Tuna Fishing Capacity: Conservation and Socio-economics: 15–18 March 2004, Madrid, Spain. FAO, p. 235.

Chang, W., Smyth, D., 1972. Stability and instability of is-lm equilibrium. Oxford Economic Papers, 372–384.

Charles, A., 1983. Optimal fisheries investment under uncertainty. Canadian Journal of Fisheries and Aquatic Sciences 40 (12), 2080–2091.

Charnov, E. L., 1976. Optimal foraging: The marginal value theorem. Theor. Pop. Biol. 9, 129–136.

Chen, W., Satyanarayana, C., 2007. Applied graph theory: graphs and electrical networks. Systems, Man and Cybernetics, IEEE Transactions on 8 (5), 418.

Cheung, W., Lam, V., Sarmiento, J., Kearney, K., Watson, R., Zeller, D., Pauly, D., 2010. Large-scale redistribution of maximum fisheries catch potential in the global ocean under climate change. Global Change Biology 16 (1), 24–35.

Christensen, P. W., 2008. An Introduction to Structural Optimization. Kluwer.

Christensen, V., Maclean, J., 2011. Ecosystem approaches to fisheries: a global perspective. Cambridge University Press.

Christensen, V., Walters, C. J., 2004. Ecopath with Ecosim: methods, capabilities and limitations. Ecol. Model. 172 (2–4), 109–139.

Clements, F., 1936. Nature and structure of the climax. The Journal of Ecology 24 (1), 252–284.

Colyvan, M., Ginzburg, L., 2003. The Galilean turn in population ecology. Biology and Philosophy 18 (3), 401–414.

Connell, J., 1978. Diversity in tropical rain forests and coral reefs. Science 199 (4335), 1302–1310.

Cook, K. S., Whitmeyer, J. M., 1992. Two approaches to social structure: Exchange theory and network analysis. Annual Review of Sociology.

Cooke, S., Cowx, I., 2004. The role of recreational fishing in global fish crises. BioScience 54 (9), 857–859.

Costanza, R., Andrade, F., Antunes, P., van den Belt, M., Boersma, D., Boesch, D., Catarino, F., Hanna, S., Limburg, K., Low, B., et al., 1998. Principles for sustainable governance of the oceans. Science 281 (5374), 198–199.

Cottle, R., Pang, J.-S., Stone, R. E., for Industrial, S., Applied, M., 2009. The linear complementarity problem. Classics in applied mathematics. SIAM, Philadelphia.

Cuddington, K., 2001. The "balance of nature" metaphor and equilibrium in population ecology. Biol. Phil. 16 (4), 463–479.

Cunningham, S., Greboval, D., 2001. Managing fishing capacity: A review of policy and technical issues. Vol. 409. Food & Agriculture Organization of the UN (FAO).

Curran, S., Kumar, A., Lutz, W., Williams, M., 2002. Interactions between coastal and marine ecosystems and human population systems: perspectives on how consumption mediates this interaction. Ambio: A Journal of the Human Environment 31 (4), 264–268.

Cury, P., Bakun, A., Crawford, R., Jarre, A., Quiñones, R., Shannon, L., Verheye, H., 2000. Small pelagics in upwelling systems: patterns of interaction and structural changes in wasp-waist ecosystems. ICES Journal of Marine Science: Journal du Conseil 57 (3), 603–618.

Cury, P., Cayré, P., 2001. Hunting became a secondary activity 2000 years ago: marine fishing did the same in 2021. Fish and Fisheries 2 (2), 162–169.

Cury, P., Shannon, L., Shin, Y., 2003. The functioning of marine ecosystems: a fisheries perspective. Responsible fisheries in the marine ecosystem, 103–123.

Dafermos, S., 1980a. Traffic equilibrium and variational inequalities. Transportation science 14 (1), 42–54.

Dafermos, S., 1980b. Traffic equilibrium and variational inequalities. Transportation science 14, 42–54.

Daniele, P., 2006. Dynamic networks and evolutionary variational inequalities. Edward Elgar Publishing.

De Angelis, D. L., Waterhouse, J., 1987. Equilibrium and non-equilibrium concepts in ecological models. Ecol. Monog. 57 (1), 1–21.

Denny, M., 2011. How the ocean works: an introduction to oceanography. Princeton University Press.

Dietz, T., Ostrom, E., Stern, P., 2003. The struggle to govern the commons. science 302 (5652), 1907–1912.

Donzelli, F., 1997. Pareto's Mechanical Dream. Dipartimento de Economia Politica e Aziendale, Università di Milano.

Dorogovtsev, S., Mendes, J., 2002. Evolution of networks. Advances in Physics 51 (4), 1079–1187.

Dunne, J., John, J., Adcroft, A., Griffies, S., Hallberg, R., Shevliakova, E., Stouffer, R., Cooke, W., Dunne, K., Harrison, M., et al., 2012. Gfdl's esm2 global coupled climate-carbon earth system models part i: Physical formulation and baseline simulation characteristics. Journal of Climate.

Easley, D., Kleinberg, J., 2010. Networks, crowds, and markets. Cambridge Univ. Press.

Eco, U., 2009. Vertige de la liste. Flammarion Québec.

Erdos, P., Rényi, A., 1960. On the evolution of random graphs. Akad. Kiadó.

Evans, T., Plato, A., 2008. Network rewiring models. Networks and Heterogeneous Media 3 (2), 221.

Facchinei, F., Fischer, A., Piccialli, V., 2007. On generalized Nash games and variational inequalities. Operations Research Letters 35 (2), 159–164.

Facchinei, F., Pang, J.-S., 2003. Finite-Dimensional Variational Inequalities and Complementary Problems. Vol. I and II. Springer-Verlag, New York.

FAO, 2012. The state of the world fisheries and aquaculture. Food and Agriculture Organization of the United Nations; Rome.

FAO, 2013. Fishstat software and database. Website, `http://www.fao.org/fishery/statistics/software/fishstat/en`.

Farrell, J., 1988. Communication, coordination and Nash equilibrium. Economics Letters 27 (3), 209–214.

Fath, B., 2004a. Distributed control in ecological networks. Ecological Modelling 179 (2), 235–245.

Fath, B. D., 2004b. Network analysis applied to large-scale cyber-ecosystems. Ecol. Model. 171 (4), 329–337.

Fawcett, T., Higginson, A., 2012. Heavy use of equations impedes communication among biologists. Proceedings of the National Academy of Sciences 109 (29), 11735–11739.

Fisher, F., 1989. Disequilibrium foundations of equilibrium economics. Vol. 6. Cambridge University Press.

Forrester, J., 1973. Principles of systems. Wright-Allen Press Cambridge, MA.

Fortilus, J., 2012. Modélisation bio-économique des pêcheries thonières: mise en place d'aires marines protégées en haute mer de l'océan indien. Ph.D. thesis, Université de Nantes.

Fox-Genovese, E., Fox, G., 1976. The origins of physiocracy: Economic revolution and social order in eighteenth-century France. Cornell University Press Ithaca & London.

Freeman, L. C., 2004. The development of social network analysis : a study in the sociology of science. North Charleston, Vancouver, BC.

Fréon, P., Misund, O., 1999. Dynamics of pelagic fish distribution and behaviour: effects on fisheries and stock assessment. Blackwell Science. Oxford.

Fretwell, S. D., Lucas, J. H. L., 1970a. On territorial behavior and other factors influencing habitat distribution in birds. Acta Biotheor. 19, 16–36.

Fretwell, S. D., Lucas, J. H. L., 1970b. On territorial behavior and other factors influencing habitat distribution in birds. Acta Biotheoretica 19, 16–36.

Frisch, R., 1936. On the notion of equilibrium and disequilibrium. The Review of Economic Studies 3 (2), 100–105.

Fromentin, J., 2003. The east atlantic and mediterranean bluefin tuna stock management: uncertainties and alternatives. Scientia Marina 67 (S1), 51–62.

Fromentin, J., Powers, J., 2005. Atlantic bluefin tuna: population dynamics, ecology, fisheries and management. Fish and Fisheries 6 (4), 281–306.

Galaz, V., Biermann, F., Folke, C., Nilsson, M., Olsson, P., 2012. Global environmental governance and planetary boundaries: An introduction. Ecological Economics.

Galeotti, A., Goyal, S., Jackson, M., Vega-Redondo, F., Yariv, L., 2010. Network games. The review of economic studies 77 (1), 218–244.

Game, E., Grantham, H., Hobday, A., Pressey, R., Lombard, A., Beckley, L., Gjerde, K., Bustamante, R., Possingham, H., Richardson, A., 2009. Pelagic protected areas: the missing dimension in ocean conservation. Trends in Ecology & Evolution 24 (7), 360–369.

Garrison, R., 2000. Time and money: The macroeconomics of capital structure. Routledge.

Gause, G., 1936. The Struggle for Existence. Williams and Wilkins, Baltimore.

Ginsburgh, V., Keyzer, M., 2002. The structure of applied general equilibrium models. MIT Press.

Ginzburg, L., Colyvan, M., 2004. Ecological orbits: how planets move and populations grow. Oxford University Press, USA.

Giske, J., Huse, G., Fiksen, O., 1998. Modelling spatial dynamics of fish. Reviews in Fish Biology and Fisheries 8 (1), 57–91.

Gonzalez, E. A., 2007. Introductory graph theory for electrical and electronics engineers. IEEE multidisciplinary engineering education magazine 2 (2).

Gordon, H., 1954. The economic theory of a common-property resource: the fishery. The Journal of Political Economy 62 (2), 124–142.

Goyal, S., 2007. Connections: an introduction to the economics of networks. Princeton University Press.

Grafton, R., Hannesson, R., Shallard, B., Sykes, D., Terry, J., 2010. The economics of allocation in tuna regional fisheries management organizations. Conservation and Management of Transnational Tuna Fisheries, 155–162.

Granovetter, M., 1973. The strength of weak ties. American journal of sociology, 1360–1380.

Gribkovskaia, I., Halskau Sr, Ø., Laporte, G., 2007. The bridges of Königsberg-A historical perspective. Networks 49 (3), 199–203.

Grimm, V., Berger, U., DeAngelis, D., Polhill, J., Giske, J., Railsback, S., 2010. The odd protocol: A review and first update. Ecological Modelling.

Grimm, V., Revilla, E., Berger, U., Jeltsch, F., Mooij, W., Railsback, S., Thulke, H., Weiner, J., Wiegand, T., DeAngelis, D., 2005a. Pattern-oriented modeling of agent-based complex systems: lessons from ecology. science 310 (5750), 987.

Grimm, V., Revilla, E., Berger, U., Jeltsch, F., Mooij, W. M., Railsback, S. F., Thulke, H.-H., Weiner, J., Wiegand, T., L., D. D., 2005b. Pattern-oriented modeling of agent-based complex systems: lessons from ecology. Science 310, 987–991.

Grimm, V., et al., 1998. To be, or to be essentially the same: the self-identity of ecological units'. Trends in ecology & evolution 13 (8), 298.

Gross, J. L., Yellen, J., 2004. Handbook of graph theory. Discrete mathematics and its applications. CRC Press, Boca Raton.

Hadjisavvas, N., Komlósi, S., Schaible, S. (Eds.), 2005. Handbook of Generalized Convexity and Generalized Monotonicity. Springer, New York.

Halsey, T., 2000. Diffusion-limited aggregation: a model for pattern formation. Physics Today 53 (11), 36–41.

Hannesson, R., 2007. Do buyback programs make sense? Fisheries Buybacks, 55–66.

Hannon, B., 1973. The structure of ecosystems. Journal of theoretical biology 41 (3), 535–546.

Harary, F., Norman, R. Z., 1953. Graph theory as a mathematical model in social science. Research Center Group Dynamics series,. Ann Arbor.

Harcourt, B., 2011. The illusion of free markets: Punishment and the myth of natural order. Harvard Univ. Pr.

Harris, P., 2011. Ethics and Global Environmental Policy: Cosmopolitan Conceptions of Climate Change. Edward Elgar Publishing.

Hayek, F., 1980. Individualism and economic order. University of Chicago Press.

Hays, G., Richardson, A., Robinson, C., 2005. Climate change and marine plankton. Trends in Ecology & Evolution 20 (6), 337–344.

Hempel, L., 1996. Environmental governance: the global challenge. Island Pr.

Hicks, J., 1937. Mr. Keynes and the "classics"; a suggested interpretation. Econometrica: Journal of the Econometric Society, 147–159.

Hjort, J., 1926. Fluctuations in the year classes of important food fishes. Journal du Conseil 1 (1), 5–38.

Hugie, D., Dill, L., 1994. Fish and game: a game theoretic approach to habitat selection by predators and prey*. Journal of Fish Biology 45, 151–169.

Hutchinson, G., 1953. The concept of pattern in ecology. Proceedings of the Academy of Natural Sciences of Philadelphia 105, 1–12.

Hutchinson, G., 1961. The paradox of the plankton. Am. Nat. 95, 137–145.

Hyrenbach, K., Forney, K., Dayton, P., 2000. Marine protected areas and ocean basin management. Aquatic Conservation: Marine and Freshwater Ecosystems 10 (6), 437–458.

Ihmels, J., Bergmann, S., Gerami-Nejad, M., Yanai, I., McClellan, M., Berman, J., Barkai, N., 2005. Rewiring of the yeast transcriptional network through the evolution of motif usage. Science 309 (5736), 938.

Ingrao, B., Israel, G., 1990. The invisible hand: economic equilibrium in the history of science. MIT Press.

Istratescu, V., 2001. Fixed point theory an introduction. Vol. 7. Kluwer Academic Print on Demand.

Jackson, M., 2010. Social and economic networks. Princeton University Press.

Janssen, M., Bodin, Ö., Anderies, J., Elmqvist, T., Ernstson, H., McAllister, R., Olsson, P., Ryan, P., 2006. Toward a network perspective of the study of resilience in social-ecological systems. Ecology and Society 11 (1), 15.

Jax, K., Jones, C., Pickett, S., 1998. The self-identity of ecological units. Oikos, 253–264.

Jennings, S., Kaiser, M., Reynolds, J., 2001. Marine fisheries ecology. Wiley-Blackwell.

Jeon, Y., Reid, C., Squires, D., 2008. Is there a global market for tuna? Policy implications for tropical tuna fisheries. Ocean Development & International Law 39 (1), 32–50.

Jiménez-Toribio, R., Guillotreau, P., Mongruel, R., 2010. Global integration of European tuna markets. Progress in Oceanography 86 (1–2), 166–175.

Jofré, A., Rockafellar, R., Wets, R., 2007. Variational inequalities and economic equilibrium. Mathematics of Operations Research 32 (1), 32–50.

Jordan, F., Liu, W.-C., Wyatt, T., 2005. Topological constraints on the dynamics of wasp-waist ecosystems. J. Mar. Syst. 57 (3-4), 250–263.

Jordan, F., Scheuring, I., 2004. Network ecology: topological constraints on ecosystem dynamics. Physics of Life Reviews 1 (3), 139–172.

Jorgensen, S., Nielsen, S., Mejer, H., 1995. Emergy, environ, exergy and ecological modelling. Ecological modelling 77 (2-3), 99–109.

Jorgensen, S. E., 2007. Ecological network theory. Elsevier.

Julius, A., Zavlanos, M., Boyd, S., Pappas, G., 2009. Genetic network identification using convex programming. Systems Biology, IET 3 (3), 155–166.

Junker, B., Schreiber, F., ebrary, I., 2008. Analysis of biological networks. Wiley Online Library.

Kaiser, M., ennings, S., Attrill, M., 2005. Marine ecology: processes, systems, and impacts. Oxford University Press Oxford, UK.

Kakutani, S., 1941. A generalization of Brouwers fixed point theorem. Duke Math. Journal 8, 457–459.

Kaldor, N., 1972. The irrelevance of equilibrium economics. The Economic Journal, 1237–1255.

Kasperski, S., Holland, D., 2013. Income diversification and risk for fishermen. Proceedings of the National Academy of Sciences 110 (6), 2076–2081.

King, R., 1993. Will the new keynesian macroeconomics resurrect the is-lm model? The Journal of Economic Perspectives, 67–82.

Knight, F., 1934. The nature of economic science in some recent discussion. The American Economic Review 24 (2), 225–238.

Koen-Alonso, M., Yodzis, P., 2005. Multispecies modelling of some components of the marine community of northern and central patagonia, argentina. Canadian Journal of Fisheries and Aquatic Sciences 62 (7), 1490–1512.

Konnov, I., 2001. Combined Relaxation Methods for Variational Inequalities. Vol. 495 of Lecture Notes in Economics and Mathematical Systems. Springer-Verlag, Berlin, Heidelberg, New York.

Koomey, J., Holdren, J., 2008. Turning numbers into knowledge: mastering the art of problem solving. Analytics Pr.

Kramer-Schadt, S., Revilla, E., Wiegand, T., Grimm, V., 2007. Patterns for parameters in simulation models. Ecol. Model. 204 (3-4), 553–556.

Krause, A. E., Frank, K. A., Mason, D. M., Ulanowicz, R. E., Taylor, W. W., 2003. Compartments revealed in food-web structure. Nature 426 (6964), 282–285.

Lang, S., 1986. A first course in calculus, 5th Edition. Undergraduate texts in mathematics. Springer-Verlag, New York.

Latham, L., 2006. Network flow analysis algorithms. Ecological Modelling 192 (3), 586–600.

Lee, G. M., Tam, N. N., Dong, N. D., 2005. Quadratic programming and affine variational inequalities. Vol. 78 of Non convex Optimization and Its Applications. Springer.

Leichenko, R., O'Brien, K. L., 2008. Environmental Change and Globalization. Double Exposures. Oxford University Press.

Leontief, W., 1951. The Structure of the American Economy, 1919–1939: An Emprical Application of Equilibrium Analysis. Oxford University Press.

Levin, S., 1970. Community equilibria and stability, and an extension of the competitive exclusion principle. American Naturalist, 413–423.

Libralato, S., Christensen, V., Pauly, D., 2006. A method for identifying keystone species in food web models. Ecol. Model. 195 (3-4), 153–171.

Liebowitz, S., Margolis, S., 1994. Network externality: An uncommon tragedy. The Journal of Economic Perspectives 8 (2), 133–150.

Lindeman, R., 1942. The trophic-dynamic aspect of ecology. Ecology 23 (4), 399–417.

Link, J., 2010. Ecosystem-based fisheries management: confronting tradeoffs. Cambridge University Press.

MacArthur, R. H., Pianka, E. R., 1966. On the optimal use of a patchy environment. American Naturalist 100, 603–609.

Machlup, F., 1959. Statics and dynamics: kaleidoscopic words. Southern Economic Journal, 91–110.

Majkowski, J., 2007. Global fishery resources of tuna and tuna-like species. FAO.

Mankiw, N., 2011. Principles of microeconomics. South-Western Pub.

Mann, K., 1993. Physical oceanography, food chains, and fish stocks: a review. ICES Journal of Marine Science: Journal du Conseil 50 (2), 105–119.

Mann, K., Lazier, J., 1996. Dynamics of marine ecosystems. Vol. 389. Blackwell Science Malden, Massachusetts, USA.

Margalef, R., 1997. Turbulence and marine life. Scientia marina 61, 109–123.

May, R., et al., 2006. Network structure and the biology of populations. Trends in Ecology and Evolution 21 (7), 394–399.

May, R. M., 2008. Network structure and the biology of populations. Trends Ecol. Evol. In Press, Corrected Proof.

McCann, K., Rooney, N., 2009. The more food webs change, the more they stay the same. Philosophical Transactions of the Royal Society B: Biological Sciences 364 (1524), 1789.

McGill, B., Enquist, B., Weiher, E., Westoby, M., 2006. Rebuilding community ecology from functional traits. Trends in Ecology & Evolution 21 (4), 178–185.

Meadows, D., Meadows, D., Randers, J., Behrens, W., of, R. C., 1972. The limits to growth. Vol. 205. Universe Books New York.

Ménard, F., Fonteneau, A., Gaertner, D., Nordstrom, V., Stéquert, B., Marchal, E., 2000. Exploitation of small tunas by a purse-seine fishery with fish aggregating devices and their feeding ecology in an eastern tropical atlantic ecosystem. ICES Journal of Marine Science: Journal du Conseil 57 (3), 525.

Mendes, J., Dorogovtsev, S., Ioffe, A., 2003. Evolution of networks: From biological nets to the internet and WWW.

Merino, G., Barange, M., Mullon, C., 2010. Climate variability and change scenarios for a marine commodity: Modelling small pelagic fish, fisheries and fishmeal in a globalized market. Journal of Marine Systems 81 (1–2), 196–205.

Miller, R., Blair, P., 1985. Input-output analysis: foundations and extensions. Vol. 328. Prentice-Hall Englewood Cliffs, NJ.

Mills, C., Carlton, J., 1998. Rationale for a system of international reserves for the open ocean. Conservation biology 12 (1), 244–247.

Milo, R., Shen-Orr, S., Itzkovitz, S., Kashtan, N., Chklovskii, D., Alon, U., 2002. Network motifs: simple building blocks of complex networks. Science 298 (5594), 824.

Mirowski, P., 1984. Physics and the marginalist revolution. Cambridge Journal of Economics 8 (4), 361.

Mittlebach, G., 2002. Fish foraging and habitat choice: a theoretical perspective. In: Hart, P. J. B., Reynolds, J. (Eds.), Handbook of Fish and Fisheries. Oxford Blackwell Science, p. 251–266.

Miyake, M., Guillotreau, P., Sun, C., Ishimura, G., 2010. Recent developments in the tuna industry: stocks, fisheries, management, processing, trade and markets. Food and Agriculture Organization of the United Nations.

Montoya, J. M., Pimm, S., Solé, R. V., 2006. Ecological networks and their fragility. Nature 442, 259–264.

Montoya, J. M., Rodriguez, M. A., Hawkins, B. A., 2003. Food web complexity and higher-level ecosystem services. Ecology Letters 6 (7), 587–593.

Moody, A., Houston, A., McNamara, J., 1996. Ideal free distributions under predation risk. Behavioral Ecology and Sociobiology 38 (2), 131–143.

Mullon, C., Mittaine, J., Thébaud, O., Péron, G., Merino, G., Barange, M., 2009a. Modeling the global fishmeal and fish oil markets. Natural Resource Modeling 22 (4), 564–609.

Mullon, C., Nagurney, A., 2012. A game theoretical approach to the vertical coexistence of small and big fish. Ecological Modelling 240, 41–48.

Mullon, C., Shin, Y., Cury, P., 2009b. Neats: a network economics approach to trophic systems. Ecological Modelling 220 (21), 3033–3045.

Myers, R., Worm, B., 2003. Rapid worldwide depletion of predatory fish communities. Nature 423 (6937), 280–283.

Myerson, R., 1978. Refinements of the Nash equilibrium concept. International Journal of Game Theory 7 (2), 73–80.

Myerson, R., 1999. Nash equilibrium and the history of economic theory. Journal of Economic Literature, 1067–1082.

Nagurney, A., 1993. Network Economics. A variational inequality approach. Kluwer Academic Publishers, Dordrecht.

Nagurney, A., 1999. Network economics : a variational inequality approach. Advances in computational economics. Kluwer Academic, Boston; London.

Nagurney, A., Zhang, D., 1996. Projected dynamical systems and variational inequalities with applications. Kluwer Academic Publishers.

Nahin, P., 2007. Chases and escapes: the mathematics of pursuit and evasion. Princeton Univ. Pr.

Neira, S., Arancibia, H., 2004. Trophic interactions and community structure in the upwelling system off central chile (33-39âs). J. Exp. Mar. Biol. Ecol. 312 (2), 349–366.

Neira, S., Arancibia, H., Cubillos, L., 2004. Comparative analysis of trophic structure of commercial fishery species off central chile in 1992 and 1998. Ecol. Model. 172 (2-4), 233–248.

Neuhoff, K., Barquin, J., Boots, M., Ehrenmann, A., Hobbs, B., Rijkers, F., Vazquez, M., 2005. Network-constrained Cournot models of liberalized electricity markets: the devil is in the details. Energy Economics 27 (3), 495–525.

Nielsen, M., 1999. E.U. seafood markets—integration and demand. Tech. rep., SJFI.

Nøstbakken, L., Thebaud, O., and Sørensen, L. C. (2011). Investment behaviour and capacity adjustment in fisheries: a survey of the literature. Marine Resource Economics, 26(2), 95–117.

Odum, H. T., Odum, E. P., 2000. The energetic basis for valuation of ecosystem services. Ecosystems 3 (1), 21–23.

Ostrom, E., 2008. The challenge of common-pool resources. Environment: Science and Policy for Sustainable Development 50 (4), 8–21.

Ostrom, E., Burger, J., Field, C., Norgaard, R., Policansky, D., 1999. Revisiting the commons: local lessons, global challenges. science 284 (5412), 278–282.

Padgett, J. F., McLean, P. D., 2006. Economic credit and elite transformation in renaissance Florence. SFI.

Pan, J., Nagurney, A., 1994. Using markov chains to model human migration in a network equilibrium framework. Mathematical and computer modelling 19 (11), 31–39.

Park, J., Newman, M., 2004. Statistical mechanics of networks. Physical Review E 70 (6), 66–117.

Pascual, M., Dunne, J., 2005. Ecological networks: linking structure to dynamics in food webs. Oxford University Press, USA.

Pascual, M., Dunne, J., 2006. Ecological networks: linking structure to dynamics in food webs. Oxford University Press, USA.

Pauly, D., 2007. The sea around us project: Documenting and communicating global fisheries impacts on marine ecosystems. AMBIO: a Journal of the Human Environment 36 (4), 290–295.

Pauly, D., 2010. Five Easy Pieces: How Fishing Impacts Marine Ecosystems. Island Press.

Pauly, D., Alder, J., Bennett, E., Christensen, V., Tyedmers, P., Watson, R., 2003. The future for fisheries. Science 302 (5649), 1359–1361.

Pauly, D., Christensen, V., Guénette, S., Pitcher, T., Sumaila, U., Walters, C., Watson, R., Zeller, D., 2002. Towards sustainability in world fisheries. Nature 418 (6898), 689–695.

Pauly, D., Christensen, V., Walters, C., 2000. Ecopath, Ecosim, and Ecospace as tools for evaluating ecosystem impact of fisheries. ICES Journal of Marine Science: Journal du Conseil 57 (3), 697.

Pauly, D., Watson, R., Alder, J., 2005. Global trends in world fisheries: impacts on marine ecosystems and food security. Philosophical Transactions of the Royal Society B: Biological Sciences 360 (1453), 5–12.

Pennock, D., Flake, G., Lawrence, S., Glover, E., Giles, C., 2002. Winners don't take all: Characterizing the competition for links on the web. Proceedings of the National Academy of Sciences 99 (8), 5207.

Pereira, H., Leadley, P., Proença, V., Alkemade, R., Scharlemann, J., Fernandez-Manjarrés, J., Araújo, M., Balvanera, P., Biggs, R., Cheung, W., et al., 2010. Scenarios for global biodiversity in the 21st century. Science 330 (6010), 1496–1501.

Perry, A., Low, P., Ellis, J., Reynolds, J., 2005. Climate change and distribution shifts in marine fishes. Science 308 (5730), 1912–1915.

Perry, I, R., Ommer, R., 2010. Introduction: Coping with global change in marine social-ecological systems. Marine Policy 34 (4), 739–741.

Pickard, G., Emery, W., 1990. Descriptive physical oceanography: an introduction. Elsevier.

Pikitch, E., Santora, C., Babcock, E., Bakun, A., Bonfil, R., Conover, D., Dayton, P., Doukakis, P., Fluharty, D., Heneman, B., et al., 2004. Ecosystem-based fishery management. Science 305 (5682), 346–347.

Pimm, S., 1991. The balance of nature: ecological issues in the conservation of species and communities. University of Chicago Press.

Pimm, S. L., 1979. The structure of food webs. Theor. Pop. Biol. 16 (2), 144–158.

Pitcher, T., 2001. Fish schooling: Implications for pattern in the oceans and impacts on human fisheries. In: Steele, J., Turekian, K., Thorpe, S. (Eds.), Encyclopedia of Ocean Sciences. Academic Press, pp. 975–987.

Platt, T., Sathyendranath, S., Fuentes-Yaco, C., 2007. Biological oceanography and fisheries management: perspective after 10 years. ICES Journal of Marine Science: Journal du Conseil 64 (5), 863–869.

Polovina, J., 1984. Model of a coral reef ecosystem. Coral reefs 3 (1), 1–11.

Poulard, J., Blanchard, F., 2005. The impact of climate change on the fish community structure of the eastern continental shelf of the bay of biscay. ICES Journal of Marine Science: Journal du Conseil 62 (7), 1436–1443.

Power, M. E., Tilman, D., Estes, J. A., Menge, B. A., Bond, W. J., Mills, S. L., Daily, G., Castilla, J. C., Lubchenco, J., Paine, R. T., 1996. Challenges in the quest for keystones. BioScience 46 (8), 609–620.

Prance, G. T. (Ed.), 1982. The Biological model of Diversification in the Tropics. Columbia U.P.

Quirin, G., Sethi, S., 1977. Market feedbacks and the limits to growth. INFOR 15 (1), 1–21.

Ragan, M., 2009. Trees and networks before and after Darwin. Biology Direct 4 (1), 43.

Railsback, S., 2001. Getting results: The pattern-oriented approach to analyzing natural systems with individual-based models. Natural Resource Modeling 14 (3), 465–475.

Randers, J., 1980. Elements of the system dynamics method. MIT Press Cambridge, MA.

Ravier, C., Fromentin, J., 2001. Long-term fluctuations in the eastern atlantic and mediterranean bluefin tuna population. ICES Journal of Marine Science: Journal du Conseil 58 (6), 1299–1317.

Reid, C., Squires, D., Jeon, Y., Rodwell, L., Clarke, R., 2003. An analysis of fishing capacity in the western and central pacific ocean tuna fishery and management implications* 1. Marine Policy 27 (6), 449–469.

Rockafellar, R., 1990. Nonsmooth analysis and parametric optimization. Methods of Nonconvex Analysis, 137–151.

Rogers, C., 1980. A less strange version of milnor's proof of brouwer's fixed-point theorem. The American Mathematical Monthly 87 (7), 525–527.

Rossi, V., López, C., Sudre, J., Hernández-García, E., Garçon, V., et al., 2008. Comparative study of mixing and biological activity of the benguela and canary upwelling systems. Geophys. Res. Lett 35, L11602.

Rykiel, E., 2006. Towards a definition of ecological disturbance. Australian Journal of Ecology 10 (3), 361–365.

Safina, C., Klinger, D., 2008. Collapse of bluefin tuna in the western atlantic. Conservation Biology 22 (2), 243–246.

Samuelson, P., 1941. The stability of equilibrium: comparative statics and dynamics. Econometrica: Journal of the Econometric Society, 97–120.

Samuelson, P., 1943. Dynamics, statics, and the stationary state. The Review of Economics and Statistics 25 (1), 58–68.

Samuelson, P. A., 1952. Spatial price equilibrium and linear programming. American Economic Review 42, 283–303.

Sander, L., 2000. Diffusion-limited aggregation: a kinetic critical phenomenon? Contemporary Physics 41 (4), 203–218.

Schaefer, M., 1957. Some considerations of population dynamics and economics in relation to the management of the commercial marine fisheries. Journal of the Fisheries Board of Canada 14 (5), 669–681.

Scheffer, M., Carpenter, S. R., 2003. Catastrophic regime shifts in ecosystems: linking theory to observation. Trends Ecol. Evol. 18 (12), 648–656.

Schwartzlose, R. A., Alheit, J., Bakun, A., Baumgartner, T. R., 1999. Worldwide large-scale fluctuations of sardine and anchovy populations. S. Afr. J. Mar. Sci. 21, 289–347.

Scovazzi, T., 2004. Marine protected areas on the high seas: some legal and policy considerations. Int'l J. Marine & Coastal L. 19, 1.

Serrano, M., Boguná, M., 2003. Topology of the world trade web. Physical Review E 68 (1), 015–101.

Shannon, L., Neira, S., Taylor, M., 2008. Comparing internal and external drivers in the southern Benguela and the southern and northern Humboldt upwelling ecosystems. African Journal of Marine Science 30 (1), 63–84.

Sinclair, M., 1988. Marine populations. an essay on population regulation and speciation. Washington Sea Grant, Seattle, Washington (USA).

Solodov, M., Tseng, P., 1996. Modified projection-type methods for monotone variational inequalities. SIAM Journal on Control and Optimization 34 (5), 1814–1830.

Sousa, W., 1979. Disturbance in marine intertidal boulder fields: the nonequilibrium maintenance of species diversity. Ecology 60 (6), 1225–1239.

Steele, J., Thorpe, S., Turekian, K. K., 2009. Encyclopedia of Ocean Sciences. Academic Press, London.

Steenblik, R., Munro, G., 1998. International work on fishing subsidies: an update. The Impact of Government Financial Transfers on Fisheries Management, Resource Sustainability, and International Trade, 17–19.

Strömberg, K., Smyth, T., Allen, J., Pitois, S., O'Brien, T., 2009. Estimation of global zooplankton biomass from satellite ocean colour. Journal of Marine Systems 78 (1), 18–27.

Suh, S., 2005. Theory of materials and energy flow analysis in ecology and economics. Ecological modelling 189 (3-4), 251–269.

Sumaila, U., Khan, A., Dyck, A., Watson, R., Munro, G., Tydemers, P., Pauly, D., 2010. A bottom-up re-estimation of global fisheries subsidies. Journal of Bioeconomics 12 (3), 201–225.

Summerhayes, V., Elton, C., 1923. Contributions to the ecology of Spitsbergen and Bear Island. University Press.

Tang, J., Scellato, S., Musolesi, M., Mascolo, C., Latora, V., 2010. Small-world behavior in time-varying graphs. Physical Review E 81 (5), 55–101.

Todd, M., 1976. The computation of fixed points and applications. Vol. 1. Springer-Verlag Berlin.

Tufte, E., 2003. The cognitive style of PowerPoint. Graphics Press Cheshire, Conn.

Tufte, E., Howard, G., 1983. The visual display of quantitative information. Graphics press Cheshire, CT.

Turner, M., 1989. Landscape ecology: the effect of pattern on process. Annual review of ecology and systematics 20, 171–197.

Turner, M., Romme, W., Gardner, R., O'Neill, R., Kratz, T., 1993. A revised concept of landscape equilibrium: disturbance and stability on scaled landscapes. Landscape Ecology 8 (3), 213–227.

Tyedmers, P., 2004. Fisheries and energy use. Encyclopedia of energy 2, 683–693.

Tyedmers, P., Watson, R., Pauly, D., 2005. Fueling global fishing fleets. Ambio, 635–638.

Ulanowicz, R., 1997. Ecology, the ascendent perspective: Robert E. Ulanowicz. Columbia University Press.

Ulanowicz, R., 1999. Life after Newton: an ecological metaphysic. BioSystems 50 (2), 127–142.

United Nations Organization, 2013. Comtrade database. Website, `http://comtrade.un.org`.

Van Notten, P., Rotmans, J., van Asselt, M., Rothman, D., 2003. An updated scenario typology. Futures 35 (5), 423–443.

Vandermeer, J., Evans, M. A., Foster, P., Hook, T., Reiskind, M., Wund, M., 2002. Increased competition may promote species coexistence. PNAS 13, 8731–8736.

Vicsek, T., 1992. Fractal growth phenomena. World Scientific Pub Co Inc.

Von Neumann, J., Morgenstern, O., 1972. Theory of games and economic behaviour. Princeton University Press.

Vroey, M., Hoover, K., 2004. Introduction: Seven decades of the is-lm model. History of Political Economy 36 (5), 1–11.

Walras, L., 1874. Eléments dEconomie Politique Pure. Corbaz.

Ward, J., Mace, P., Thunberg, E., 2005. The relationship of fish harvesting capacity to excess capacity and overcapacity. Marine Resource Economics 19, 525–29.

Wardrop, J. G., 1952. Some theoretical aspects of road traffic research. In: Proceedings, Institute of Civil Engineers. Vol. 1. pp. 325–378.

Watts, A., 2001. A dynamic model of network formation. Games and Economic Behavior 34 (2), 331–341.

Watts, D. J., 2004a. Six degrees. Norton.

Watts, D. J., 2004b. Small worlds. Princeton University Press.

Watts, D. J., Strogatz, S. H., 1998. Collective dynamics of 'small-world' networks. Nature 393 (6684), 440–442, times Cited: 4325.

Wilhite, A., 2006. Economic activity on fixed networks. Handbook of computational economics 2, 1013–1045.

Winfree, A., 2001. The geometry of biological time. Vol. 12. Springer Verlag.

Worm, B., Hilborn, R., Baum, J., Branch, T., Collie, J., Costello, C., Fogarty, M., Fulton, E., Hutchings, J., Jennings, S., et al., 2009. Rebuilding global fisheries. science 325 (5940), 578.

Wulff, F., Field, J., Mann, K., 1989. Network analysis in marine ecology: methods and applications. Vol. 32. American Geophysical Union.

Young, T., 2003. Developing a legal strategy for high seas marine protected areas. Legal Background Paper. WWF/IUCN High Seas Marine Protected Areas Project.

Index